Communications in Computer and Information Science 649

Commenced Publication in 2007
Founding and Former Series Editors:
Alfredo Cuzzocrea, Dominik Ślęzak, and Xiaokang Yang

More information about this series at http://www.springer.com/series/7899

Axel-Cyrille Ngonga Ngomo · Petr Křemen (Eds.)

Knowledge Engineering and Semantic Web

7th International Conference, KESW 2016
Prague, Czech Republic, September 21–23, 2016
Proceedings

 Springer

Editors
Axel-Cyrille Ngonga Ngomo
Leipzig University
Leipzig
Germany

Petr Křemen
Czech Technical University in Prague
Prague
Czech Republic

ISSN 1865-0929   ISSN 1865-0937   (electronic)
Communications in Computer and Information Science
ISBN 978-3-319-45879-3   ISBN 978-3-319-45880-9   (eBook)
DOI 10.1007/978-3-319-45880-9

Library of Congress Control Number: 2016949634

Printed on acid-free paper

This Springer imprint is published by Springer Nature
The registered company is Springer International Publishing AG Switzerland

Preface

These proceedings contain the papers accepted for oral presentation at the 7th International Conference on Knowledge Engineering and Semantic Web (KESW 2016). The conference was held in Prague, Czech Republic, during September 21–23, 2016.

The principal mission of the KESW conference series is to provide a discussion forum for the community of researchers currently underrepresented at the major International Semantic Web Conference (ISWC) and Extended Semantic Web Conference (ESWC). This mostly includes researchers from Eastern and Northern Europe, Russia, and former Soviet republics. This year, the conference was held in Prague to catalyze discussions between the traditional KESW community and the European research community.

As in previous years, KESW 2016 aimed at helping the community to get used to the common international standards for academic conferences in computer science. To this end, KESW featured a peer reviewing process in which every paper was reviewed in a rigorous but constructive way by at least three members of the Program Committee. As before, the PC was international, representing countries ranging from the USA to Japan and Germany.

We received a total of 53 submissions. The strict reviewing policies have resulted in the acceptance of 17 full research papers. This translates into an acceptance rate of 32 %. Additional 9 papers (17 %) were accepted for short presentation and have also been given space in these proceedings. The authors represent mainly EU countries, including Germany, Spain, and the Czech Republic as well as various parts of Russia.

KESW 2016 continued the tradition of inviting established researches for keynote presentations. We are grateful to Lynda Hardman (CWI, Netherlands), Axel Polleres (WU Vienna, Austria), Steffen Staab (University of Koblenz, Germany), and Filip Železný (FEE CTU, Czech Republic) for their insightful talks. The program also included posters and position paper presentations to help attendees, especially younger researchers, discuss preliminary ideas and promising PhD topics.

We thank Dmitry Mouromtsev and Pavel Klinov, who helped us immensely during conference preparation. Next, we would like to thank both organizing institutions, Czech Technical University in Prague and ITMO University, for their support. Next, we would like to express our thanks to this year's sponsors, namely Datlowe, s.r.o., STI Innsbruck, and metaphacts GmbH – without their support the event would hardly be possible. We would also like to thank the hardworking PC as well as our publicity chairs, Martin Ledvinka and Maxim Kolchin, for their reliable and quick work. Last but not least, we would like to thank the Action M Agency, particularly Milena Zeithamlová, for their reliable administrative support.

July 2016

Petr Křemen
Axel-Cyrille Ngonga Ngomo

Organizing Committee

General Chair

Petr Křemen   FEE CTU in Prague, Czech Republic

Program Chair

Axel-Cyrille Ngonga Ngomo   Institute for Applied Informatics, Germany

Publicity Chairs

Martin Ledvinka   FEE CTU in Prague, Czech Republic
Maxim Kolchin   ITMO University, Russia

Program Committee

Alessandro Adamou	KMI, The Open University, UK
Long Cheng	Technische Universität Dresden, Germany
Evangelia Daskalaki	ICS-FORTH, Greece
Jeremy Debattista	University of Bonn, Germany
Chiara Del Vescovo	British Broadcasting Corporation, UK
Elena Demidova	University of Southampton, UK
Ivan Ermilov	Universität Leipzig, Germany
Irini Fundulaki	ICS-FORTH, Greece
Ujwal Gadiraju	L3S Research Center, Germany
Kleanthi Georgala	Universität Leipzig, Germany
Peter Haase	metaphacts GmbH, Germany
Ali Hasnain	Digital Enterprise Research Institute, Ireland
Martin Homola	Comenius University Bratislava, Slovakia
Konrad Höffner	Universität Leipzig, Germany
Dmitry Ignatov	National Research University Higher School of Economics, Russia
Vladimir Ivanov	Kazan Federal University, Russia
Valentina Ivanova	Linköping University, Sweden
Natalya Keberle	Zaporizhzhya National University, Ukraine
Evgeny Kharlamov	University of Oxford, UK
Jakub Klímek	FIT CTU in Prague, Czech Republic
Pavel Klinov	Complexible Inc., USA
Boris Konev	University of Liverpool, UK

Roman Kontchakov Birkbeck	University of London, UK
Liubov Kovriguina	NRU ITMO, Russia
Dmitry Kudryavtsev	Saint Petersburg State Polytechnical University, Russia
Christoph Lange	University of Bonn, Germany
Steffen Lohmann	Fraunhofer IAIS, Germany
Nicolas Matentzoglu	University of Manchester, UK
Dmitry Mouromtsev	NRU ITMO, Russia
Elena Mozzherina	Saint Petersburg State University, Russia
Rafael Pealoza	Free University of Bozen-Bolzano, Italy
Denis Ponomaryov	A.P. Ershov Institute of Informatics Systems, Russia
Héctor Pérez-Urbina	Google, USA
Mariano Rodríguez Muro	IBM Research, USA
Yuliya Rubtsova	A.P. Ershov Institute of Informatics Systems, Russia
Muhammad Saleem	Universität Leizpig, Germany
Tzanina Saveta	ICS-FORTH, Greece
Marvin Schiller	Universität Ulm, Germany
Daria Stepanova	Technical University of Vienna, Austria
Lauren Stuart	Purdue University, USA
Julia Taylor	Purdue University, USA
Ioan Toma	STI Innsbruck, Austria
Dmitry Tsarkov	The University of Manchester, UK
Joerg Unbehauen	Universität Leipzig, Germany
Dmitry Ustalov	Ural Federal University, Russia
Amrapali Zaveri	Stanford University, USA
Dmitriy Zheleznyakov	University of Oxford, UK

Additional Reviewers

Callahan, Alison	Gottschalk, Simon	Uhliarik, Ivor
Del Corro, Luciano	Grangel, Irlan	Vahdati, Sahar
Déraspe, Maxime	Kozlov, Artem	Zhukova, Nataly
Gossen, Gerhard	Siu, Amy	

·

Organizers

CTU in Prague

ITMO University

Sponsors

Datlowe

STI Innsbruck

Metaphacts

Contents

Data Management

Applications

Ontologies

Multi-viewpoint Ontologies
for Decision-Making Support

Sergey Gorshkov[1(✉)], Stanislav Kralin[2],
and Maxim Miroshnichenko[1]

[1] TriniData, Mashinnaya 40-21, 620089 Ekaterinburg, Russia
{serge,miroshnichenko}@trinidata.ru
[2] Ekaterinburg, Russia
stanislav.kralin@gmail.com

Abstract. Considering multiple viewpoints is often required when building ontologies for decision-making support systems. The notion of subjective context is useful for designing such a systems. We review the evolution of the subjectivity representation in the knowledge engineering, then choose an appropriate definition of the context for our application. This allows formulating the functional requirements for a multi-viewpoint decision-making support system and choosing the technical way of context representation. We propose a method of ontological representation of multiple viewpoints using named graphs as a response to these requirements. Decision-making support in the socio-economic realms is an especially valuable application for multi-viewpoint ontologies. We consider a demonstration use case, including software implementation. The inference rules may be used in such applications both for making conclusions within every particular context, or transferring knowledge between them. We present a set of sample rules for our demonstration use case and discuss the results achieved.

Keywords: Multi-viewpoint ontology · Decision making support · Context modeling

1 Introduction

It is often necessary to reflect different standpoints concerning the same objects and operation when using Semantic Web technologies for building decision-making support systems.

The development of a software application usually includes implicit construction of the uniform model for a fragment of reality. This model reflects only one view of it, or combines several views cleared of contradictions. Meanwhile every process in the socio-economic realm includes a number of independent subjects. Each of them possess a unique set of opinions. Sometimes it is possible to neglect this diversity, but tasks of the other classes require their careful consideration. Among the examples of

---

S. Kralin—Independent researcher.

© Springer International Publishing Switzerland 2016
A.-C. Ngonga Ngomo and P. Křemen (Eds.): KESW 2016, CCIS 649, pp. 3–17, 2016.
DOI: 10.1007/978-3-319-45880-9_1

such tasks are the socio-economic systems management, administration of legislation, management of the complex projects performed by the groups of contractors.

In this work, we will choose the way of representation of multiple viewpoints in ontologies using available functionality of OWL and SPARQL, implementation of the inference on these models, and verification of the practical applicability of the results achieved. For the last task, we will consider a demonstration prototype of the decision-making support system. The practical goal of this prototype is the assistance in political decisions making by representing opinions of various social groups on the different variants of the possible developments of situation, reproduction of the way of thinking of the people in these groups. The analysis of this information allows revealing the possible influence of the decisions on the citizen's attitude, helps balance all the interests, and shows a set of possible ways of leveling negative effects of unpopular actions.

On the side, we have split this article into several blocks reflecting viewpoints of the demonstration system's end-user, ontologist/analyst formalizing the task and building a model for it, and the programmer implementing the solution. Each block starts with the wording like "From the analyst's point of view". We hope this will structure our text, show the benefits of considering multiple viewpoints even at the meta-level of modeling and software engineering, and demonstrate how the different people are seeing the same things.

2 Related Works

From the analyst's point of view, the description logic lying in the base of the simplest OWL variants does not offer some features, which may be found in the more common logic. It is designed for the ease of determining truthfulness of the statements (and provides decidability and low computation complexity), but does not offer tools for working with the "truth-value gluts", particularly rooted in the distinction of the judgements of different subjects. Such situations are common in our life; however, the relativity of truth is generally counterintuitive for the human way of thinking, and our mind tends to search for the "absolute truth" it presume to exist even in purely subjective debates. The ontologists also may be misled in this way, which makes it harder to grasp and model the multi-subject situations.

There are two strategies of dealing with "truth-value gluts" in common logic: integrating and isolating ones. The examples of the integrating strategy are the paraconsistent and non-monotonic logical systems. There are also the extralogical frameworks which may serve for this purpose, such as Dung's one [9]. However, we do not have a task of direct resolution of contradictions; we want to assert them and to model how these contradictory judgements are producing different conclusions in various subjective contexts.

The examples of the isolating strategy are the multi-modal epistemic and doxastic logics going back to Hintikka [10], and McCarthy's notion of *context* formulated with the first-order tools [15]. Both multi-modal and first-order approaches cannot be directly translated into description logic. The tools of the first-order approaches lie beyond the limits of the *guarded fragment*, which contains description logics [1, 20].

The modal approaches transform the description logic, which is a multi-modal system by its nature, into the many-dimensional modal system [14], which leads to the rise of computational complexity or even undecidability.

Actually, there are some contextual [12], epistemic [2] and even paraconsistent description logics of limited expressivity, which are rather far from implementation in OWL, although their features are considered important for using with description logics as a modeling language [2].

An overview of multi-modal, first-order and other approaches to the subjectivity modeling was given by Sowa [21]. We are not considering there the attempts of modeling subjectivity directly in semantics, and more or less technical semantics allowing "relativistic" interpretations, such as Kripke's *possible worlds* [13], Hintikka's *model sets* [11], and Barwise and Perry's *situations* [3].

Among the reviewed approaches, J. McCarthy' notion of a *context* [21] is important for us as a starting point for our implementation. A context is a just one version of the reality description. This definition is important for us, as it implies existence of several descriptions of the same fragment of reality, which may be contradictory and equally authoritative. Some of the further developers of this theory has emphasized, for example, the linkage of a context with a subject or a group of subjects sharing it [19].

Montague has shown that it is necessary to consider the points of reference for establishing truthfulness of the statements. A set of meaningful objects is distinguished for each point of reference, and the value (intension) and extension of each predicate is defined [16]. We will solve our task basing on this approach.

The context's content in our implementation is a usual OWL model. We build a meta-level over this model, describing the relations of the subjects and contexts, and methods of their interaction. These methods may also include the rules of "translation" from one contextual "language" to another.

From the programmer's point of view, even the OWL Full standard does not offer an explicit method for creating meta-models described above. The researchers facing necessity of reflecting several viewpoints in ontology are finding their own ways to do this. For example, A. Boer, considering the ontological representation of legislative information, has introduced the notion of subjective betterness, which is used for selection of appropriate positions of regulation [4]. The betterness is an OWL relation, defined for the pairs of situations considered by a legislator. Situations are represented as OWL individuals. The LKIF ontology, designed with participation of A. Boer, is representing the concepts like "believes", "states" as OWL relations. The authors are proposing the mode of modalities representations, which does not break the OWL limits, avoids emergence of extra objects, but has a limited expressiveness.

The common method for modalities formalization is the reification. It implies the generation of the special entities for each statement, belonging to the classes like "Belief" [19]. These entities are linking subjects and the opinions. Let us consider an example: "John believes that Penguins are living in Arctic". We have to create an individual #John_sBelief (see Fig. 1) belonging to the class #Belief, which has relation #hasBearer (all identifiers are arbitrary and given just for readability) with the individual #John, and relation #hasBelievedFact with the individual #ArcticPenguins. The #ArcticPenguins, in turn, is an instance of the class #AnimalArea, having relation #hasSpecies with the #Penguins and relation #hasArea with the region #Arctic

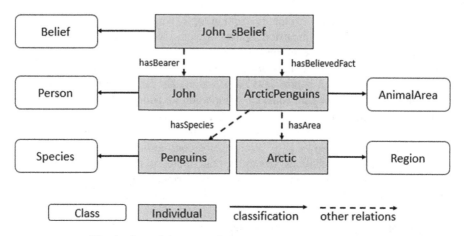

Fig. 1. One of the ways of representing beliefs in ontology

(of course, other structures for expressing this fact are also possible). This method also allows remaining within the limits of the OWL standard. It has its own opportunities and drawbacks, among which is the creation of "extra" entities such as #John_sBelief.

On the side, let us note that someone may argue that it is not necessary to represent false assertions (such as "Penguins are living in Arctic") in ontologies. There are two reasons to do that: (a) many assertions cannot be surely proved or refuted at all, (b) sometimes pragmatics dictates to do that. Imagine John is your boss and discussion about penguins may cost you too much.

The C-OWL is an extension of OWL, which allows description of the rules of reference between different ontologies elements. Each of these ontologies is representing a specific context [5]. C-OWL solves the task of knowledge transformation between different subjective contexts, "translation" from one language to another, but does not explicitly defines the subjects, context owners. It does not allow representation of contradictory opinions on the same situations expressed in the same terms by the different subjects, does not provide the ability to distinguish logical rules they use. P-SOC-OWL ontology extends C-OWL ideas, resolving some of these problems, and introducing means for representing subjective opinions on the same objects [17]. This ontology even allows expressing the confidence degree of the subject in some statement.

Djakhdjakha et al. are offering a rather mature approach to the same task in [7] and the following series of papers. The authors are considering multi-viewpoint ontologies using the notions of a global and local context, global and local role, individual, subsumption rules and bridge rules. This approach implements means of expressing alignments and contradictions between points of view. Some particular questions are well developed, such as classification of individuals from different points of view [8].

Let us note that the prevailing approach to the contexts formalization in OWL is the segregation of each context into a separate ontology, and linking them with various means. This way is not the only possible. An OWL ontology may be considered as a theory of some subject matter area, with several points of view built in [18]. We will design our implementation basing on the idea of using named graphs for context representation [6].

3 Motivation

From the analyst's point of view, the modeled system often shall be regarded in a number of points of view, each one matching some subject or group. Our task is to propose a method of implementation of the next functional requirements using OWL standard features:

1. To provide ability of description of subjects or their groups in a model, and their distinct points of view (contexts).
2. Each viewpoint may contain its special terms (notions) for expressing some concepts (ideas), which may be shared between viewpoints, or be specific for some of them.
3. In each viewpoint, a distinct set of facts about some individuals may be expressed. If the individuals are shared between contexts, some facts on them may be contradictory. The important special case is the representation of different regards on the composition of some object.
4. Each viewpoint may contain its own user-defined inference rules.
5. Arbitrary rules of knowledge transformation between viewpoints may be defined, as well as the rules for synthesis of knowledge from different viewpoints. The special case is the situation when some notions from different viewpoints are referring to the same concept. Then it is necessary to define rules of "translation from one conceptual language to another".

From the programmer's point of view, we have to meet the following technical requirements:

1. The whole model shall be stored in a single ontology.
2. We shall avoid reification (introduction of the extra individuals expressing things such as "[someone] has opinion").
3. We shall provide a method for accessing particular viewpoints as well as the whole ontology through a SPARQL endpoint.

From the end user's point of view, we need a sample use case, for which the listed requirements are actual. Let us consider a hypothetical decision-making support software, which is used by the top executive of some municipality. As it is an elective office, the politician wants to know how some variants of his or her actions will affect the public opinion. Imagine that some company wants to obtain permission for construction of the industrial facility in the city. In our simplified example, let us consider three point of view on this question:

- Of the citizens aware of economy problems: for them, a new facility is a source of workplaces.
- Of the citizens aware of ecology problems: for them, a new facility is a source of pollution.
- Of the state and municipal employee: for them, a new facility is just an object, which is classified by regulations and which shall be compliant with its requirements.

We shall consider the city's top official as the person who is able to switch between various points of view to make the most pragmatic decision. In our sample use case, the politician wants to have an IT system, which will model the way of thinking of the people of the various social groups, predict their reaction on the approval or denial of the facility construction, and estimate parameters of the affected interests.

4   Implementation

From the analyst's point of view, our conceptual model will have the next structure. First, there are subjects and groups of them, having their points of view. Each subject or group has its own interests, which may be fulfilled or ignored as a result of the political decisions. Collective subjects has their own metrics, which can be measured in sociological surveys: amount of people involved, distribution of subjective perception of the importance of the interest within group etc.

Let us consider classification of the individual object representing the planned facility (we will name it "Acme Manufacturing") in various viewpoints. This will illustrate implementation of functional requirement #2 from the above list (Table 1).

Table 1. Classification of the acme manufacturing individual

Citizens (ecology)	Citizens (economy)	State employee
Industrial facility	Industrial facility	Object of technical regulation
Source of pollution		Source of pollution; object of ecological regulation
	Source of workplaces	Source of workplaces; object of fiscal regulation

Figure 2 shows how the relations of various classes and properties with the different viewpoints reflects the perception of the situation by the subjects.

Thereby, various points of view may use different notions (terms) for denotation of the same (or mismatching) concepts and individuals.

Different subjects are viewing an object's structure differently. For example, for the ecology-aware citizens it is important that the facility include the boiler working on the hydrocarbon fuel. State employee are viewing object's structure in a much more complex way. The schema of the ontology fragment representing this situation is shown on the Fig. 3. This is how we implement functional requirement #3 from the above list.

Various subjects are making inference using different rules. For example, for the citizens the valid rule is "An industrial facility is a source of pollution, if it includes a hydrocarbon-powered boiler". In the same time, for the state employee, who are using regulations as a source of inference, the valid rule may sound like "An industrial facility is a source of pollution, if it emits a dangerous substance X in the concentration Y". The both rules are assigning the same class name ("source of pollution"), but the conditions when it is applicable are different. What is a source of pollution for one

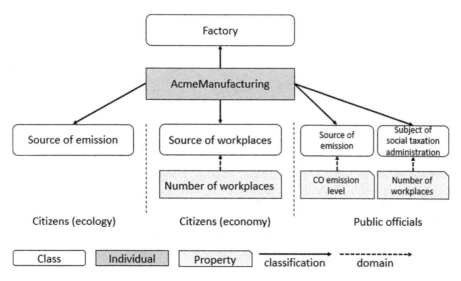

Fig. 2. Classification and properties of the object in different viewpoints

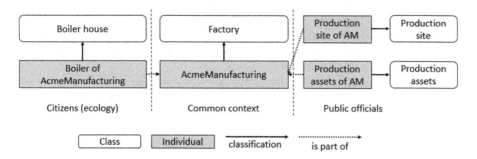

Fig. 3. Decomposition of the object in different viewpoints

subject is not necessary a source of pollution for another, and both of them have their reasons. Therefore, we have implemented the functional requirement #4.

We will follow the next way of implementations:

– Describe the subjects and their interests in the model.
– Describe the individual object (facility), common for all the viewpoints (although the subjects are seeing it differently).
– Formulate the rules of inference for various viewpoints, reflecting the way of thinking of the people of various social groups.
– Represent the possible variants of the management decisions. The rules are formulated according to the social group's opinions analysis, and are allowing to infer how each decision affects the interests of the subjects.
– The system's end user, a politician, can explore the variants of management decision, and see which interests are violated in each case, and by which reason.

The Fig. 4 shows overall ontology structure:

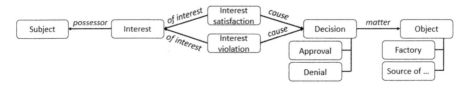

Fig. 4. Ontology structure for the use case

This diagram tells that every subject may possess some interest(s), which may be satisfied of violated by some decisions. Every decision is either an approval or a denial of a specific action over some object. Of course, this ontology is very sketchy; in the real use, the model shall look much more complicated.

From the programmer's point of view, the implementation will include a SPARQL endpoint, in which the above listed information will be stored. The following software components will work with it: an inference engine, an ontology editor, and a user interface for model exploration (Knowledge Management System, KMS).

We represent subjects or the groups of them as the individuals belonging to the class #Subject. All the content of the endpoint is split onto named graphs, each of them representing a particular point of view. One of the graphs represents a "consensus" (or "common") point of view, containing the facts that are true for all the subjects considered. The other facts are distributed between the named graphs according to the viewpoints in which they are asserted.

For example, we have the individual #EcologyCitizens representing the group of subjects (and thus belonging to the class #Subject), and the #PublicOfficials individual. The next triples represent the fact that ecology-aware citizens consider Acme Manufacturing as a source of pollution, while public officials are treating it as just an object of ecological regulation. Both parties agree that it is an industrial facility (Table 2):

Table 2. Triples representing classification of the #AcmeMfg individual

RDF triple			Graph
Entity ("subject")	Predicate	Object	
#AcmeMfg	rdf:type	#PollutionSource	#EcologyCitizens
#AcmeMfg	rdf:type	#EcoRegulationObject	#PublicOfficials
#AcmeMfg	rdf:type	#IndustrialFacility	

We see that the name of the graph matches the identifier of the group of subjects possessing the point of view, or it is empty (or has some default value) for the "consensus" one. Our engine expects to find the graphs named after the designated subjects. When we need to obtain information valid for particular point of view, our engine rewrites SPARQL query using graph name(s). The engine functions below the application layer of the system. For example, user opens #AcmeMfg individual, and the application has to display the list of classes to which it belongs. The application issues a simple query like:

```
SELECT ?type WHERE {#AcmeMfg rdf:type ?type }
```

The engine seamlessly rewrites it into a more complicated query:

```
SELECT ?type ?graph WHERE {GRAPH ?graph {#AcmeMfg rdf:type ?
type } }
```

The engine filters the result upon receiving. For example, it will retain only the triples from #EcologyCitizens or default graph, if the chosen viewpoint is #EcologyCitizens. As an alternative, graph name may be explicitly specified in the query. Our ontology editor (Onto.pro) is working by the same way, assigning entered assertions to the chosen point(s) of view.

This approach is more convenient than, for example, using OWL annotations for assigning facts to the viewpoints. It is more laconic, allows specifying easily the point of view in SPARQL queries and/or filter query results according to the points of view, and even combining data from different viewpoints in a single query (in this case query rewriting becomes a little more complicated, but still can be handled automatically). By the same reasons, it is more convenient than separating various viewpoints onto different ontologies. Less changes in SPARQL queries (or no changes at all, for the front-end application) comparing to the single viewpoint ontologies means better performance and flexibility. In addition, it isolates the application layer from the viewpoints implementation when needed.

In the model exploration interface (we use our software ArchiGraph.KMS for this) we provide ability of choosing point of view of some subject. After that, the user sees only the facts that are valid for the chosen viewpoint and/or the "consensus" one. The software filters the information according to the named graphs to which it belongs to achieve this.

Let us show some examples of inference rules used in our sample case (Table 3). The last rule illustrates fulfillment of the functional requirement #5 of our list.

In our implementation, we use the SPARQL rules tagged by the appropriate viewpoint(s) in their metadata. Our engine follows the idea of SPIN rules, but has the original implementation, which adds some extra features to them. Each triple emerged as a result of the rule is annotated with the reference to the rule which has produced it.

5   Use Case Demonstration

From the analyst's point of view, work with the system include the next steps:

1. Create the model structure and fill it with the initial fact information. In our case, it shall include data describing the facility such as the number of workplaces, emission parameters.
2. Create two variants of the management decision: construction approval and denial.
3. Define the inference rules. In the real use cases, they must be based on the public opinion surveys and reflect the people's way of thinking as close as it is possible.
4. Apply the inference rules. They produces, particularly, the classification of modeled objects from the particular points of view, and conclusions which interests will be fulfilled and violated by each variant of the decision.

Table 3. Examples of the inference rules

Condition (if...)	Conclusion (then...)	Source viewpoint	Target viewpoint	Comment
?X is a facility. ?X has number of workplaces > 0	?X is the subject of fiscal administration. ?X is a source of workplaces	Consensus	State employee	Making particular conclusions (individuals classification) according to the common facts
?X is a facility. ?X has number of workplaces > 50	?X is a source of workplaces	Consensus	Citizens (economy)	Conditions leading to the same conclusion may differ in various viewpoints
?X is an approval of ?Y project. ?X is a source of workplaces	A new object ?Z is the fulfillment of the interest "Workplaces creation" for the economy-aware citizens caused by ?Z	Citizens (economy)	Citizens (economy)	Making complex inference within one viewpoint
?X includes a hydrocarbon-powered boiler	?X is a source of pollution	Citizens (ecology)	Citizens (ecology)	Reflecting the people's way of thinking
?X has CO emission level > 0	?X is a source of pollution	State employee	Citizens (ecology)	Knowledge transfer from one viewpoint to another

From the end user's point of view, the system is primarily a GUI that allows performing certain actions, so we will pay attention to the visual representation in this section. Lack of GUI features is often a problem that eliminates any possible benefits of the intellectual systems for the end user. There is the scenario of use of our system.

1. The decision maker uses KMS interface to explore the conclusions made by the system. He starts with the page presenting information on each possible decision. For example, at the page of the Acme Manufacturing project approval (shown on the Fig. 5) he will see that this decision will violate the interests of one group of citizens and satisfy the interests of another:
2. Turning to the interest violation or satisfaction page, the user sees the links to the interest and the social group which possess it. The link to the inference rule, which has produced this fact, is also present, as shown on the Fig. 6.

The objects pointing on the current one: [ungroup]

• Cause (2)   [-]

• Interest violation of EcologyConcernedCitizens by ApprovalOfProject
• Satisfaction of EconomyConcernedCitizens by ApprovalOfProject

Object relations schema   [hide]

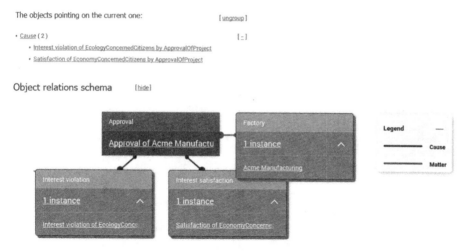

Fig. 5. Fragment of the management solution page in the KMS interface

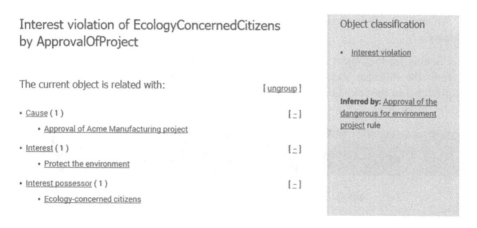

Fig. 6. Fragment of the violated interest page in the KMS interface

Clicking on the rule, the user sees the description given by the analyst during its construction. For example, it may be information on the public survey with the link to its result. The user may discover that the reason of the opinion that Acme Manufacturing facility is a source of pollution is the conviction of some part of people that firing hydrocarbon fuel has a negative effect on ecology. This allows considering the conclusions not as restrictions, which are limiting the range of possible decisions, but as a direction for working on the social consensus achievement. Disproof of particular cognitive patterns may take away some objections against proposed decisions. In our sample case it may be the work for ensuring and proving facility's ecological safety, or the proposal of compensational actions.

3. Turning to the interest page (presented at Fig. 7), the user sees its description and parameters of the social group possessing it. Exploring all the possible variants of the decision by this way, the politician will obtain the basis for decision-making.

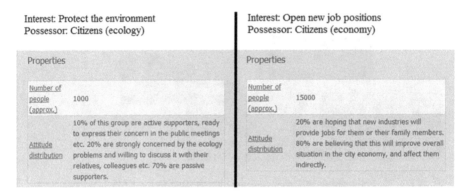

Fig. 7. Parameters of the social group's interest

An applied software may provide a convenient interface for comparison of decision variants.

4. Clicking on the object name (Acme Manufacturing), the user sees its description from the different social group's points of view. The special switch for this is provided in the interface, as shown on the Fig. 8:

Fig. 8. Point of view switch in the KMS interface

The Fig. 9 shows properties and classification of the object with the different switch positions:

KMS also takes into consideration the chosen point of view when constructing search queries, offering to use only the terms from the specific viewpoint. Therefore, our user may freely switch between different points of view while working with the system, see different statements on the same objects, use different terminology for denotation of the same entities. In other use cases, some users may be assigned to particular points of view, then they will see information in the system only in their own terminology, structure, details level.

In our sample case, the politician may make the next conclusions after using the system:

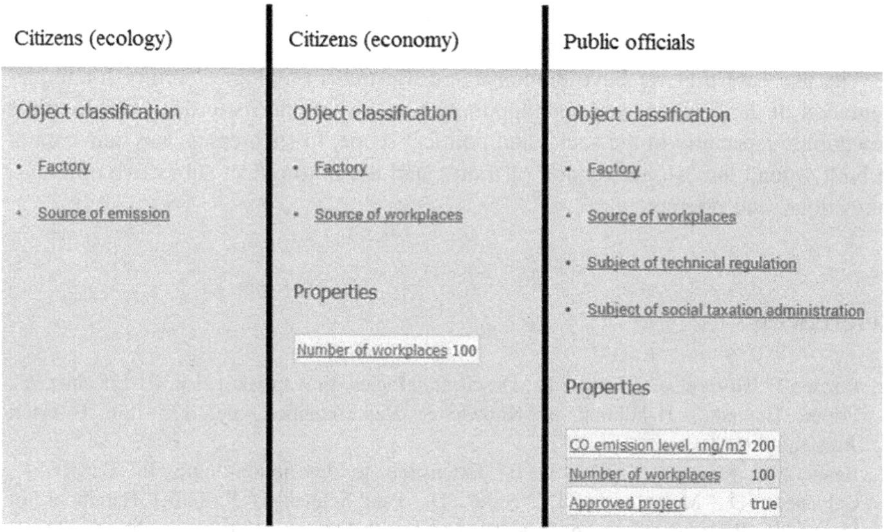

Citizens (ecology)	Citizens (economy)	Public officials
Object classification	Object classification	Object classification
• Factory	• Factory	• Factory
• Source of emission	• Source of workplaces	• Source of workplaces
		• Subject of technical regulation
	Properties	• Subject of social taxation administration
	Number of workplaces 100	Properties
		CO emission level, mg/m3 200
		Number of workplaces 100
		Approved project true

Fig. 9. Object classification and parameters from the various points of view

1. The facility construction shall be approved, as the number of people interested in new workplaces significantly exceeds the number of protesters against pollution. The new facility does not exceeds emission limits defined by regulations.
2. The facility's owners has to make some efforts on the facility's image as the ecologically harmless object. They shall do some compensatory actions – for example, develop new parks. This shall be included as conditions into the investment agreement.

Let us emphasize that the final conclusions shall be made by the politician, not the system itself (this is inacceptable from the legal and ethical point of view).

The main opportunity of using semantic model in the considered use case is its flexibility: the facts expressed in ontology may and has to be permanently updated when a new information becomes known, as well as the inference rules. The conclusions made by the system will be updated accordingly, so the system will always contain the actual image of the modeled scope of reality according to the most actual information on it.

6 Conclusions

We have formulated a general set of the requirements for the ontology-powered IT system, considering different points of view of various subjects. We have combined several methodological approaches, rooted in the logical theories, and technologies for its programmatic implementation. This allowed us to reach our goal using features of the existing versions of OWL and SPARQL standards, and to demonstrate working with the multiple points of view in the sample application. The key component of the solution, which provides its rationality and functionality, is the use of the named graphs for representation of the viewpoints of various subjects.

The use case that we have demonstrated illustrates ontologies' utility for implementation of the decision-making support systems, which needs to deal with multiple viewpoints, especially in the social and political scope. In such cases, a system cannot be built around the "single version of truth", and needs to reflect subjective opinions, convictions, and interests.

References

1. Baader, F., Horrocks, I., Sattler, U.: Description logics. In: van Harmelen, F., Lifschitz, V., Porter, B. (eds.) Handbook of Knowledge Representation, pp. 135–180. Elsevier, Amsterdam (2008)
2. Baader, F., Kusters, R., Wolter, F.: Extensions to description logic. In: Baader, F., Calvanese, D., McGuinness, D., Nardi, D., Patel-Schneider, P. (eds.) Handbook of Knowledge Representation, pp. 219–261. Cambridge University Press, Cambridge (2003)
3. Barwise, J., Perry, J.: Situations and Attitudes. MIT Press, Cambridge (1983)
4. Boer, A.: Legal Theory, Sources of Law and the Semantic Web. IOS Press, Amsterdam (2009)
5. Bouquet, P., Giunchiglia, F., van Harmelen, F., Serafini, L., Stuckenschmidt, H.: C-OWL: contextualizing ontologies. In: Fensel, D., Sycara, K., Mylopoulos, J. (eds.) ISWC 2003. LNCS, vol. 2870, pp. 164–179. Springer, Heidelberg (2003)
6. Carroll, J., Bizer, C., Hayes, P., Stickler, P.: Named graphs, provenance and trust. In: Proceedings of the 14th International Conference on World Wide Web (2005)
7. Djakhdjakha, L., Hemam, M., Boufaida, Z.: Foundations on multi-viewpoints ontology alignment In: Proceedings of 4th International Conference on Web and Information Technologies, Algeria
8. Djezzar, M., Boufaida, Z.: Ontological classification of individuals: a multi-viewpoint approach. Int. J. Reasoning-Based Intell. Syst. 7(3/4), 276–285 (2015)
9. Dung, P.M.: On the acceptability of arguments and its fundamental role in non-monotonic reasoning, logic programming and n-person games. Artif. Intell. 77(2), 321–357 (1995)
10. Hintikka, J.: Knowledge and Belief: An Introduction to the Logic of the Two Notions: Contemporary Philosophy. Cornell University Press, Ithaca (1962)
11. Hintikka, J.: Modality and quantification. Theoria 27(3), 119–128 (1961)
12. Klarman, S.: Reasoning with contexts in description logics. Ph.D. thesis, Vu University Amsterdam (2013)
13. Kripke, S.: Semantical analysis of modal logic I. Normal propositional calculi. Zeitschrift fur mathematicshe Logik und Grundlagen der Mathematik 9(56), 67–96 (1963)
14. Kurucz, A., Wolter, F., Zakharyaschev, M., Gabbay, D.: Many-Dimensional Modal Logics: Theory and Applications: Studies in Logic and the Foundations of Mathematic. Elsevier Science, Amsterdam (2003)
15. McCarthy, J.: Notes on formalizing context. In: Proceedings of the 13th International Joint Conference of Artificial Intelligence – Volume I, IJCAI 1993, pp. 555–560. Morgan Kaufmann Publishers (1993)
16. Montague, R.: Pragmatics and intensional logic. In: Davidson, D., Harman, G. (eds.) Semantics of Natural Language, vol. 42, pp. 142–168. Springer, Heidelberg (1970)

17. Nickles, M., Cobos, R.: Social contexts and the probabilistic fusion and ranking of opinions: towards a social semantics for the semantic web In: Proceedings of the Second International Conference on Uncertainty Reasoning for the Semantic Web – Volume 218 (2006)
18. Obrst, L.: Ontological architectures. In: Poli, R., Healy, M., Kameas, A. (eds.) Theory and Applications of Ontology. Springer, Heidelberg (2010)
19. Obrst, L., Nichols, D.: Context and ontologies: contextual indexing of ontological expressions. In: AAAI 2005 Workshop on Context and Ontologies, AAAI 2005, Pittsburgh, PA (2005)
20. Sattler, U., Calvanese, D., Molitor, R.: The description logic handbook, pp. 137–177. Cambridge University Press (2003)
21. Sowa, J.: Knowledge Representation: Logical, Philosophical and Computational Foundations. Brooks/Cole, Boston (2000)

Ontological Anti-patterns in Aviation Safety Event Models

Jana Ahmad[(✉)] and Petr Křemen

Faculty of Electrical Engineering,
Czech Technical University in Prague, Prague, Czech Republic
{jana.ahmad,petr.kremen}@fel.cvut.cz

Abstract. Last years, there has been growing interest in developing high quality models to ensure interoperability of applications as well as proper understanding among a community. To improve productivity of model designers as well as to improve quality of resulting models, proper theoretical foundations and tool support is necessary. This paper discusses some types of conceptual modeling anti-patterns that lead to error- prone modeling decisions, and describes a reproducible solution to a general anti-pattern detection problem during conceptual design. Novel contribution of this paper is the definition of new ontological anti-patterns, we observed during our work in design and model ontologies in the domain of Aviation Safety. The approach is illustrated on models designed by means of OntoUML, an ontology-founded UML profile based on the Unified Foundational Ontology (UFO).

1 Introduction

In recent years, more and more attention is paid to the quality of conceptual models. Conceptual models should be sufficiently expressive to represent positive instances. Furthermore, elected conceptual modeling language (CML) must be intuitive and expressive enough to help modelers to produce flexible, reusable and intended models. In this paper, we are interested in studying and presenting anti-patterns. By an anti-pattern we understand a syntactically correct description in the particular CML that leads to an unintended model (i.e. it leads to construct excess, see Chap. 2 in [1]). In the scope of this paper, the considered CML is the OntoUML language defined in [1], as a representant of a suitable well-founded general purpose conceptual modeling language. The goal of this paper is to elaborate semantic anti-patterns identified during usage of OntoUML for modeling the aviation safety domain.

In recent decades, in the aviation safety community, there is an increasing amount of requirements especially relating to technology and performance. They understand the challenges facing industry and government today in challenging technologies, applications and procedures for the integration, analysis and presentation of business information to support better business decision-making. Aviation helps the global economy to grow. Thus, an increasing attention has been paid to the development of Aviation Safety, i.e. to improve aviation safety

A.-C. Ngonga Ngomo and P. Křemen (Eds.): KESW 2016, CCIS 649, pp. 18–30, 2016.
DOI: 10.1007/978-3-319-45880-9_2

systems, to enhance quality of safety data, and finally to avoid accidents and incidents in this domain. In the aviation safety community, there is a big effort to adopt techniques such as Business Intelligence on a national and international scale [4].

We came across the idea of this paper during our cooperation with the Civil Aviation Authority of the Czech Republic and several Czech aviation organizations including Prague airport or Air Navigation Services of the Czech Republic, within two national projects focused on IT support for aviation safety. Particular contributions of this paper include extension of UML class diagrams for representing UFO-B models, as well as novel ontological anti-patterns detected in those models.

Section 2 presents related work. In Sect. 3 we briefly define the notion of Unified Foundational Ontology (UFO) [1], which is a basis for OntoUML (one of Ontologically Well-Founded Conceptual Modeling languages). Section 4 discusses the possibility to extend OntoUML, in order to define and evaluate event and process based models. In Sect. 5 the use case and methodology used in our research is presented. Section 6 discusses ontological anti-patterns with their negative consequences [5–8]. In Sect. 7 we extend OntoUML towards perdurant modeling by introducing the notion of an Event and define anti-patterns in UFO-B models, identified during our work in design and model ontologies in the domain of Aviation Safety (AS). Finally, Sect. 8 shows some final considerations of this work.

2   Related Work

The main commonly used conceptual modeling languages are Nijssen's Information Analysis Method (niam), which provides a powerful grammar for generating conceptual schema diagrams NIAM [14], ER [15], UML which is a modeling language typically used in the field of software engineering, [16], and OWL [17]. Unified Foundational Ontology (UFO) started as a joint effort of Giancarlo Guizzardi and Gerd Wagner in 2001 to unify the General Ontological Language, which was a predecessor of the General Formal Ontology (GFO) proposed by Herre et al. [21] and the top-level ontology of universals underlying widely-used OntoClean methodology proposed by Guarino and Welty [13]. The main contribution of OntoClean was the beginning of a formal foundation for ontological analysis [13]. OntoUML is an ontologically well-founded UML extension based on the Unified Foundational Ontology (UFO) [2,11]. An ontology pattern OP is a modeling solution to solve a recurrent ontology design problem. It is a template that represents a schema for specific design solutions. An ontology design pattern ODP consists of a set of prototypical ontology entities that constitute the abstract form of a pattern, and of a set of metadata about its use cases, motivations, provenance, the pros and cons of its application, the links to other patterns, etc. Design solutions based on ODPs encode ontology entities that apply, specialize, or instantiate the prototypical entities defined by the schema [22].

3   UFO and OntoUML

We use OntoUML to illustrate anti-pattern detection problems. OntoUML is based on the Unified Foundational Ontology (UFO), which is a top-level ontology aimed at specifications of domain ontologies and languages. It can be used for evaluating business modeling methods and providing real-world semantics for their modeling constructs. In general, it aims at developing theories, methodologies and engineering tools with the goal of advancing conceptual modeling as a theoretically sound discipline but also one that has concrete and measurable practical implications [3]. UFO is divided into three layers:

– UFO-A: Object and Trope model part (An Ontology of Endurants) [1].
– UFO-B: Event and Process model part (An Ontology of Perdurants) [11] see Fig. 2.
– UFO-C: Social and Agent model part (An Ontology of Intentional and Social Entities) [18].

As a top-level ontology, UFO distinguishes individuals and universals. Several categories of universals are applied in the UML metamodel for class diagrams in order to turn it into a well-founded conceptual modeling language named OntoUML. Selected universals, their relationships and restrictions are represented as UML stereotypes and OCL constraints. OntoUML has been successfully applied in different domains (e.g., Telecommunication, Business, Management, IT Governance, Services, etc.) see Fig. 1 (taken from [2]).

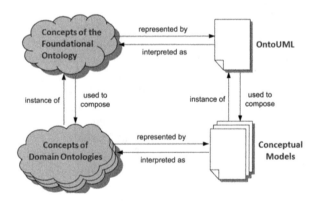

Fig. 1. Relation between concepts of domain ontologies and concepts of foundational ontology and OntoUML

4   OntoUML Perdurant Extension

The OntoUML metamodel uses ontological distinctions among the categories of object types (Kind, Subkind and Roles), trope types (Relator) and relations (formal and material relations) [1]. During our work in modelling Aviation Safety

Events, we recognize that, there is persistent need to extend OntoUML meta-model to use perdurant individuals and universals (Event) types. Events are individuals that may be composed of temporal parts. They happen in time in the sense that they may extend in time accumulating temporal parts [12]. The ontology proposed in [1] accounts for a descriptive commonsensical view of reality, focused on structural (as opposed to dynamic) aspects. Taking UML, as a general-purpose CML, we believe that extending UML class diagrams with perdurants is beneficial for the sake of properly visualizing structural relationships of perdurants. Sequence diagrams, on the other hand might be used for visualization of temporal dependencies, that are however not handled in our approach and will be considered for future work.

To model an event system, we have to describe event types and object types, that event depends on them (their participants) in order to exist see Fig. 2. For any event type, we have to specify the state changes of objects and the follow-up events caused by the occurrence of an event of that type [12]. Conceptual process models are based on the event types, by OntoUML Class Diagrams. Almost all of the basic type concepts that participant in process are supported, the three categories of snapshot objects, event types and situations are not supported. Consequently, we have to add them to the Onto-UML profile in the form of the class stereotypes snapshot objects, event type and situations.

Extending OntoUML editor tools and adding new stereotypes makes it possible to define event based models, in order to help modelers to evaluate and improve the models produced using OntoUML conceptual modeling languages. The Object Constraint Language (OCL) is an expression language that helps modelers to formulate constraints in the context of a given UML model. OCL is used to specify invariants attached to classes, pre and post conditions of operations, and guards for state transitions [20]. But currently, it is important to specify OCL constraints over the dynamic behavior of a UML model, i.e., consecutiveness of states and state transitions as well as time-bounded constraints. However, it is essential to specify such constraints to guarantee correct system behavior, e.g., for modeling real-time systems [20]. OCL temporal retraction helps modelers to specify required occurrences of actions, events and states. Temporal constraints are a means to declaratively describe properties of components and the relation between them in event or temporal systems. The first constraint below states that endurant universal cannot be subclass of a perdurant universal; The second and third constraints specify that events should be connected to their participants via an existential dependence relation (entailing an immutability constraint in the association end connected to the types representing each participant), in (Participant of) relation the source must be Object (2nd constraint) and the target should be Event (3rd constraint).

- allParents()-> select(x | x.oclIsKindOf(ObjectClass))-> isEmpty().
- participating().oclIsKindOf(ObjectClass).
- participated().oclIsKindOf(Event).

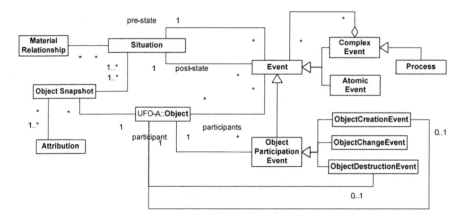

Fig. 2. UFO-B, an ontology of events

5   Use Case and Methodology

In this paper we study ontological anti-patterns identified during building ontologies to model and analyze the domain of aviation safety, and add new (Event) stereotypes and (Participant of) relation that connects events to their participants to OntoUML, in order to raise the quality of model and to add the possibility for modeling process and event in Aviation Safety reality. In these efforts, to increase the awareness of analytical methods and tools in aviation community for safety analysis in aviation, our strategy is to build and design ontological conceptual models in the domain of aviation safety, analyze safety events that lead to incidents or accidents, and explain factors, that contribute to these safety events, in order to transfer impressible safe to incredibly safe. Within the projects we developed the aviation safety ontology and several domain specific ontologies for describing safety issues in specific organizations of the aviation industry. Thus, in this paper we consider the following domains and corresponding conceptual models developed with the OntoUML methodology:

– A Conceptual Model representing the domain of aviation safety. It defines general well understood concepts in Aviation domain such as Aircraft, Flight, Agents and etc. (more detail in [23]). See Fig. 3.
– A Conceptual Model that describes Ramp Error Decision Aid (REDA) Contributing Factors ontology. It describes conditions that contribute to a ramp system failure which lead to other event [9].
– A Conceptual Model that describes DSA (one of our partners in the project) is an organization providing various aviation services, e.g. flight school, medical flights, rescue flights, maintenance [19].

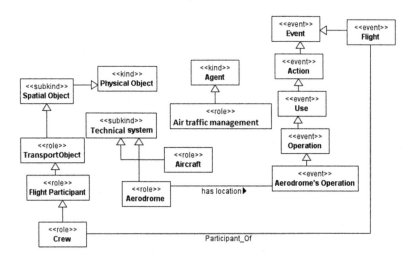

Fig. 3. Excerpt from the aviation safety ontology

6 Ontological Anti-patterns in UFO-A

A modeling language prevents the presentation of syntactically non valid state of affairs, but it cannot guarantee to have only desired instances [6]. According to [7], this is because the admissibility of domain specific state of affairs depends on domain specific rule not on the ontological one. To explain this situation, we studied the occurrence of these undesired consequences in our ontological conceptual models. This section presents some examples of ontological anti-pattern happened in UFO-A concepts (structural concepts), these types of anti-patterns are discussed in details in [5–8]. Figures 4 and 5 are taken from Aviation Safety ontology models, which represent the basic vocabularies, concepts and the relations between concepts in aviation safety domain (ASD). Figure 4 represents Relation Between Overlapping Subtypes (RBOS) anti-pattern that defined in [5,6], it happens in a model having two potentially overlapping (i.e., non-disjoint) types T1 and T2 whose principle of identity is provided by a common Kind ST, and such that T1 and T2 are related through a formal relation R. According to the domain conceptualization, a Person can be a Student and a person can be an Instructor that teaches the Student. However, the same person cannot play both roles.

Figure 5 represents a binary relation between overlapping types (BinOver) anti-pattern that explained in [6], which Occurs when an association of any given meta-type connects two types that constitute an overlapping set. If this is the case, it means that the same individual may eventually instantiate both ends of the relationship.

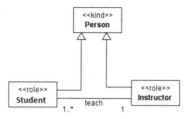

Fig. 4. Example of RBOS anti-pattern

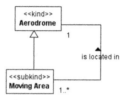

Fig. 5. Example of BinOver anti-pattern

7   Ontological Anti-patterns in UFO-B

Our research corresponds to the evaluation of models made during the development of aviation safety ontologies, and perform the evaluation of models in OntoUML editor tools (e.g., ontouml-lightweight-editor OLED) [10]. In order to make this evaluation, we need a mapping of the original models in UML to OntoUML class diagram syntax. Because most of our work depends on UFO-B, which is not supported by OntoUML profile, we are interested in mapping UFO-B concepts to OntoUML syntax and applying anti-patterns detection algorithms to event types [11]. In this case, we can test the event model, detect undesired instances then terminate them. Within the research we could add new Event stereotype into OLED to represent UFO-B based concepts and relations. This section presents some parts of ontologies based on UFO-B models, and discusses some types of semantic anti-patterns related to those models.

7.1   Aviation Safety Ontology for Maintenance Organizations

In national aviation authorities, there will be a regulatory requirement to do reactive events and factors investigation in all aircraft maintenance organizations around the world. Thus we built ontologies (management ontology) to represent and model factors that are considered a part of management system, to define what may cause or contribute to incidents in order to directly reduce or eliminate the contributing factors to error or damage events. As a part of this domain, we designed an ontological model of the Ramp Error Decision Aid (REDA) Contributing Factors, while REDA is designed to investigate events caused by worker performance and that occurred during the receiving, unloading, servicing, maintaining, uploading, and dispatching of commercial aircraft at an airport [9]. REDA Contributing Factor is used to describe conditions that contribute to a ramp system failure which lead to other event. For example, if "Incomplete or vague written communication" event happened it might become a factor of another event (e.g., ramp system failure). This Event model contains similar situation of Imprecise abstraction (ImpAbs) anti-pattern in UFO-A concepts, that defined by guizzardi in [5,6]. In perspective to [5] an association R characterizes the logical anti-pattern named Imprecise Abstraction (ImpAbs) if at least one of the following holds see Fig. 6:

- R's source end upper bound multiplicity is equal or greater than 2 and the Class connected to it has 2 or more subtypes.
- R's target end upper bound multiplicity is equal or greater than 2 and the Class connected to it has 2 or more subtypes.

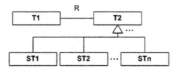

Fig. 6. Imprecise abstraction anti-pattern

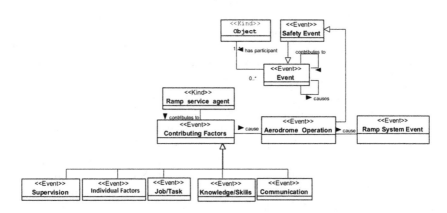

Fig. 7. Imprecise abstraction anti-pattern in REDA

Figure 7 shows REDA contributing factors (source) may cause some aerodrome operation events (target), this source has many subtypes, that may also cause aerodrome operation events, which in turn cause another event type called ramp system event.

7.2 Aviation Safety Ontology for High Risk Organizations

In recent years, an increasing attention has been paid to the development of the hazard taxonomy with the potential to be used throughout the aviation

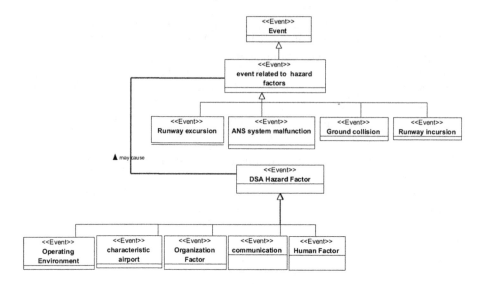

Fig. 8. Excerpt from DSA model

industry. This involves necessity for the approach to the problem, which does not exclude, in contrary it strongly supports finding a contributors or root causes that lead to the failure or accident realization [19]. As part of our work we designed models, which represent DSA ontology, see Fig. 8, in order to define hazard factors and events or problem related to these factors. We intend to detect semantic anti-patterns and avoid their occurrence in particular OntoUML event based models. However, OntoUML does not support event stereotypes that are important to evaluate and analyze our aviation safety events models. For example, during analyzing the results and instances, we faced some ambiguous states and undesired instances, and those unintended patterns are not defined in the catalogue of ontological anti-patterns in [5], that lead us to propose new type of semantic anti-pattern called (Imprecise Intersection). Imprecise Intersection anti-patterns occurs in a model having at least two Event types E1, E2 and every type has at least one joint subevent type SE, two relations R1, R2 connect this subevent type to different Event types (e.g., E5, E4), or to same type (e.g., E5) see Fig. 9.

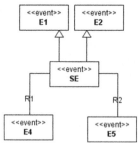

Fig. 9. Imprecise intersection anti-pattern

To exemplify this anti-pattern type, consider the following fragment of the DSA ontology, which contains Imprecise Intersection anti-pattern depicted in Fig. 10. The ontology searches for potential hazards from each of the defined categories, the most relevant categories are: characteristics airport, communication, human factor, operating environment, organization factor and Service factor, in order to reach correct comprehension of the analyzed event. Detected hazard defined in any category does not exclude hazards from the other categories, in other words it tries to find potential causes of event by deeper analysis and determination of the events chain. According to this model, psychological action-procedure violation 103010400 event, which considers in Hazard taxonomy both human and organization factor event type, it causes both (Push-back or taxi interference) event if it is related to organization factors and

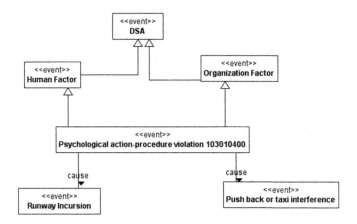

Fig. 10. Imprecise intersection anti-pattern in DSA

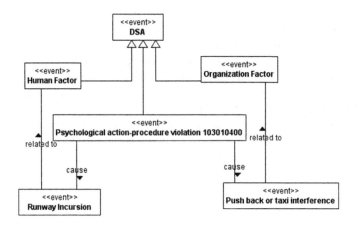

Fig. 11. Imprecise intersection anti-pattern solution

(Runway incursions) event if it is related to human factors, which may lead to undesired consequences and unintended instances because of this Imprecise Intersection situation. To solve this undesired consequences, we propose to remove this intersection and add new two associations.

Figure 11 proposes to delete this intersection between event types and define new two associations to define the intended consequences, as solution to this type of anti-pattern.

In this model when for example psychological action-procedure violation 103010400 event happens, either human or organization factors are responsible for its happen. Thus regarding to (Participant of) relation that connect event to its participants, psychological action-procedure violation 103010400 event has human and organization participants see Fig. 12.

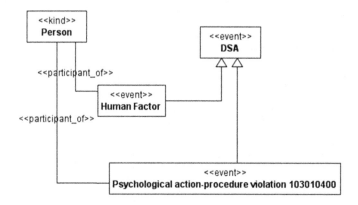

Fig. 12. Imprecise intersection anti-pattern solution

8   Final Consideration

In this paper, we discuss how ontological anti-patterns cause undesired consequences in ontological conceptual models, especially in Aviation Safety event models, in order to help designer to produce intended model, increase the accuracy of conceptual models and improve a conceptual model assessment tool based on Unified Modeling Language (UML) that assumes a well-founded conceptual modeling language named OntoUML. We present new ontological anti-patterns happened in UFO-B models called imprecise intersection anti-pattern.

We also add (Event) stereotype and (Participant of) relation that connects events to their participants to OLED to have the possibility for modeling UFO-B concepts based models and present some OCL restrictions regarding to perdurant concepts.

Acknowledgments. This work was supported by grant No. GA 16-09713S Efficient Exploration of Linked Data Cloud of the Grant Agency of the Czech Republic and by grant No. SGS16/229/OHK3/3T/13 Supporting ontological data quality in information systems of the Czech Technical University in Prague.

References

1. Guizzardi, G. Ontological Foundations for Structural Conceptual Models, Ph.D. thesis (CUM LAUDE), University of Twente, the Netherlands. Published as the book Ontological Foundations for Structural Conceptual Models, Telematica Instituut Fundamental Research Series No. 15. ISBN 90-75176-81-3 ISSN 1388-1795; No. 015; CTIT Ph.D. thesis, ISSN 1381-3617; No. 05–74
2. Carraretto, R., A modeling infrastructure for OntoUML, B.Sc. thesis, Federal University of Esprito Santo (2010)
3. Guizzardi, G., Wagner, G., Almeida, J.P.A., Guizzardi, R.S.S.: Towards ontological foundations for conceptual modeling: the unified foundational ontology (UFO) story. Appl. Ontol. **10**(3–4), 259–271 (2015)
4. http://eccairsportal.jrc.ec.europa.eu. Accessed 4 Oct 2016
5. Sales, T.P., Guizzardi, G.: Ontological anti-patterns: empirically uncovered error-prone structures in ontology-driven conceptual models. Data Knowl. Eng. **99**, 72–104 (2015)
6. Guizzardi, G.: Formal ontology, patterns, anti-patterns for next-generation conceptual modeling. In: KDIR 2015, p. 9 (2015)
7. Guizzardi, G.: Ontological patterns, anti-patterns and pattern languages for next-generation conceptual modeling. In: Yu, E., Dobbie, G., Jarke, M., Purao, S. (eds.) ER 2014. LNCS, vol. 8824, pp. 13–27. Springer, Heidelberg (2014)
8. Guizzardi, G., Sales, T.P.: Detection, simulation and elimination of semantic anti-patterns in ontology-driven conceptual models. In: Yu, E., Dobbie, G., Jarke, M., Purao, S. (eds.) ER 2014. LNCS, vol. 8824, pp. 363–376. Springer, Heidelberg (2014)
9. Ramp Error Decision Aid (REDA) Users Guide (2013)
10. Sales, T.P., Guizzardi, G., Almeida, J.P.A.: OntoUML lightweight editor: a model-based environment to build, evaluate and implement reference ontologies. In: EDOC Workshops, pp. 144–147 (2015)
11. Guizzardi, G., Wagner, G., de Almeida Falbo, R., Guizzardi, R.S.S., Almeida, J.P.A.: Towards ontological foundations for the conceptual modeling of events. In: Ng, W., Storey, V.C., Trujillo, J.C. (eds.) ER 2013. LNCS, vol. 8217, pp. 327–341. Springer, Heidelberg (2013)
12. Guizzardi, G., Wagner, G.: Towards an ontological foundation of discrete event simulation. In: Winter Simulation Conference 2010, pp. 652–664 (2010)
13. Guarino, N., Welty, C.A.: An overview of OntoClean. Laboratory for Applied Ontology (ISTC-CNR) Polo Tecnologico, Via Solteri 38, 38100 Trento, ITALY guarino@isib.cnr.it. IBM Watson Research Center 19 Skyline Dr., Hawthorne, NY 10532, USA
14. Weber, R., Zhang., Y.: An Ontological Evaluation of NIAMs Grammar for Conceptual Schema (1991)
15. Wand, Y., Storey, V.C., Weber, R.: An ontological analysis of the relationship construct in conceptual modeling. ACM Trans. Dtatabase Syst. **24**(4), 494–528 (1999)
16. Evermann, J., Wand, Y.: Towards ontologically based semantics for UML constructs. In: Kunii, H.S., Jajodia, S., Sølvberg, A. (eds.) ER 2001. LNCS, vol. 2224, p. 354. Springer, Heidelberg (2001)
17. Brea, P., Wand, Y.: Analyzing OWL using a philosophy-based ontology. In: Formal Ontology in Information Systems. IOS press, Amsterdam (2004)

18. Julio, C.N., de Almeida Falbo, R., Almeida, J.P.A., Guizzardi, G., Pires, L.F., van Sinderen, M., Guarino, N.: Towards a commitment-based reference ontology for services. In: EDOC 2013. pp. 175–184 (2013)
19. http://www.dsa.cz/. Accessed 4 Dec 2016
20. Flake, S., Mueller, W.: A UML Profile for Real-Time Constraints with the OCL
21. Herre, H., Heller, B., Burek, P., Hoehndorf, R., Loebe, F., Michalek, H.: General Formal Ontology (GFO) **8** (July 2006)
22. Gangemi, A.: Ontology design patterns for semantic web content. In: Gil, Y., Motta, E., Benjamins, V.R., Musen, M.A. (eds.) ISWC 2005. LNCS, vol. 3729, pp. 262–276. Springer, Heidelberg (2005)
23. http://www.inbas.cz/ontologie. Accessed 20 June 2016

User-Driven Ontology Population from Linked Data Sources

Panagiotis Mitzias, Marina Riga, Efstratios Kontopoulos[✉], Thanos G. Stavropoulos,
Stelios Andreadis, Georgios Meditskos, and Ioannis Kompatsiaris

Information Technologies Institute, Thessaloniki, Greece
{pmitzias,mriga,skontopo,athstavr,andreadisst,
gmeditsk,ikom}@iti.gr

Abstract. In order for ontology-based applications to be deployed in real-life scenarios, significant volumes of data are required to populate the underlying models. Populating ontologies manually is a time-consuming and error-prone task and, thus, research has shifted its attention to automatic ontology population methodologies. However, the majority of the proposed approaches and tools focus on analysing natural language text and often neglect other more appropriate sources of information, such as the already structured and semantically rich sets of Linked Data. The paper presents PROPheT, a novel ontology population tool for retrieving instances from Linked Data sources and subsequently inserting them into an OWL ontology. The tool, to the best of our knowledge, offers entirely novel ontology population functionality to a great extent and has already been positively received according to user evaluation.

Keywords: Ontologies · OWL · Ontology population · Linked data · DBpedia

1 Introduction

The rapidly increasing interest in building ontologies derives from their capability of representing knowledge in a structured and uniform way [1]. However, in order for semantically rich ontology-based applications to be deployed at an enterprise level, significant volumes of data are also required to populate the underlying models.

Populating ontologies with knowledge manually is a time-consuming and error-prone task. As a result, research has shifted attention to automating this process, introducing *ontology population*, which refers to a set of methodologies for automatically identifying and adding new instances of concepts from an external source into an ontology [2]. Ontology population does not affect the concept hierarchies and non-taxonomic relations in the ontology, leaving the structure of the ontology itself unmodified. What is affected in essence are the realisations of concepts (i.e. individuals) and the relations in the domain.

The majority of proposed tools for ontology population are aimed at NLP applications, which typically extract knowledge from natural language text and offer significant advantages over traditional export formats [3]. However, other sources of information

© Springer International Publishing Switzerland 2016
A.-C. Ngonga Ngomo and P. Křemen (Eds.): KESW 2016, CCIS 649, pp. 31–41, 2016.
DOI: 10.1007/978-3-319-45880-9_3

are very often neglected, like e.g. *Linked Data* [4], which are already more structured and semantically rich in comparison to free text.

Towards this direction, this paper presents *PROPheT* (*PERICLES*[1] *Ontology Population Tool*), a novel instance extraction engine for locating realisations of concepts (i.e. instances) in a Linked Data source, filtering them and subsequently inserting them into an OWL ontology. To the best of our knowledge and as described in the next section, no other tool currently exists that can offer the extent of functionality delivered by PROPheT.

The rest of the paper is as follows: Sect. 2 provides an overview of related work paradigms. Section 3 presents the proposed ontology population tool, focusing on its architecture, core components and ontology population capabilities, followed by a case study that better illustrates PROPheT's functionality. Section 5 features a user evaluation of the tool, followed by a brief discussion its limitations and directions for future work.

2   Related Work

Below is a brief account of ontology learning and population tools found in literature, mostly varying in the source of knowledge, functionality and degree of automation. For example, *DB2OWL* [5] automatically generates ontologies from relational database schemas. A mapping process detects particular cases for conceptual elements in the database and accordingly converts database components to the corresponding ontology components. The approach presented in [6] is aimed at a more efficient Linked Data consumption by proposing a learning process that can automatically construct a simple mid-level ontology, linking related ontology predicates in different data sets. Data collection is performed using SPARQL querying to the *Linking Open Data* (*LOD*) cloud. In [7], a workflow is proposed including the development of a domain ontology, the mapping between predicates in the latter and in DBpedia or other LOD sources and, finally, ontology population using SPARQL queries against DBpedia and other LOD sets. Other proposed tools conduct ontology population with knowledge derived from text documents [8, 9], spreadsheets [10] and XML files [11].

Most of the tools suggested in literature build an ontology from scratch, while our approach is aimed at enriching an existing ontology with relevant instances and property values, without any restriction to thematic domains. To the best of our knowledge, no other ontology population tool can instantiate new concepts from a LOD source so flexibly, regardless the domain of interest or the content of the source. PROPheT's flexibility lies in the fact that any kind of LOD with a served endpoint can be handled by the tool as an external source of knowledge for extracting concepts of interest and populating them to corresponding resources into the domain ontology.

Concerning the degree of automation, there is no implemented tool that carries out the whole population process automatically. PROPheT may be considered as a semi-automatic, user-driven system with different degrees of automation in various

---

[1] The tool has been developed in the context of the PERICLES FP7 project: http://www.pericles-project.eu/.

tasks: the interaction of the software with the LOD endpoint for performing the search process is done automatically, while the final selection of the instances to be populated in the ontology and the mapping of properties needs to be done manually. The user-driven mapping process enables the dynamic and proper definition of matching elements between source and target ontologies.

3 PROPheT Overview

PROPheT is a sophisticated GUI-equipped instance extraction engine for searching instantiations of concepts in a SPARQL-served LOD source, filtering them and subsequently populating them into a local model. It offers three types of instance extraction-related functionalities, along with user-driven mapping of datatype properties. It is flexible enough to work with any OWL domain ontology and any RDF LOD set that is available via a SPARQL endpoint. PROPheT, along with documentation, is freely available at: http://mklab.iti.gr/project/prophet-ontology-populator. A detailed description of PROPheT's architecture and functionality is given subsequently.

3.1 Architecture

PROPheT's overall architecture is illustrated in Fig. 1. The front-end was implemented in Python along with PyQt[2], while specialised Python APIs (RDFLib[3], SPARQL-Wrapper[4]) in the back-end are deployed for handling local and remote models (ontologies) and their content. Additionally, an SQLite database was set up in the back-end for storing dynamic data (e.g. settings, user preferences) that are created during the tool's operation.

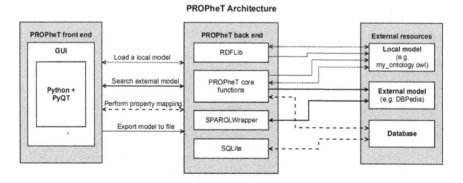

Fig. 1. PROPheT overall architecture.

---

[2] https://riverbankcomputing.com/software/pyqt.
[3] https://github.com/RDFLib/rdflib.
[4] https://github.com/RDFLib/sparqlwrapper.

The arrows in the figure display key PROPheT processes; varying arrow types indicate the corresponding back-end components and external resources employed for each process.

3.2   Core Components

Below is a list of PROPheT's core components (see also workflow in Fig. 2):

- *My Model* (*MM*) is the ontology model to be populated with new instances. The ontology must comply with a specific format (.owl, .rdf, .ttl) and can reside at a local or remote location.
- *External Model* (*EM*) is the source from which new instances will be retrieved. This source should be facilitated by a SPARQL endpoint[5], so that PROPheT will be able to query directly the knowledge base via SPARQL.
- *Extraction module* (search mechanism): The current version of PROPheT features class-based and instance-based population; for more details see next subsection.
- *Mapping module*: Allows the user to match MM and EM datatype properties. When populating new instances, the module instantiates MM datatype properties with values derived from "similar" user-defined EM datatype properties.
- *Storage module* (database) stores preferences relevant to PROPheT utilities and functions, like (i) the total number of derived instances from EM, (ii) MM and EM sources, (iii) user-defined mappings, (iv) known namespaces, (v) handling of general ontology properties.
- *Export module*: A mechanism for storing the already processed/populated MM in a local file, in some of the most popular ontology file formats (*.owl, .rdf, .ttl, .nt*).

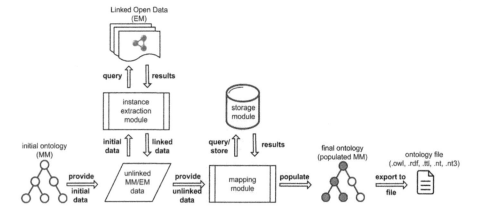

Fig. 2.   PROPheT main workflow.

5 List of SPARQL endpoints available at https://www.w3.org/wiki/SparqlEndpoints.

3.3   Ontology Population

PROPheT offers the following types of instance extraction-related functionalities:

Class-Based Populating. This method enables the user to populate MM with new instances "from-scratch", by searching for specific types of instances in EM via entering the exact type of class, e.g. *dbo:Artist*[6]. In order to be used efficiently, the user needs to know the structure of EM or where the class type belongs in the hierarchy of the EM ontology. PROPheT submits SPARQL queries to the EM endpoint and retrieves a list of instances that belong to the specified class. The user may then select the instance(s) that he/she wishes to import (populate) under an existing MM class.

In order for PROPheT to proceed with the defined instantiation(s) for MM, a user-driven ontology mapping is performed. A list of all unique datatype properties (*owl:DatatypeProperty*) for the selected instance(s) is given to the user in order to consistently define their mapping into existing datatype properties in MM. The mapping process is further described in Sect. 3.4.

Instance-Based Populating. Detecting and importing new instances via an instance-based search may be done in two different ways:

1. *Search by existing instance* - The user may select an instance already existing in MM and query the endpoint for similar instances. In particular, PROPheT performs an *rdfs:label*-based search and finds EM classes that include an instance with the desired label. The user may then select the classes of his/her interest, view their extension (i.e. set of instances) and choose which of them to import into MM.
2. *Search by instance label* - Similarly, a direct *rdfs:label*-based search is performed, with exact or partial match of the input text. However, in this case the user needs to type the desired label[7] and the search will result in a set of instances with this label, rather than similar instances of the same type (class).

In both instance-based mechanisms, the software presents a list of search results (instances) which may be selected by the user and entered into MM. As described above, a mapping process needs to take place, in order for the tool to also import the datatype properties of instances and their corresponding values.

Enrich Existing Instance. This type of functionality gives the user the ability to enrich an already existing instance within MM with properties and values derived from other instances in EM that have the same *rdfs:label* with the existing instance. Through this method, PROPheT performs an *rdfs:label*-based search for instances in EM that include the desired label. The derived instances may belong to one or more different classes in EM and the software tracks and presents the different *rdf:type* property declarations defined for these instances. Based on the content and semantics of the derived instances, the user may decide which property-value pair(s) he/she wishes to import into MM for the initially selected instance. Similarly to the aforementioned functions, an ontology

---

[6] *dbo* is the prefix for a specific DBpedia URI, that is http://dbpedia.org/ontology/.

[7] The search process offers options such as *Exact match*, *Contains term* and case sensitivity.

mapping process should be performed in order for the new properties and values to be added to the existing instance.

3.4 Mapping Classes and Properties

While populating a model, the user is required to manually define mappings between MM and EM classes and properties. In particular, the user needs to specify the MM class where instances extracted by any of the three above processes will be imported. Additionally, a mapping between EM and MM datatype properties is considered fundamental in order to insert datatype property values to the latter. For example, the user might define that the EM property *dbo:birthDate* corresponds to the MM property *:dateOf-Birth*. Once defined by the user, PROPheT memorizes such mappings and offers suggestions when the need for the same mappings reappears. The local model may also be semantically enriched by optionally affiliating EM and MM properties via relation *owl:equivalentProperty* and classes via *owl:sameAs* or *rdfs:seeAlso*. Additional associations, like e.g. *skos:narrower* and *skos:borader* from SKOS [12] are considered for the next version of the tool.

4   Case Study

This section illustrates PROPheT's functionality through specific examples that involve extracting information from two different LOD sources: DBpedia and LinkedMDB[8]. Consider a case where the user wants to populate an existing ontology containing information regarding artists (*ex:Artist*) and artworks (*ex:Artwork*). More specifically, let's suppose the user wishes to add the movie *"The Godfather"* into *ex:Artwork*. In order to use PROPheT, the user loads his/her model and registers LinkedMDB as the current EM source.

Since the name of the movie is specified a priori, he/she could search for instances through the *"Search by Instance Label"* method: when typing the movie title only one result is retrieved that corresponds to the entry of this movie in LinkedMDB[9]. Information retrieved for that entry is limited, due to the fact that PROPheT handles only datatype properties and their values, and not the attached object properties; on the other hand, LinkedMDB contains mostly object properties that connect instances of different types of classes.

Continuing the case study, suppose that the user wishes now to take advantage of information stored in other LOD sources, like for example in DBpedia. In this case, he/she has to modify the currently selected EM in PROPheT to DBpedia, and then to apply the *"Enrich existing instance"* method. PROPheT submits a SPARQL query to DBpedia and retrieves a set of instances that may belong to different classes, but they all share the same *rdfs:label* with the newly populated instance in MM. At that point, the user may select any pair(s) of datatype properties/values he/she wants to add to the MM

---

[8] http://www.linkedmdb.org/.

[9] The exact entry in LinkedMDB is http://data.linkedmdb.org/resource/film/43338.

instance of *"The Godfather"* movie. After manually mapping the relevant EM and MM properties, the data is inserted into the corresponding fields in the MM ontology. An indicative screenshot of the populated ontology in PROPheT thus far, can be seen in Fig. 3.

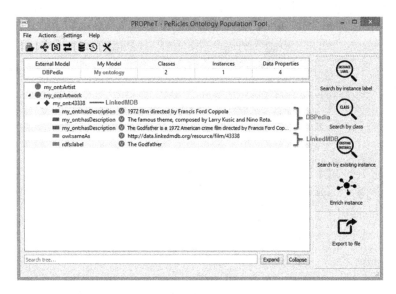

Fig. 3. PROPheT's main window with newly populated instance in class *ex:Artwork*; indicative values in properties are presented, retrieved from DBpedia and LinkedMDB.

In case the user wishes to search for and retrieve similar resources that belong to a specific class, then he/she can employ PROPheT's methods *"Search by Class"* or *"Search by Existing Instance"* for any selected endpoint. For instance, if the former method is selected, the user has to type the exact *prefix:class_name* that he/she is interested in, e.g. *dbo:Film* for DBpedia, or *movie:film* for LinkedMDB. A set of instances will be retrieved, and the user may then proceed with the selection and mapping process as described previously.

If, on the other hand, *"Search by Existing Instance"* is selected, PROPheT will take into account the predicate-object pair (*<rdfs:label 'The Godfather'>*) of the newly populated instance (*ex:43338*[10]) to search for alternative classes that contain instances with the same label. When applying this search in DBpedia, numerous classes are retrieved, specified either as internal classes via DBpedia URIs or as external classes via URIs originating from adopted ontologies (e.g. *YAGO*[11], *SKOS*[12], *Wikidata*[13], etc.).

An interesting aspect is the diversity of types of the retrieved instances, that range from more generic (e.g. *skos:Concept, dbo:Work*) to more specialized concepts (e.g. *yago:*

---

[10] Name of instance was retrieved from instance of *"The Godfather"* movie in LinkedMDB.
[11] www.mpi-inf.mpg.de/YAGO/.
[12] https://www.w3.org/2008/05/skos.
[13] https://meta.wikimedia.org/wiki/Wikidata/Development/RDF.

1970sCrimeFilms, *yago:1972Films*, *yago:FilmsBasedOnNovels*, *wikidata:Q11424*, etc.).
The user may select one or more classes from which instances will be retrieved and proceed
with the selection of instances to be populated (see Fig. 4) and with the mapping process.
Consequently, with this method we achieve integrating instances from *different* DBpedia
classes into *one single* MM class.

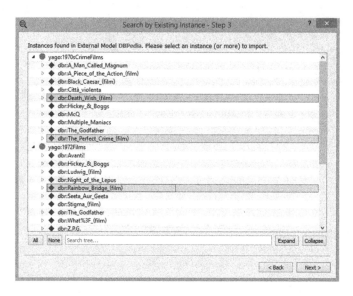

Fig. 4. Selection of instances from different classes (declared via *rdf:type*) where the newly
populated instance belongs to.

5   User Evaluation

The tool's user evaluation involved 15 participants, aged above twenty, with a computer
science background (either B.Sc., M.Sc. or Ph.D.), most of them (~80 %) familiar and
experienced with Semantic Web technologies (RDF, OWL). The trial was performed
remotely at the convenience of each participant, as they were provided with a link to
download the tool and access the online user guide. Initially, users had to fill in demo-
graphic details and perform a set of predefined tasks in the form, and answer based on
the outcomes.

As the tasks required full grasp of the tool's function, even when consulting the user
guide, the answers were used simply to validate a user's eligibility, which turned out to
be 100 %. Rating the system with respect to different aspects, in a scale from 1 to 7,
where 1–3, 4, 5–7 are considered a positive, neutral and negative answer respectively,
yielded positive results in all domains, shown on Fig. 5.

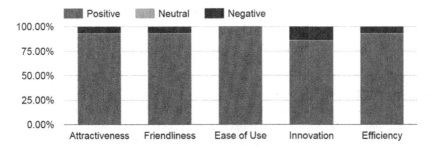

Fig. 5. User evaluation results.

Due to space restrictions and the currently small user sample size, a more extended evaluation, using a universal questionnaire, (e.g. SUS [13]), is planned as future work.

6 Limitations and Future Work

The type of populated information is significant for the completeness of an ontology population tool. Instances in ontologies may contain values for both *object properties* (*owl:ObjectProperty*) that interrelate individuals, and *datatype properties* (*owl:DatatypeProperty*) that relate individuals to literal values. In PROPheT's current version, population is limited to datatype properties only; relationships between instances are significantly more complex since each property is limited by domain and range declarations and the manual alignment between MM and EM properties would most probably lead to inconsistent results.

Furthermore, the current version of the tool does not handle direct or indirect imports of modular ontologies. Our aim in the future versions is to handle the MM ontology as an extended version of an ontology-graph, where declarations of triples of imported ontologies will be combined with those triples that are defined only in the domain ontology.

A further limitation is that PROPheT does not exploit the structure of the ontology that the user wants to populate: an alignment of the internal ontology with the external ones could save lots of effort and increase the usefulness of the system.

Last, the problem of instance redundancy (i.e. two or more instances in the ontology refer to the same real object) is handled by PROPheT in a way that instances with the same name-identifier cannot be populated multiple times in the ontology, i.e. values of populated data properties are linked to one instance. As future work, PROPheT may encompass more complex handling mechanisms, such as heuristics or machine learning methods to identify similar resources.

7 Conclusions

The process of ontology population is a non-trivial but also laborious task, while the relevant proposed approaches are typically focused on analysing natural language text, often overlooking other sources of more structured information, like e.g. Linked Data. This paper presents PROPheT, a novel user-driven ontology population tool for semi-automatically retrieving instances from a Linked Data source and inserting them into an OWL ontology. Through the use of embedded wizards, the user may apply advanced class-based and instance-based queries to the LOD endpoint, without any precondition in knowing technical details of the applied queries or of the SPARQL query language's syntax. The embedded mapping process enables the dynamic and proper definition of matching elements (i.e. classes and properties) between source and target (populated) ontologies. The extent of functionality and flexibility offered by the tool cannot be matched by any other software currently found in literature, making PROPheT a truly novel system for populating ontologies.

Acknowledgments. This project has received funding from the European Union's Seventh Framework Programme for research, technological development and demonstration under grant agreement no. 601138.

References

1. Stephan, G.S., Pascal, H.S., Andreas, A.S.: Knowledge representation and ontologies. In: Studer, R., Grimm, S., Abecker, A. (eds.) Semantic Web Services: Concepts, Technologies, and Applications, pp. 51–105. Springer, Heidelberg (2007)
2. Buitelaar, P., Cimiano, P.: Ontology Learning and Population: Bridging the Gap Between Text and Knowledge, vol. 167. Ios Press, Amsterdam (2008)
3. Petasis, G., Karkaletsis, V., Paliouras, G., Krithara, A., Zavitsanos, E.: Ontology population and enrichment: state of the art. In: Paliouras, G., Spyropoulos, C.D., Tsatsaronis, G. (eds.) Multimedia Information Extraction. LNCS, vol. 6050, pp. 134–166. Springer, Heidelberg (2011)
4. Bizer, C., Heath, T., Idehen, K., Berners-Lee, T.: Linked data on the web (LDOW2008). In: Proceedings of 17th International Conference on World Wide Web, pp. 1265–1266. ACM, April 2008
5. Ghawi, R., Cullot, N.: Database-to-ontology mapping generation for semantic interoperability. In: VLDB 2007 Conference, VLDB Endowment, Vienna, Austria, pp. 1–8. ACM (2007)
6. Zhao, L., Ichise, R.: Mid-ontology learning from linked data. In: Pan, J.Z., Chen, H., Kim, H.-G., Li, J., Wu, Z., Horrocks, I., Mizoguchi, R., Wu, Z. (eds.) JIST 2011. LNCS, vol. 7185, pp. 112–127. Springer, Heidelberg (2012)
7. Gavankar, C., Kulkarni, A., Fang Li, Y., Ramakrishnan, G.: Enriching an academic knowledge base using linked open data. In: Proceedings of Workshop on Speech and Language Processing Tools in Education in 24th International Conference on Computational Linguistics, pp. 51–60 (2012)
8. Maynard, D., Funk, A., Peters, W.: SPRAT: a tool for automatic semantic pattern-based ontology population. In: International Conference for Digital Libraries and the Semantic Web, Trento, Italy (2009)
9. Velardi, P., Navigli, R., Missikoff, M.: Integrated approach for web ontology learning and engineering. IEEE Comput. **35**(11), 60–63 (2002)

10. Han, L., Finin, T.W., Parr, C.S., Sachs, J., Joshi, A.: RDF123: from spreadsheets to RDF. In: Sheth, A.P., Staab, S., Dean, M., Paolucci, M., Maynard, D., Finin, T., Thirunarayan, K. (eds.) ISWC 2008. LNCS, vol. 5318, pp. 451–466. Springer, Heidelberg (2008)
11. Modica, G.A., Gal, A., Jamil, H.M.: The use of machine-generated ontologies in dynamic information seeking. In: Batini, C., Giunchiglia, F., Giorgini, P., Mecella, M. (eds.) CoopIS 2001. LNCS, vol. 2172, pp. 433–447. Springer, Heidelberg (2001)
12. Miles, A., Bechhofer, S.: SKOS simple knowledge organization system reference. In: W3C recommendation, 18, W3C (2009)
13. Brooke, J.: SUS-a quick and dirty usability scale. Usability Eval. Indus. **189**(194), 4–7 (1996)

Ontology for Performance Control in Service-Oriented System with Composite Services

Maksim Khegai[✉], Dmitrii Zubok, Tatiana Kharchenko, and Alexandr Maiatin

ITMO University, Saint Petersburg, Russia
MaxHegai@rambler.ru,
{zubok,kharchenko}@mail.ifmo.ru, mavr.mkk@gmail.com
http://www.ifmo.ru/

Abstract. Providing high performance to large systems based on service-oriented architecture is a difficult issue. Such systems are composed of a big number of interacting composite services, each consisting of one or several applications. To process jobs that income to such a system collaboration between several applications is needed and processing time will be influenced by choice of a set of applications, resources that they have and time consumed to exchange data between them. For effective hardware resources utilization virtualization technologies are used. Applications that implement services functionality are placed in virtual machines, deployed in a number of physical servers. One of main advantages of a service-oriented architecture is scalability that leads to frequent changes in applications set, their placement in virtual machines and resources available to them. To provide high performance jobs queuing is needed to choose optimal set and order of applications for processing. Efficiency of jobs queuing algorithms highly depends on up-to-date information about every object in a system: applications, virtual machines, physical servers and telecommunications. That, because of inconsistency in configuration may become difficult. One of the proven methods of choosing a set of interacting services to process a complex job is use of ontologies. In this paper an extension to this method is proposed to increase performance of a system. Ontology that describes not only functional abilities of services but also information about their current performance and communicative abilities is described.

Keywords: Ontology · Performance optimization · Service-oriented architecture · Queuing

1 Introduction

Service-oriented architecture for servers becomes more and more popular. Their modular approach, when servers components are distributed and loosely coupled, provides wide possibilities to systems scalability and control. At the same time

© Springer International Publishing Switzerland 2016
A.-C. Ngonga Ngomo and P. Křemen (Eds.): KESW 2016, CCIS 649, pp. 42–55, 2016.
DOI: 10.1007/978-3-319-45880-9_4

services implementations are encapsulated, hidden from any other component. One of the main issues to be solved when using solutions with service-oriented architecture is providing scalability. Ability to add system servers, services or available functionality is necessary to reduce down time during maintenance, thus reducing costs. Scalability also allows to increase computing power of a system without needing to drastically change configuration. To achieve it, hyper convergent systems are often used. Those are module systems that represent a set of servers, data storage systems, network devices and control components. When using such systems there is no need to check compatibility of different elements with each other.

Services, added to a system, may interact with each other. In this case to process a job that came to one service, other services may be needed. It should be noted that a system may have a lot of services that have abilities to process a job. Choosing a proper path, that is a set of applications to process a composite job, especially in case of frequent change of available services, is a difficult process, although it is very well researched. However most of proposed solutions study choice of applications set based on their functional abilities but performance of a service-oriented system is not considered. On the other hand, a system that has a big amount of loosely coupled services is a difficult case from a performance control point of view, and choice of a set to process each composite job may heavily influence overall performance of a system. There are a lot of approaches to services description, involved in such systems, for example WSDL, but as of late ontologies are widely used. WSDL based approach, as of now, is the most popular approach and is recommended by W3C organization since 2007. WSDL is an XML based language for web-services interfaces and their functionality description in a convenient machine-readable way. WSDL file includes a description of how a service may be called, which parameters it includes and which data structures it returns.

Ontologies were proposed as an abstraction of descriptive languages such as WSDL. Their usage to services description is studied in [4]. In this work ontologies are used as a WSDL substitute, since the latest is not sufficient and demands each developer to fully understand inner structure of a server. To solve this ontologies and semantic web were used. Presented in [3,5,6] ontologies do not describe parameters such as performance or service placement in a system. This information is crucial for choosing an optimal path.

Ontologies are actively used for computing systems infrastructure description. For example in [2] a testing ontology, that includes all basic components of computing systems infrastructure, is presented, however it doesn't allow to provide information about current performance of components. In case of service-oriented architecture used for cloud infrastructure, it changes constantly and doesn't allow reflecting changes in program components or changes in connections between them, which is typical for cloud-based systems with dynamic control of resources allocation.

A research of ontologies with fuzzy link between objects for choosing services in clouds exists. It solves a multi-criteria task, involving weights of different service parameters, including performance parameters [7]. However, in this case an issue of defining a single service is overviewed and interacting services, that are used to build a sequence of services when a single element of a sequence can heavily influence performance, is not studied.

Ontologies for composite services description also find their reflection in newer studies. The study [1] is a good example. It proposes a use of ontology that allows to build a sequence of services for issue solving. But this study doesn't research problems of service-oriented systems performance, concentrating on services choice according to business requirements. However, independent satisfaction of business requirements of separate services cannot guarantee optimal performance of a system. Another important aspect that is not studied in those works is a problem of increasing performance of systems that a build upon interacting services. In this case one of factors, influencing a choice of applications set to process a composite job, is a minimization of average processing time of a job.

However, while solving this issue other difficulties occur. This is due to flaws in service-oriented architecture:

1. Set of applications and computing servers is not constant and changes with time
2. When services are composed of a few applications, a number of factors that influence performance increases
3. Average processing time is not always predictable
4. Systems overload moment is not always predictable

Nowadays service-oriented solutions are placed in cloud infrastructure. This, in particular, allows to provide more efficient distribution of hardware resources when number of services and their load frequently changes.

Section 2 of this paper analyzes a typical service-oriented architecture from a performance control point of view. Section 3 describes an ontology for services description with all required to performance control parameters and proposes its use in service-oriented platforms. Section 4 presents a use case of a proposed approach.

2   Typical Model of a System with Service-Oriented Architecture and Composite Services

A system with service-oriented architecture, from a software point of view, is a set of applications. Each service may be implemented by one or several interacting applications. Applications may be implemented with use of different technologies, include many existing components (libraries, DBCS, frameworks, etc.) and demand different environment for their functionality (operating systems, interpretators, developer environments etc.). An important issue in this case is effective utilization of hardware resources. Different requirements to environment requires services to be deployed on separate servers and unpredictable (and sometimes insignificant) need in resources in different services leads to inefficient usage of hardware resources. A solution has been found in virtual servers consolidation. In this case a system has virtual machines, implementing different environments, with deployed service or services.

A system has a few physical servers with a few virtual machines in them. In those machines services that process jobs are placed. A service may include one single application or several interacting applications. In the second case incoming job may be sent to be processed by other services. It should be noted that with this architecture a lot of situations, when processing leads to a competition for resources between services, occur. In one virtual machine services compete for its memory and total CPU time that was allocated to this machine. Services in different virtual machines compete for CPU time of host and its disk subsystems. And, finally, services that were deployed in different hosts, compete for general telecommunication resources. In this case a solution to the issue of choosing optimal services combination to process a job is non-trivial. A typical model is represented in Fig. 1.

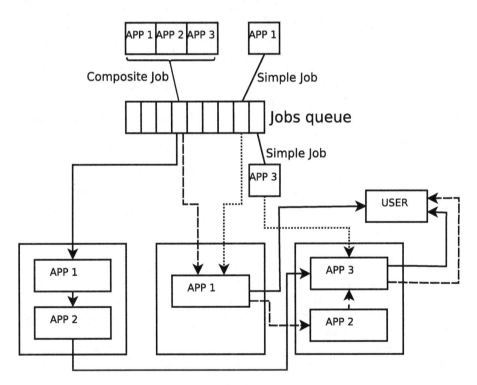

Fig. 1. Generic model of a system

A set of jobs incomes to a system. Since there are several computational servers that have a number of applications that are able to process a job, a problem of finding an optimal server occurs. If there is a job that requires a few applications for its processing, the system must decide a required sequence.

Required applications may be deployed in different servers and there may be several similar applications, so there will be several different paths that this job may follow. Another difficulty is characterized by a fact that applications may process several jobs in parallel or one single job when it is possible (for example when a job needs data from a data base and processing of data that is not stored in the base).

A choice of optimal path is also influenced by current performance of servers that include those applications. This performance value is not constant and depends on a number of parameters. First, on a number of jobs that are being processed at the moment. Every new job decreases performance because of overhead costs when processing streams are being switched in terms of preemptive multitasking. The performance value may have a threshold. For example, if a number of simultaneously processing jobs exceeds threshold value, a RAM overflow may occur, which will lead to swapping and will drastically decrease performance. Second, virtual machines with applications are deployed on the same physical server and compete for its resources, such as CPU time and access to I/O ports. In result, to evaluate probable performance of a single application we need to consider not only current load on applications deployed in the same virtual machine, but also load on other virtual machines, deployed in the same physical server. Third, different physical servers use same telecommunication channels and hardware. It leads to necessity of current load on telecommunication channels and hardware, caused by jobs distribution.

Thus, performance of the system constantly changes with time. Main factors that influence it are:

1. Number of simultaneously processed jobs in the system and in a separate application.
2. Placement of applications in virtual machines and virtual machines in physical servers considering telecommunication resources for data exchanging between systems objects.

Solution to performance optimization for the described system requires solution of a number of issues

1. Monitoring of performance and available resources of each server, virtual machine and application without decrease in overall performance.
2. Quick search of a path for job processing that includes a set of instances of applications that implement required services functionality.
3. Storing knowledge about systems objects and efficiency of decisions in path choosing for different jobs type, that were made before.

The first issue was solved and the solution was presented in [9]. There to keep systems performance when receiving up-to-date data about services performance, a set of intellectual agents was proposed. Those agents monitor performance and send the data to a main controlling agent. The second and the third issue a use of ontology-based knowledge base is proposed. This paper will not study path finding algorithms but instead concentrate on ways to provide controller with up-to-date information required to make those decisions.

Required for decision making information is characterized by few features. First, information sources (systems objects) set is constantly changing. Second, parameters set, that describe each object, is non-constant. Third, Information gathering channels are non-constant. To provide means of gathering and storing information about system an ontology-based knowledge base may be used. This will allow for efficient information accumulation in conditions of constantly changing sources and data structures. The knowledge base will be an informational core of the system. Creation, deletion or change of state of each object will be reflected in the knowledge base, which will be a source of information for performance control algorithms.

3   Ontology Approach to Services Description

To achieve scalability and keep controllability in situation of constantly changing composition and characteristics of objects, the system needs a description of components. Each service, application or server must be connected to each other and components themselves must be able to be easily connected to the model. Ontologies provide means to describe such components. Use of ontologies is justified by information not having well-defined structure. During its activity, new components with uncertain attributes will inevitably appear, which have an impact on interaction between elements and will define ability to control them.

A system with the described architecture may be counted as context-dependant. Use of ontologies for creating scalable context-dependant models is well researched in [8]. To build a basic ontology that will become a base for the knowledge base, we need to determine main concepts and domain specific ontologies representing knowledge of different application domains.

To determine main concepts we need to describe connections between main objects of the system. The system is a cloud with several consolidated servers. These servers have a set of services, simple and composite, that process incoming jobs. Composite services represent a sequence of separate simple services that have one single application; incoming job will be processed in series by several different applications. A number of services, servers and applications is changing with time.

There is always a single main server in the system and a lot of computational servers. Main server includes controller, that distributes jobs, knowledge base, that stores ontology and its fragments, application templates, that need to be deployed in services and a sub-system that sends templates to a computational server with services.

Every computational server contains a controller, an application copies receiving subsystem and services with interpretators to process jobs. Since different services may contain different interpretators, when receiving a copy of application, receiving subsystem must detect a service with compatible interpretators. One service may include several applications, Thus, the first ontology concept is physical server. Its ontology fragment must characterize hardware resources that were allocated to it.

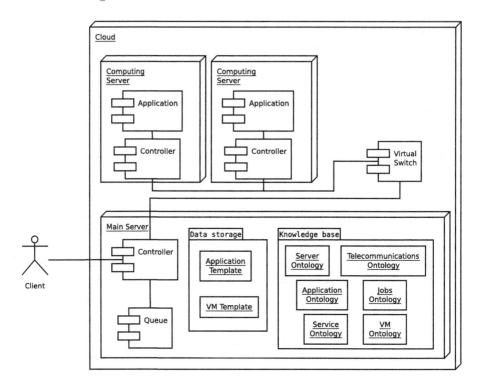

Fig. 2. System architecture

The architecture is presented in Fig. 2.

Applications that implement each service are placed in *virtual machines*. When needed, a creation of a new *virtual machine* is performed through the main server's controller. *Virtual machine* is, in general, a set of limited resources, allocated by server it is placed in. But with that a *virtual machine* has one distributed operating system and a set of system software that allows deploying a limited set of applications. This demands separating virtual machines in a separate concept. One of properties of the concept is a placement in a physical server. This provides interaction between those concepts. It should be noted that there may be many ways of organizing communication between *servers* and *virtual machines*. However all of them have a huge impact on the overall performance of the system, leading to a necessity of separating means of communication in a concept of thir own. their "Telecommunication Node".

The next pair of objects, that we need to look at, are *service* and *application*. Despite obvious similarities, they are two separate entities from the server point of view. *Service* is a functional object, providing a certain system function implementation. From the information point of view a *service* is characterized by its functional abilities and interaction interfaces. *Application* is a software entity that implements server's functions. Its characteristics are requirements to a platform it may be deployed in and characteristics that influence performance.

For example an ability for multi-streaming, working with data bases etc. *Service* and *application* are connected to each other: *application* ensures that a *service* works properly. An entity of *application* is deployed in a *virtual machine* and this property allows to connect those two concepts.

Finally, another system object that we need to separate in its own concept is a *job*. The main controller has an inner queue where all incoming jobs are placed before being distributed to appropriate virtual machines. The controller must correctly determine job's type: define with which services and in which order it should be processed, build a sequence of applications and then send it to be processed. When incoming to an application a *job* is added to processing queue of the *application*. Depending on applications features and operating systems that operates it, these queues may have different service disciplines, for example round-robin. It should be noted that these queues have threshold value that denies addition of new jobs to the queue if this value was exceeded. This happens due to lack of required resources in virtual machine. If it was found while processing a job, a sequence must be rebuilt.

Thus, the knowledge base is formed by four connected concepts:

1. Physical server
2. Virtual machine
3. Telecommunication node
4. Service
5. Applications
6. Job

Their ontology fragments form one common ontology of the system that is scalable by adding new variations of existing concepts. Figure 3 shows the ontology for the system.

The ontology and its fragments are stored in a knowledge base as RDFS triplets using EasyRDF library.

The next section will describe those concepts from the point of their parameters and their additional entities and their parameters.

3.1 Physical Server

Main entities of the concept are: "Performance", "Server Resource Quotas" and "VM".

"Performance" is characterized by parameters "Response time", "Availability", "Jobs number" and "Throughput". The existence of the entity is explained by necessity of choosing the most productive server to send jobs to.

Entity "Server Resource Quotas" describes resources allocated to a server. Parameters "HDD", "RAM" and "CPU" are responsible for available free space, RAM amount and CPU time, respectively.

"VM" describes virtual machines that were deployed on the server, and includes ontology fragments of this type.

"Server ID" is a parameter that describes an identification number of a server.

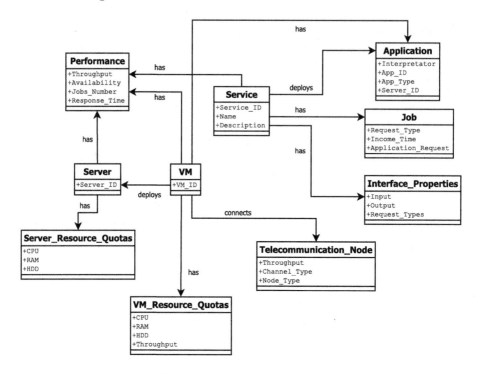

Fig. 3. Ontology of the system

3.2 Virtual Machine

Virtual machines concept work with entities "Application", "Telecommunication Node" and "VM Resource Quotas".

Entity "Application" includes ontology fragments of every application that was deployed in the virtual machine.

Entity "VM Resource Quotas" describes quotas for resources allocated to the VM. Parameters "HDD", "RAM", "Throughput" and "CPU" are responsible for available free space, RAM amount, network activity and CPU time, respectively.

"VM ID" is an identification number of a virtual machine.

3.3 Service

Service concept works with entities "Application", "Job" and "Interface properties".

"Job" describes jobs type that the server can process. "Application" includes ontology fragments of this type, that describe applications able to process a job. Types of jobs and applications are the same, since service can receive jobs only of a type that an application, deployed inside it, can process.

"Performance" is not involved in the work of service directly but is an important part of optimal path finding algorithms. Parameters of the entity are: "Throughput", "Availability", "Jobs number", "Response time".

"Interface properties" entity describes which input and output interfaces does a service use and also, request types a service can process.

Main entities of the concept are: "Job", "Application", "Performance", "Interface properties".

3.4 Telecommunication Node

The concept describes a telecommunication node and, since it is related to virtual machines, typically represents a physical or virtual switch. There are three parameters "Throughput", "Channel type" and "Node Type".

The concept is connected to concepts "VM" and "Server" so they could send and receive messages from each other.

3.5 Application

Application concept describes applications that process jobs and is characterized by four parameters: application type, intepretator, application ID and server ID. The last parameter describes which server an application is placed on, thus connecting "Service" and "Server" concepts.

3.6 Job

Job concept describes jobs that are incoming to a service to be processed. It is characterized by three parameters: job type, applications sequence that must process the job and time when a job was received.

4 Test Case

A practical use of this approach may be demonstrated on the following scenario. Consider a system with a few *physical servers*. In this system an OpenStack platform is deployed. Its advantage for the proposed approach is in already available RabbitMQ framework that will interact with each systems component. This allows implementing information exchanging mechanism when servers are subscribed. Components that are created during the work: services, applications, virtual machines, are subscribed to changes in knowledge base, sending their fragment of the ontology that describes their parameters. Before being deleted, components unsubscribe from changes and information about them is deleted. The system contains storage with applications templates, providing a way to deploy applications copies in *virtual machines* that satisfy requirements (supported operating system, required memory amount, free hard drive space, interpretators, DBCS, etc.). The system also has virtual machine templates storage with pre installed operating systems and other components, necessary for applications functionality. It provides flexibility in control of number and placement of applications in the system. There is also a main server, deployed in one of virtual machines. It contains a knowledge base with information about systems components, their current performance and components of decision making algorithms. Those algorithms control performance and manage incoming jobs queue.

Performance control strategy includes several ways of controlling the system:

1. Queuing
2. Virtual machines migration
3. Management of applications number in separate services

Appearance of a new *physical server* in the system leads to appearance of a new ontology fragment that describes its parameters. The same happens when a new *virtual machine* appears, and also a template is created and added to the templates storage. When a new *service* appears, another ontology fragment is added to the knowledge base, containing its properties, its applications properties. Deployment of a new *application* or *virtual machine*, their destruction, and migration of *application* to different *virtual machine* or *virtual machine* to another *physical server* leads to a change in appropriate ontology fragments. Thus, the knowledge base always contains information about not only deployed applications and virtual machines, but also systems potential abilities to deploy new applications or virtual machines.

A new *job* that a system received is, first, analyzed. A sequence of different *services* that may process the job is built, and their parallel or serial way of usage is determined. This is due to dependencies between different parts of a *job* that is being processed in different *services*. The next step is to determine if there are services that can process the *job* and each *server* is listed. There may be a situation when there are no *services* with such applications but templates base contains templates for them and the knowledge base has information about it. In this case a decision is made if an application must be deployed and in which way. This influences performance of the system. Depending on information in the knowledge base, about current load, parameters and current performance of available virtual machines, a decision to deploy a copy of the application or to create a new virtual machine may be made. The new virtual machine then will contain a copy of application. The sequence diagram of this case is presented in Fig. 4.

If all *applications* are deployed in every required *service*, the algorithm goes into next stage optimal path finding. The easiest solution may be found in immediate sending of knowledge from composite job to applications according to a number of free services. If all or a part of required applications is overloaded and there is a reason to suspect a decrease in performance after sending a *job*, another control algorithm from the listed ones is used.

Queuing allows solving short-term scheduling. To implement this type of control, we need information about the current number of jobs that are being processed at the moment. A forecasting algorithm may decide to delay a job in the queue and send it to a certain application after some time, when load on it will decrease. The algorithm was studied in [10]. Size of queue and time limit for the job to stay in queue are calculated at this stage.

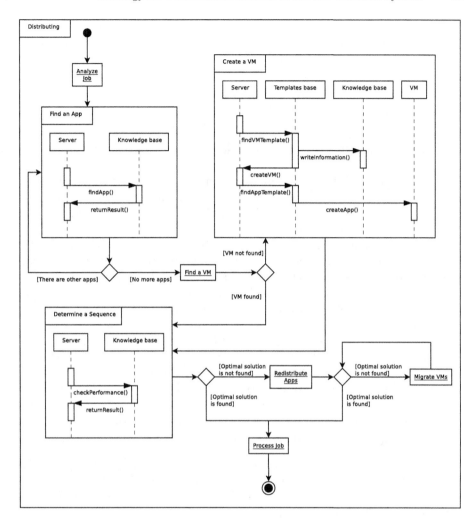

Fig. 4. Interaction overview diagram of test case

Virtual machines migration is used when, according to information in the knowledge base, there are free resources in a physical server. And a jobs stream intensity is forecasted. In this case a virtual machine with application that is required, is migrated to another physical server and current overload to free resources ratio allows it to receive a new job to be processed, without decreasing performance level.

Applications number control is effective when change of incoming jobs stream leads to uneven load on virtual machines. A decision to deploy application copies in existing virtual machines that support them and have free resources or to create such virtual machines is made. In the last case a placement in free physical servers is also controlled.

The ontology, described in the previous section, provides means to receive a required up-to-date information for decision making in scope of listed control types. In background algorithms for separate systems components demand evaluating are working. Their job is to destroy applications or virtual machines that are not needed at the moment, to free resources. This allows to quickly deploy required applications and virtual machines.

5   Conclusion and Future Work

The presented study demonstrates an approach to information providing for performance control in service-oriented system with composite services. The main goal of presented ontology and scenario is to show a possibility and way to build a knowledge base that ensures that scalability is provided. It is also ensures that information is up-to-date in case of complex hierarchical architecture with changing composition of its components. Main concepts of the ontology represent system control objects for cloud-based infrastructure with service-oriented architecture: physical servers, virtual machines, services and applications. The solution assumes that all objects are able to scale the ontology by fragments of information, characterizing their current state. Use of such knowledge base allows to implement difficult control strategies, applied to different system objects in real time. Modern cloud computing platforms have tools to provide data exchange between their components. They allow use of the proposed approach for gathering and storing up-to-date data in systems with different scale and purpose.

Development of the study will be in improving decision making algorithms that use data accumulated in the knowledge base, and improving control strategies of service-oriented systems with composite services for machine learning.

Acknowledgements. This work was partially financially supported by the Government of Russian Federation, Grant 074-U01. The presented result is also a part of the research carried out within the project funded by grant #15-07-09229 A of the Russian Foundation for Basic Research.

References

1. vom Brocke, J., Braccini, A.M., Sonnenberg, C., Spagnoletti, P.: Living IT infrastructures an ontology-based approach to aligning IT infrastructure capacity and business needs. Int. J. Acc. Inf. Syst. **15**, 274 (2014)
2. Ghijsen, M., van der Ham, J., Grosso, P., Dumitru, C., Zhu, H., Zhao, Z., De Laat, C.: A semantic-web approach for modeling computing infrastructures. Comput. Electr. Eng. **39**, 2553–2565 (2013)
3. Li, W., Zhong, Y., Wang, X., Cao, Y.: Resource virtualization and service selection in cloud logistics. J. Netw. Comput. Appl. **36**, 1696–1704 (2013)
4. Liu, M., Shen, W., Haob, Q., Yana, J.: An weighted ontology-based semantic similarity algorithm for web service. Expert Syst. Appl. **36**, 12480–12490 (2009)
5. Medjahed, B., Bouguettaya, A., Elmagarmid, A.K.: Composing web services on the semantic web. Int. J. Very Large Data Bases **12**, 333–351 (2003)

6. Nguyen, T., Loke, S.W., Torabi, T., Lu, H.: On the practicalities of place-based virtual communities: ontology-based querying, application architecture, and performance. Expert Syst. Appl. **41**, 2859–2873 (2014)

7. Sun, L., Ma, J., Zhang, Y., Dong, H., Hussain, F.K.: Cloud-FuSeR: fuzzy ontology and MCDM based cloud service selection. Future Gener. Comput. Syst. **57**, 42–55 (2016)

8. Teslya, N., Smirnov, A., Levashova, T., Shilov, N.: Ontology for resource self-organisation in cyber-physical-social systems. In: Klinov, P., Mouromtsev, D. (eds.) KESW 2014. CCIS, vol. 468, pp. 184–195. Springer, Heidelberg (2014)

9. Zubok, D.A., Maiatin, A.V., Khegai, M.V.: Ontology-based approach in the scheduling of jobs processed by applications running in virtual environments. In: Klinov, P., Mouromtsev, D. (eds.) Knowledge Engineering and the Semantic Web, vol. 518, pp. 273–282. Springer, Heidelberg (2015)

10. Zubok, D.A., Maiatin, A.V., Kiryushkina, V.E., Khegai, M.V.: Functional model of a software system with random time horizon, pp. 259–266 (2015)

Privacy in Online Social Networks:
An Ontological Model for Self-Presentation

Javed Ahmed[1,2(✉)]

[1] CIRSFID, University of Bologna, Bologna, Italy
[2] CSC, University of Luxembourg, Luxembourg, Luxembourg
shahanijaved@gmail.com

Abstract. Online Social Networks (OSNs) have become an important part of daily digital interactions for more than half billion users around the world. Unconstrained by physical spaces, the OSNs offer to web users new interesting means to communicate, interact, and socialize. The OSNs exhibit many of the characteristics of human societies in terms of forming relationships and how those relationships are used for personal information disclosure. However, current OSNs lack an effective mechanism to represent social relationships of the users that leads to undesirable consequences of leakage of users' personal information to unintended audiences. We propose an ontological model to represent diverse social relationships and manage self-presentation of social web users. This model is inspired from most influential social theories about self-presentation and tie strength. This model regulates personal information disclosure on the basis of social role and relationship quality between the users. We also present results of our user study, which demonstrates that relationship quality plays vital role to control personal information disclosure in social web, and quality of relationship between users can be easily inferred from user interaction patterns in online social networks.

Keywords: Online social networks · Privacy · Ontology · Audience segregation · Tie strength

1 Introduction

Internet has become an inevitable part of lives of people today. Nearly half of the users who have access to Internet are members of some online social networking site[1]. Online social networks (Facebook and Google+) are top most visited sites on Internet,[2] and fourth most popular activity of Internet users.[3] Online social networks are one of the most popular fora for self-presentation and user interactions. These social networking sites promote the vision of human centric web and

---

[1] PewResearchCenter http://www.pewglobal.org/2010/12/15/global-publics-embrace-social-networking/.

[2] Alexa http://www.alexa.com/topsites.

[3] Nielsen http://blog.nielsen.com/nielsenwire/wp-content/uploads/2009/03/nielsen-globalfaces-mar09.pdf.

© Springer International Publishing Switzerland 2016
A.-C. Ngonga Ngomo and P. Křemen (Eds.): KESW 2016, CCIS 649, pp. 56–70, 2016.
DOI: 10.1007/978-3-319-45880-9_5

resulted in a fundamental shift in the status of end-users. An individual end-user becomes content creator and manager instead of just being content consumer. Today, for every single piece of data shared on OSNs, the uploader must decide which of his friends should be able to access the data. People spend an unprecedented amount of time interacting with social networking sites and uploading large amount of personal information. An exponential growth in usage of online social networks created a myriad of privacy concerns. As a result, the issue of privacy in online social networks has received significant attention in both the research community [1–4] and the main stream media[4,5].

The current online social networks provide multitude of privacy settings to manage access to uploaded content. Facebook is well known for providing detailed privacy settings. However, the privacy setting interface is too complicated to most of the normal users. The current interface has limited visual feedback and promotes a poor mental model of how the settings affect the profile visibility [5]. Even after modifying settings, users can experience difficulty in ensuring that their settings match the actual desired outcome. Madejski [2] shows that privacy settings for uploaded content are often incorrect, failing to match users' expectations. A number of papers report that users have trouble with existing extensive privacy settings. The vast majority of users do not utilize privacy settings to customize their accessibility [1,3,4].

In order to help users share content selectively with their friends, friend-list feature is another approach. Each friend-list contains a subset of a users' friends and then allow a user to share content only with members of the friend-list. Unfortunately, the usefulness of this feature is overshadowed by the cognitive burden that is placed on users. It is responsibility of users to populate their lists and maintain the appropriateness of these lists over time. The relationships in everyday life evolve with time and friend-list feature do not offer any mechanism to deal with this evolution. So the appropriateness of these lists is questionable with passage of time. As a result, it is unsurprising that many users do not use the friend-list feature. Recently, smart-list feature is introduced to overcome this problem. The smart-list is generated automatically based on profile attributes of users. The majority of users do not provide detailed attribute information necessary, so the smart list only contains a subset of the correct users. It is also important to note that smart list does not take into consideration relationship strength, but only function on profile similarity attributes [6,7]

Despite of the multitude of privacy controls, current online social networks fail to provide an effective mechanism to manage access to uploaded content of the users. The main reason for this failure is shortcoming of the online social networks to represent diverse social relationships. Online social networks carry problematic assumptions in their implicit design of representing social relationships. All friends are created equal that means they have access to same identity,

---

[4] Facebook Facelifts Its Privacy Policy http://gadgets.ndtv.com/social-networking/news/facebook-facelifts-its-privacy-policy-226824.

[5] Do Social Networks Bring the End of Privacy? http://www.scientificamerican.com/article/do-social-networks-bring/.

and same social context of the user. In real life people play diverse roles and disclose their personal information according to the role. Each individual has several role based identities to preserve the contextual integrity of the information which is being disclosed. The notion of privacy as contextual integrity is compromised by online social networks. Most online social networks employ "friendship" as the only type of bidirectional relationship. The friendships is binary, static, and symmetric relationship of equal value between all the directly connected users which provide only a coarse indication of the nature of the relationship. In reality social relationships are of varying tie strength (how close two individual are to one another), dynamic (change over time), and asymmetric in nature (one person pays attention to another, it does not mean the latter will reciprocate). It is challenging task to model dynamism, asymmetry, and relational strength in user relationships in contemporary online social networks. This is the motivation for our research work.

The main question for this research is how to represent diverse social relationships of the users in online social networks. More specifically, we want to explore whether a users' interaction patterns with his friends can be used as a basis for inferring relationship strength between users and control personal information disclosure for the user. To answer these questions, we conducted a user study with online social network users to support our following hypotheses about relationship strength and personal information disclosure.

H1. Personal information disclosure depends on relationship strength among the users.
H2. Relationship strength depends on frequency of social interactions among the users.
H3. Choice of the interaction type for communication with friends depends on relationship strength.

Our approach is to develop theoretical framework for social web privacy initially, then we present results of user study to support our hypotheses about link between personal information disclosure, relationship strength and interaction pattern of the users. Finally, we develop a ontological model to manage self-presentation and social relationships of a user in a dynamic environment of online social networks with diverse audiences. The innovative aspect of this approach is that theoretical framework for privacy and ontological model are inspired from most influential social theories of Goffman and Granovetter about self-presentation and tie strength respectively.

2   Related Work

With emergence of the semantic web, ontologies have provided new potential for enhancing expressiveness, formal semantic, and reasoning capabilities of several approaches. Ontologies together with rules can be exploited to develop an underlying privacy platform for online social networks. In this section, we compare our proposed approach with some of the other relevant initiatives in this area.

FOAF (Friend of a Friend) [8] is one of the first semantic models to grasp social interconnections between people. Persons, their activities and relationships to other people or objects are modeled in this ontology. FOAF is light weight and very simplified model. FOAF has a "knows" property which defines a social relationship. However, representing relationship using such RDF property fail to accommodation rich context information and diverse social relationships. FOAF realm [9,10] quantifies the knows relations in the context of FOAF ontology as a trust metric, and support rules that control access of friend to resources in online social networks by stating maximum distance and minimal friendship level. RELATIONSHIP ontology[6] also model user relationships in online social networks in precise manner. This ontology specialize the "knows" property of FOAF to characterize various user relationships (personal, professional, sentimental and family). The AMO (Access Management Ontology) [11] is another approach that allows annotating the resources and modeling the access control policy. These existing ontologies do not take into consideration diverse social roles and relationship strength between users.

Elahi et al. [12] propose ontologies to represent relationships among individuals and the community in order to enforce access restrictions to the resources. Carminati et al. [13,14] propose conceptually similar, but much richer OWL ontology for modelling various aspects of online social networks. The use of semantic ontology allows the model to infer about the relationships among users and resources. The authors define three type of policies, namely, access control policy, filtering policy, and admin policy. Access control policies are positive authorization rules; filtering policies can limit someone's access to information by him/herself; and admin policies can be used to express who are authorized to define those policies. Although the authors outline an access control framework, lack of formal descriptions and implementation leaves behind many ambiguities. A more detailed approach is developed by Masoumzadeh et al. [15,16], which proposes the Ontology-based Social Network Access Control (OSNAC) model, encompassing two ontologies; the Social Networking systems Ontology (SNO), capturing the information semantics of a social network, and the Access Control Ontology (ACO), which allows for expressing access control rules on the relations among concepts in the SNO. This model takes into account intricate semantic relationships among different users, data objects, and between users and data objects. The model enables expressing much more fine grained access control policies on a social network knowledge base than already discussed by Carminati et al.

Barkhuus [17] investigates the application of contextual integrity to the consideration of privacy in HCI research. The qualitative study of Shi et al. [18] provide preliminary insights in understanding user's interpersonal privacy concerns from the perspective of contextual integrity. Lipford et al. [19] claim that failure of privacy management on online social networks reflect the nuanced and contextual nature of privacy in the offline social world. The authors argue that online social networks ought to be designed from the perspective of contextual

---

[6] RELATIONSHIP, http://vocab.org/relationship/.

integrity to preserve the privacy of their users. Kayes et al. [20,21] claim that their system (Aegis) implemented contextual integrity, whereas, their example contexts and policies reflect basic access controls akin to UNIX access controls. The Aegis system express limited forms of norms and contexts and ignore underlying principles of contextual integrity such as norms with roles and attributes. Compared to Kayes et al., we provide a much richer set of contexts and roles to model diverse aspects of user's social relationships in online social networks.

Lerone et al. [22] introduced interaction count based approach to determine relationship strength. In this approach, the authors simply take into consideration three types of interactions and count them in order to calculate relationship strength. This model is not based on semantic web approach. It is very simple and does not differentiate between interactions on the basis of their role in developing relational ties. Waqar et al. [23] extend work of Lerone et al. by applying data mining approach to calculate relationship strength for online social networks, Whereas, this data mining model is not validated on real OSNs data. Christo et al. [24] show that users tend to interact mostly with small subset of friends, often having no interactions with up to 50 % of their friends. The authors suggest a model for representing user relationships based on user interactions.

Existing research literature supports our idea that all friends should not be given equal access to user personal information, but access to personal information should be administrated based on relationship strength among online social network users. The social perspective of privacy is ignored by all the approaches discussed in this section. This is the innovative aspect of our approach that we model privacy for online social networks from social perspective. Our approach takes into consideration existing rich literature of sociology on self-presentation and tie strength. We identify various contextual roles and dimensions of relationships strength, and develop an ontological model that takes into consideration this conceptual background from sociology. This model can be useful for any future social web environment which intends to incorporate privacy in their implicit design from social perspective and wants to mimic real life information disclosure pattern.

3   Privacy in Social Web

With emergence of the social web, a new debate started about the meaning and value of privacy. According to some researchers privacy has been undermined by online social networks, even some of them claim that it no longer exists.[7] The concept of privacy is so intricate that there is no universal definition of it. In following section, we present theoretical background on privacy in social web.

3.1   Theoretical Background

Privacy on the web in general revolves mostly around information privacy. Information privacy is an individual's claim to control the terms under which personal

---

[7] Do Social Networks Bring the End of Privacy? http://www.scientificamerican.com/ article/do-social-networks-bring/.

information is acquired, disclosed or used [25]. The rise of online social networks changed dynamics of the web where users are content producer instead of content consumer and share large amount of personally sensitive information. We need to redefine privacy which suits contemporary needs of the social web. The existing literature provides some definitions highlighting various aspects of privacy problem in OSNs. According to Gurses et al. [26], the researcher tackle three type of privacy problems associated with online social networks.

Surveillance Privacy Problem: This problem arises when the personal information and social interactions of OSN users are leveraged by governments and service providers.

Social Privacy Problem: This problem emerges through the necessary renegotiation of boundaries as social interactions get mediated by OSN services.

Institutional Privacy Problem: This problems is related to users losing control and oversight over the collection and processing of their information in OSNs.

This work is focused on social privacy problem that aims to protect user personal information from other users. Palen et al. [27] suggest three boundaries that users negotiate individually or collectively in OSNs to address the issue of social privacy.

Disclosure Boundary: It manages audience for uploaded content of OSN users and deals with issue of unintended audiences.

Identity Boundary: It manages self-presentation of online social network users and deals with issue of context collapse.

Temporal Boundary: It manages personal information disclosure with reference to time and deals with issue of information persistence in social web.

The users have a scope in mind when they upload personal information in online social networks. This scope is defined by disclosure, identity, and temporal boundaries. The privacy is breached when information is moved beyond its intended scope either accidentally or maliciously. Simply a breach can occur when information is shared with a party for whom it was not intended, it can also happen when information is abused for different purpose than was intended, or when information is accessed after its intended lifetime. The analysis of various aspects of privacy are also discussed in detail in our earlier published work [28–30].

Our definition of privacy is inspired from work of Pfitzmann et al. [31]. The authors' concept of privacy is customized to suits the needs of social web users. The users face three major problems in their effort to manage their privacy in online social networks. These issues are context collapse, invisible audience, and interdependent privacy. Our definition of privacy resolves around preserving contextual integrity of the user, minimizing disclosure to personal information of the users, and enhancing ability of the user to control access to their content residing into the spaces of their friends.

Contextual Integrity: It gives a user ability to keep the audiences separate and compartmentalize their social life.

Disclosure Minimization: It gives a user ability to control personal information disclosure on the basis of quality of relationship.

User Control: It gives a user ability to control access to shared resources based on their resource centric role.

The notion of privacy as contextual integrity can be useful in addressing the problem of context collapse faced by online social network users. Goffman's audience segregation [32] is valuable concept to preserve contextual integrity. According to Goffman each individual performs multiple and possibly conflicting roles in everyday life, and it need to segregate the audience for each role, in a way that people from one audience cannot witness a role performance, that is intended for another audience and there by keeping a consistent self-presentation. Audience segregation and privacy are closely linked. Nissenbaum also argues that privacy revolves around "contextual integrity" [33].

Relationship strength plays vital role in disclosure minimization. Stronger the relationship more personal information is disclosed and weaker the relationship less personal information is disclosed. Granovetter coined the term tie strength [34]. This term is one of the most influential concepts in sociology. Tie strength is a quantifiable social network concept that measures the quality of relationships. The existing sociology literature suggests seven dimensions of tie strength such as intensity, intimacy, duration, social distance, emotional support, structural dimension, and reciprocal service. According to Petroczi et al. [35], the relationship indicators in online social networks are similar to those in offline communities. All tie strength dimensions can be easily inferred from user interaction patterns and profile similarity attributes in existing online social networks.

User control is handled by interpersonal boundary regulation. Altman discusses the concept of interpersonal privacy which give user more control to regulate his interpersonal boundary which is dealing with both individual and collaborative boundary [36]. Online social network users cannot only upload a content into their own space, but also upload content into the spaces of their friends. This gives rise to the phenomenon which is termed as interdependent privacy. Photo Tagging is very common example of interdependent privacy. Current online social networks provide simple access control mechanisms allowing users to govern access to information contained in their own spaces. Unfortunately, users have no control over data residing outside their spaces [37]. We define privacy in online social networks from social perspective and it is inspired from social theories of Goffman, Granovetter and Altman.

3.2 Privacy Perspective of Social Web Users

We conducted a user study to investigate the attitude of social web users towards privacy. The survey was designed to examine privacy concerns and user interaction patterns in online social networks. An online questionnaire was distributed via numerous university mailing lists and postings in popular OSN groups.

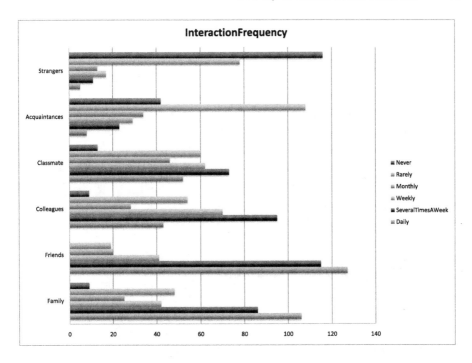

Fig. 1. Interaction frequency pattern

The survey targeted Facebook and Google+ users. The response were collected from May to August 2015, with overall gross sampling consisting of 334 participants. After deleting responses that were unusable a final net sample of 323 participants was obtained out of which 245 were male and 81 females. The vast majority of the participants belongs to age-group between 20 to 40 years with only few exceptions. The most of the participants are active online social network users either constantly logged into their accounts or check their account several times per day. Some of the relevant results of this study are presented in this section. According to the results 65 % participants added more than 200 people in their friend network, and 26 % participant also added strangers to their network, Whereas, only 3 % participants are interested to share their personal information with strangers added in their friend network. As per the results shown in Fig. 1, the vast majority of participants interact with friends and family on daily basis and their interaction with colleagues and classmates is on weekly basis, whereas their interaction pattern with acquaintances and stranger is rarely and never. The results in Fig. 2 shows that preferred interaction of OSNs users with strong ties is messaging, posting, commenting, and chatting. The participants' preferred interaction with weak ties is either liking or not-applicable. The most frequently used interaction types are messaging, liking, chatting, wishing, and posting, whereas, the least frequently used interaction types are playing games and tagging.

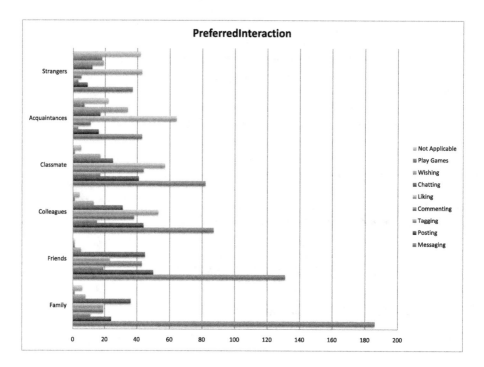

Fig. 2. Preferred interaction pattern

Our three hypotheses are supported to be true by this user study. According to results only 3 % participants agreed to share their personal information with strangers, and 12 % with acquaintances, Whereas vast majority is interested to share their personal information with family and friends 82 % and 76 % respectively. The 31 % participants are willing to share personal information with classmates, and 30 % with colleagues. These results completely support our hypothesis H1. The results also support our hypothesis H2. The vast majority interact with friends and family on daily basis, and classmates and colleagues weekly basis, whereas, their response to question regarding interaction with strangers and acquaintance was never or rarely. It is interesting to notice that participants preferred interaction for family and friends is messaging or chatting, whereas, liking is preferred interaction for acquaintances or strangers added in their friend networks. This choice of interaction type also support our hypothesis H3. While exploring scientific literature, we found that conclusion drawn by Banks et al. [38] and Ahmad et al. [23] also support our hypotheses.

4   Ontological Model for Self-Presentation

The term of ontology originated in philosophy and it is formal explicit specification of a shared conceptualization [39]. Conceptualization refers to an abstract model of phenomena in the world by having identified the relevant concepts of those phenomena. Explicit means that the type of concepts used and constraints on their use are explicitly defined. Formal refers to the fact that ontology should be machine readable. Shared reflects that ontology should capture consensual knowledge accepted by the communities. An ontology has five modeling elements: concepts, properties, relations, axioms and instances. Basically, the role of ontology is to construct a domain model using these elements. It is widely recognised that constructing a domain model or ontology is an important step in the development of knowledge based systems. In short, a good methodology for developing an ontology is needed.

4.1   Ontology Development Methodology

The methodology used to develop this ontology is Methontology [40]. Methontology is a well-structured methodology used to build ontologies from the scratch.

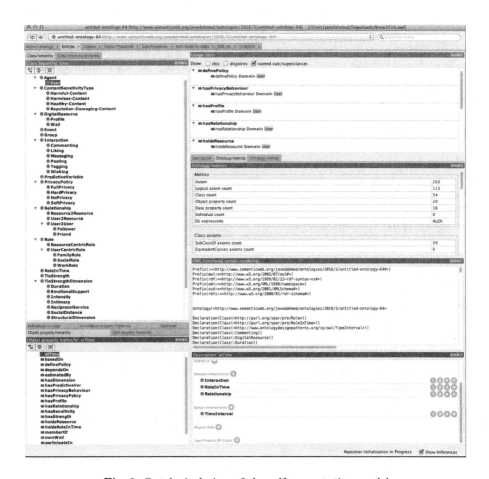

Fig. 3. Ontological view of the self-presentation model

It is based on the IEEE standard for software development. The methodology suggests three type of activities for building an ontology such as ontology management activities, ontology development-oriented activities and ontology support activities. We describe some of the ontology development oriented activities carried to build this ontological model. We developed ontology requirement specification document (ORSD) in the specification phase. In conceptualization phase, we identify the domain concepts from the state of art on privacy in computer science, engineering and social science. We developed problem scenarios to identify classes, properties and their relationships. A pool of competency questions was developed to identify functional requirements of the model. We use OWL (Web Ontology language) to represent formalization of the model. In integration phase, we reviewed existing ontologies and identify some concepts from FOAF[8], PRO[9], TimeInterval[10] and Time-indexed Value in Context (TVC)[11] ontologies for reuse. The implementation of the model is done using Protege.

4.2   Semantic Representation

We develop this model because we could not find an appropriate ontology in the literature that can capture the details of relationship strength among online social network users. The current version of the ontology consists of 54 concepts and 20 object properties. Figure 3 shows details of concepts, object properties and data properties of our ontological model. The *User* is main concept in the ontology. We extend *User* from *foaf:Agent* class and *User* is connected with *foaf:Group* via *memberOf* object property. The *User* has different kind of *Relationship* which includes *User2User*, *User2Resource*, and *Resource2Resource* relationships. The relationships are dynamic in nature and change over period of time. This concept is represented with *ti:TimeInterval* class which is connected with *Relationship* class using *tvc:atTime* object property. The relationship has *TieStrength* that is determined from *TieStrengthDimension*. There are seven different dimensions of relationship strength. These dimension have indicators in terms of *PredictiveVariable* which can be inferred from various type of user *Interaction*, and *Profile* similarity attributes. We reuse *pro:Role* and *pro:RoleInTime* classes and *pro:holdsRoleInTime* and *pro:withRole* object properties. It permits one to specify how a user has a role relating to a contextual entity, and the period of time during which that role is held. It is based on the Time-indexed Value in Context (TVC) ontology pattern. The user roles are divided into two categorize *UserCentricRole* and *ResourceCentricRole*. *UserCentricRole* are further subdivided into *SocialRole*, *FamilyRole* and *WorkRole*. The content of a user is modeled using *DigitalResource* which includes *Profile* and *Wall*. The sensitivity of user content is model using *ContentSensitivityType* and its subclasses. Object property *hasSensitivity* establish relationship between *DigitalResource*

---

[8] http://xmlns.com/foaf/spec/.

[9] http://www.sparontologies.net/ontologies/pro.

[10] http://www.ontologydesignpatterns.org/cp/owl/timeinterval.owl.

[11] http://www.essepuntato.it/2012/04/tvc.

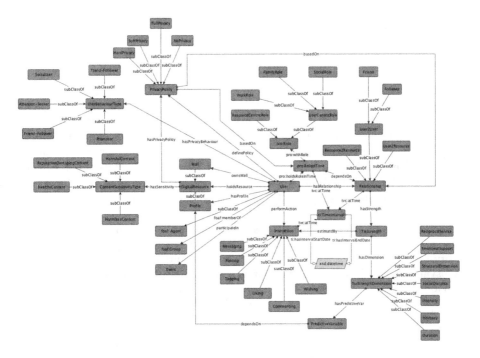

Fig. 4. Conceptual representation of ontology

and *ContentSensitivityType*. The user social interactions are represented by different subclasses of *Interaction*, which is useful for estimation of relationship strength. The object property *estimatedBy* has domain *TieStrength* and range *Interaction*. Online social networks users have different privacy requirements depending on their purpose of joining OSNs. The privacy behaviour of OSNs users is modeled using *PrivacyBehaviourType* and its subclasses. Finally, *PrivacyPolicy* and its subclasses model users' privacy in online social networks. It is based on *pro:RoleInTime* and *Relationship* and defined by *User*. Figure 4 depicts detailed diagram of the ontological model.

4.3   Evaluation

An ontology is a fairly complex structure and it is often more practical to focus on the evaluation of different aspects of the ontology. In the first phase of evaluation, we evaluate the ontological model for aspects such as vocabulary, syntax, structure and semantics. We used various reasoners for checking consistency of the model. We also used OntoClean to discover possible problematic decisions in the structure of the ontology. We also performed evaluation against three requirements coverage, granularity, and specificity of the ontology. In the second phase of evaluation, we intend to evaluate the model at assertional level by translating competency questions into SPARQL queries and retrieving data.

5 Conclusion and Future Work

In this paper, we presented an ontological model to represent diverse social relationships of users in online social networks. This model is based on well-founded social theories about self-presentation and tie strength. These social theories give insights on how to manage diverse social relationships and what are interaction patterns and personal information disclosure practices between strong and weak ties. This model suits to emerging privacy needs of social web users due to fundamental shift in their status from content consumer to content producer. Based on this model a privacy friendly online social networking environment can be developed to address existing issues of privacy. In future, we plan to build social application to evaluate applicability of our ontological model.

Acknowledgments. This research project is funded by erasmus mundus joint international doctorate in law, science and technology, and administered by CIRSFID, University of Bologna, Italy. This paper is extension of our research work published in PhD symposium of ICWE 2016. The author also acknowledges valuable comments of Leendert van der Torre, Guido Governatori, Serena Villata and Silvio Peroni.

References

1. Liu, Y., Gummadi, K.P., Krishnamurthy, B., Mislove, A.: Analyzing facebook privacy settings: user expectations vs. reality. In: Proceedings of the 2011 ACM SIGCOMM Conference on Internet Measurement Conference, pp. 61–70. ACM (2011)
2. Madejski, M., Johnson, M.L., Bellovin, S.M.: The failure of online social network privacy settings (2011)
3. Johnson, M., Egelman, S., Bellovin, S.M., Facebook, privacy: it's complicated. In: Proceedings of the Eighth Symposium on Usable Privacy and Security, p. 9. ACM (2012)
4. Netter, M., Riesner, M., Weber, M., Pernul, G.: Privacy settings in online social networks-preferences, perception, and reality. In: 2013 46th Hawaii International Conference on System Sciences (HICSS), pp. 3219–3228. IEEE (2013)
5. Akcora, C.G., Ferrari, E.: Graphical user interfaces for privacy settings. In: Alhajj, R., Rokne, J. (eds.) Encyclopedia of Social Network Analysis and Mining, pp. 648–660. Springer, New York (2014)
6. Liu, Y., Viswanath, B., Mondal, M., Gummadi, K.P., Mislove, A.: Simplifying friendlist management. In: Proceedings of the 21st International Conference Companion on World Wide Web, pp. 385–388. ACM (2012)
7. Bartel, J.W., Dewan, P.: Evolving friend lists in social networks. In: Proceedings of the 7th ACM Conference on Recommender Systems, pp. 435–438. ACM (2013)
8. Brickley, D., Miller, L.: Foaf vocabulary specification 0.98. Namespace document 9 (2012)
9. Kruk, S.R.: Foaf-realm-control your friends access to the resource. In: FOAF Workshop Proceedings, vol. 186 (2004)
10. Kruk, S.R., Grzonkowski, S., Gzella, A., Woroniecki, T., Choi, H.-C.: D-FOAF: distributed identity management with access rights delegation. In: Mizoguchi, R., Shi, Z.-Z., Giunchiglia, F. (eds.) ASWC 2006. LNCS, vol. 4185, pp. 140–154. Springer, Heidelberg (2006)

11. Buffa, M., Faron-Zucker, C.: Ontology-based access rights management. In: Guillet, F., Ritschard, G., Zighed, D.A. (eds.) Advances in Knowledge Discovery and Management. SCI, vol. 398, pp. 49–62. Springer, Heidelberg (2012)

12. Elahi, N., Chowdhury, M.M.R., Noll, J.: Semantic access control in web based communities. In: The Third International Multi-conference on Computing in the Global Information Technology, 2008. ICCGI 2008, pp. 131–136. IEEE (2008)

13. Carminati, B., Ferrari, E., Heatherly, R., Kantarcioglu, M., Thuraisingham, B.: A semantic web based framework for social network access control. In: Proceedings of the 14th ACM Symposium on Access Control Models and Technologies, pp. 177–186. ACM (2009)

14. Carminati, B., Ferrari, E., Heatherly, R., Kantarcioglu, M., Thuraisingham, B.: Semantic web-based social network access control. Comput. Secur. **30**(2), 108–115 (2011)

15. Masoumzadeh, A., Joshi, J.: OSNAC: an ontology-based access control model for social networking systems. In: 2010 IEEE Second International Conference on Social Computing (SocialCom), pp. 751–759. IEEE (2010)

16. Masoumzadeh, A., Joshi, J.: Ontology-based access control for social network systems. Int. J. Inf. Priv. Secur. Integrity **1**(1), 59–78 (2011)

17. Barkhuus, L.: The mismeasurement of privacy: using contextual integrity to reconsider privacy in HCI. In: Proceedings of the SIGCHI Conference on Human Factors in Computing Systems, pp. 367–376. ACM (2012)

18. Shi, P., Xu, H., Chen, Y.: Using contextual integrity to examine interpersonal information boundary on social network sites. In: Proceedings of the SIGCHI Conference on Human Factors in Computing Systems, pp. 35–38. ACM (2013)

19. Lipford, H.R., Hull, G., Latulipe, C., Besmer, A., Watson, J.: Visible flows: contextual integrity and the design of privacy mechanisms on social network sites. In: 2009 International Conference on Computational Science and Engineering, CSE 2009, vol. 4, pp. 985–989. IEEE (2009)

20. Kayes, I., Iamnitchi, A.: Out of the wild: on generating default policies in social ecosystems. In: ICC Workshops, pp. 204–208 (2013)

21. Kayes, I., Iamnitchi, A.: Aegis: a semantic implementation of privacy as contextual integrity in social ecosystems. In: 2013 Eleventh Annual International Conference on Privacy, Security and Trust (PST), pp. 88–97. IEEE (2013)

22. Banks, L., Shyhtsun Felix, W.: All friends are not created equal: an interaction intensity based approach to privacy in online social networks. In: International Conference on Computational Science and Engineering, 2009, CSE 2009, vol. 4, pp. 970–974. IEEE (2009)

23. Ahmad, W., Riaz, A., Johnson, H., Lavesson, N.: Predicting friendship intensity in online social networks. In: 21st International Tyrrhenian Workshop on Digital Communications (2010)

24. Wilson, C., Boe, B., Sala, A., Puttaswamy, K.P.N., Zhao, B.Y.: User interactions in social networks and their implications. In: Proceedings of the 4th ACM European Conference on Computer Systems, pp. 205–218. ACM (2009)

25. Kang, J.: Information privacy in cyberspace transactions. Stanford Law Rev. **50**, 1193–1294 (1998)

26. Gurses, S., Diaz, C.: Two tales of privacy in online social networks. IEEE Secur. Priv. **11**(3), 29–37 (2013)

27. Palen, L., Dourish, P.: Unpacking privacy for a networked world. In: Proceedings of the SIGCHI Conference on Human Factors in Computing Systems, pp. 129–136. ACM (2003)

28. Ahmed, J., Governatori, G., van der Torre, L.W.N., Villata, S.: Social interaction based audience segregation for online social networks. In: ECSI, pp. 186–197 (2014)

29. Ahmed, J.: A privacy protection model for online social networks. In: SW4LAW+ DC@ JURIX (2014)

30. Ahmed, J.: A semantic model for friend segregation in online social networks. In: Bozzon, A., Cudré-Mauroux, P., Pautasso, C. (eds.) ICWE 2016. LNCS, vol. 9671, pp. 495–500. Springer, Heidelberg (2016). doi:10.1007/978-3-319-38791-8_36

31. Borceapfitzmann, K., Pfitzmann, A., Berg, M.: Privacy 3.0: = data minimization user control contextual integrity. Information Technology Methoden und innovative Anwendungen der Informatik und Informationstechnik **53**(1), 34–40 (2011)

32. Goffman, E.: The presentation of self in everyday life [1959]. Contemporary Sociological Theory, pp. 46–61 (2012)

33. Nissenbaum, H.: Privacy as contextual integrity. Wash. Law Rev. **79**(1), 119–157 (2004)

34. Granovetter, M.S.: The strength of weak ties. Am. J. Soc. **78**, 1360–1380 (1973)

35. Petróczi, A., Nepusz, T., Bazsó, F.: Measuring tie-strength in virtual social networks. Connections **27**(2), 39–52 (2007)

36. Altman, I.: The environment, social behavior: privacy, personal space, territory, and crowding (1975)

37. Hu, H., Ahn, G.-J., Jorgensen, J.: Enabling collaborative data sharing in google+. In: 2012 IEEE Global Communications Conference (GLOBECOM), pp. 720–725. IEEE (2012)

38. Banks, L.D., Wu, S.F.: Toward a behavioral approach to privacy for online social networks. In: Bolc, L., Makowski, M., Wierzbicki, A. (eds.) SocInfo 2010. LNCS, vol. 6430, pp. 19–34. Springer, Heidelberg (2010)

39. Gruber, T.R.: A translation approach to portable ontology specifications. Knowl. Acquisition **5**(2), 199–220 (1993)

40. Fernández-López, M., Gómez-Pérez, A., Juristo, N.: Methontology: from ontological art towards ontological engineering (1997)

Design of an Ontologies for the Exchange
of Software Engineering Data
in the Aerospace Industry

Ricardo Eito-Brun[(✉)]

Universidad Carlos III de Madrid, Getafe, Spain
REITO@BIB.UC3M.ES

Abstract. The development of complex projects in the aerospace industry is based on the collaboration of geographically distributed teams and companies. In this context, the need of sharing different types of data and information is a key factor to assure the successful execution of the projects. In the case of European projects, the ECSS standards provide a normative framework that specifies, among other requirements, the different document types, information items and artifacts that need to be generated. Information integration is a must-have in aerospace projects, where different players need to collaborate and share data during the life cycle of the products about requirements, design elements, problems, etc. This paper describes the development of an OWL-based ontology to manage the different artifacts and information items requested in the European Space Agency (ESA) ECSS standards for SW development. The ECSS set of standards is the main reference in aerospace projects in Europe, and in addition to engineering and managerial requirements they provide a set of DRD (Document Requirements Documents) with the structure of the different documents and records necessary to manage projects and describe intermediate information products and final deliverables.

The proposed ontology provides the basis for building advanced information systems where the information coming from different companies and institutions can be integrated into a coherent set of related data. It also provides a conceptual framework to enable the development of interfaces and gateways between the different tools and information systems used by the different players in aerospace projects.

Keywords: Software engineering · Aerospace · Software standards · Data exchange

1 Introduction

The development of complex software-based systems in the aerospace industry normally depends on the collaboration capabilities of geographically distributed teams. The effectiveness of the work completed by distributed teams heavily depends on their ability to share data, knowledge and information. Some activity sectors have developed standards to guide teams and companies in the execution of activities and the generation of documents and records in a standardized way. Standards provide companies

© Springer International Publishing Switzerland 2016
A.-C. Ngonga Ngomo and P. Křemen (Eds.): KESW 2016, CCIS 649, pp. 71–78, 2016.
DOI: 10.1007/978-3-319-45880-9_6

with a clear guidance on the expected characteristics of the different artifacts and deliverables – both final and intermediate – that need to be generated. In the case of the European aerospace projects, the European Space Agency (ESA) maintains the ECSS set of standards that cover the whole range of processes activities related to aerospace engineering. ECSS set of standards constitutes the normative framework and specifies the features of the document types, information items and artifacts that need to be generated during the life cycle of complex aerospace projects. The ECSS standards include annexes where the features and characteristics of these document types and information items are described. These annexes, known as Data Requirements Documents (DRD), state the intended purpose, scope and structure of the documents and information items. As the DRD focus on document types, not in information items, the document types and deliverables they describe are in some cases the results of putting together and packaging for delivery different information items. According to that, an effort to information modelling as the one proposed in this paper should focus on the analysis of the contents within each DRD.

Software development is one of the critical areas within aerospace engineering. ECSS standards include independent standards for software development: ECSS-E-ST-40C that guides software engineering activities and ECSS-Q-ST-80C that refers to software quality assurance processes. Both standards must be understood as complementary documents that must be read and understood as a single unit. In addition, software development activities need to consider the requirements stated in the ECSS-M-ST-40C standard for configuration management. This paper describes the development of an OWL-based ontology to support the publishing and exchange of the different artifacts and information items requested by the ESA ECSS standards for software development. Data exchange and integration is a must-have capability in complex aerospace projects, where different actors need to share data and information about software products. This covers a wide range of data: system and software requirements, design elements, detected problems and issues, source code files and their characteristics, etc. The proposed ontology provides the basis for building advanced distributed information systems where the data generated by different companies and tools may be integrated into a coherent set. The availability of a common, conceptual framework covering the different data and information products generated during the projects' life cycle help companies share and reuse data and makes possible the development of interfaces and gateways between proprietary tools and information systems.

2   Purpose of the Ontology

The main purpose of the proposed ontology is to help companies publish and share information about the information items (requirements, design elements, source code files, test specifications, document deliverables, etc.) generated during the system life cycle. Sharing this information is considered a key factor to improve engineers' productivity and help manage one of the most complex challenges in software development: information overload. Engineers need to be aware of the availability of different data and information items that may be closely related to the activities they are working on. Missing one single piece of data may have unforeseen consequences on the result of their work. The complexity of this situation increases as a wide variety of tools is used to generate, collect and manage these data. In a standard software development working environment, engineers need to deal with different tools to manage requirements, design elements, source code, verification activities and their results, metrics, etc. The lack of standard formats and information models to share this information constitutes a handicap for software productivity and reliability.

The proposed ontology may serve as an intermediate model to support the aggregation of data coming from different companies and tools. Semantic Web technologies and data modeling languages like RDF and OWL are gaining acceptance and confidence of users in different activity sectors as a means to aggregate reusable data in a cost-efficient way. Aerospace industry can also benefit from these technologies for a better management of complex information and knowledge-intensive ecosystems.

3 Methodology

The design of the ontology has followed the Methontology methodology [1]. This is one of the most popular and recognized methodologies for building ontologies [2], and has been successfully applied in difference scenarios [3, 4]. It starts with the formulation of a set of questions related to the end users' information needs. In this particular case, typical questions that engineers need to answer include, to name a few:

- Those related to the traceability between information items to assess the impact of changes, e.g.: Which requirements may be affected by a change made in a specific source code file? Which test procedures need to be reviewed and re-executed after a change in a design element? Which interfaces may be affected by an update in a data structure?
- Those related to the quality characteristics of the source code files: which are the values of the metrics for a specific version of a source code file or file component? Which deviations from quality standards have been reported and justified for a specific source code file?

These questions are a potential source for the ontology vocabulary. Anyway, as ontologies need to be an agreed conceptualization of the target domain, it was decided to build the ontology using as the main source the text of the standards. Using the standards as the main source also provides a common, shared vocabulary that may be later mapped o the terms used by engineers. In a second step, the ontology built from the content of the standards was compared with the identified questions to assess the completeness and comprehensiveness of the identified vocabulary. The development of the ontology included a validation step to answer the competence questions with the help of the developed tool and TopBraid Composer. For the competence questions a set of SPARQL queries were built showing the ontology's capability of gathering the expected data.

The ontology has been developed from a subset of the ECSS standards, and more specifically, from the DRD annexed to them. DRD provides a description and purpose of the different document types to be generated during the development of software products. For each document type, its purpose and proposed structure and contents are provided. One interesting aspect of ECSS structure is that requirements in the standards are traced to the document type or DRD that needs to provide evidence of the fulfillment of the particular requirement. This information is quite relevant as each requirement points to the document where the requested practice, activity or evidence need to be reported. Regarding the number of DRD processed to generate the ontology, ECSS-E-ST-40C [6] includes fifteen DRD, covering project planning, requirements and interface specification, software design, user manuals, reuse file, verification and validation plans, unit and integration tests planning, software review planning, verification reports and software release. ECSS-Q-ST-80C [8] and ECSS-M-ST-40C [7] include, respectively, two and ten DRDs.

It is remarked that the resulting ontology is not a document-oriented ontology, as its purpose is not restricted to the modeling of a set of document types and their content structure: its objective is the detailed analysis of the information items aggregated in the document types, their constituent data and characteristics. This level of granularity is considered strongly relevant, as document types described in DRDs usually are thought as a means to distribute set of information items, having each item independent life cycle and characteristics that require a separate management. For example, the DRD for the Software Requirements Document (SRD) document type is aimed to serve as a container of individual requirements that may be created, updated and evolve as independent units. A working ontology should support this level of granularity and going further than a document modeling effort focused on the development of XML schemas for document encoding. This does not mean that the scope of the ontology excludes document types and document deliverables. They should also be identified as classes, but their main purpose is to serve as containers and aggregations of other information items. The final ontology incorporates the capability of managing information for different artifacts like requirements, problem reports, configuration items, design elements etc., and for the different document and reports used to publish and exchange these documents. A detailed evaluation of the use of ontologies for industrial software development can be found in [8].

4   Ontology Modeling from DRD

ECSS-E-ST40C groups document types into eight files used to organize the documentation. Document files and document types are the first items incorporated to the ontology as classes. Both documents and folders need to be typed. This is a general characteristic of most of the items in the ontologies. Two different approaches were considered for modeling types: (a) using a hierarchy of subclasses, with a separate class for each type, or (b) adding new classes for the types, e.g. TypesOfDocumentFiles, TypesOfFolders, TypesOfRequirements, etc., defined by extension. In the latter case, the different types of the items are recorded as instances of the class representing the entity types; relationships between the class representing the entities – e.g. DocumentFiles, Requirements, etc., - and the class representing their types – e.g. TypesOfDocumentFiles, TypesOfRequirements, etc. – need to be established using OWL object properties. The second choice has been systematically applied for managing types for the different entities.

From ECSS standards additional classes for information items were derived, e.g. Requirements, TestCases, DesignElements, SourceCodeFiles, ObjectCodeFiles, Modules, Classes, Functions, ConfigurationFiles, CodingRules, CodingStandards, etc. In addition, supporting classes were identified to maintain information about the context of the software system. Contextual information refers to the entities involved in the management, construction and maintenance of the software. Classes in this group correspond to concepts like contractors, customers, projects, staff, project milestones and phases, project reviews, etc. Existing ontologies may be reused to represent these concepts, as for example those developed in the context of the Nepomuk (Networked Environment for Personalized, Ontology-based Management of Unified Knowledge) project. Classes identified during document analysis were modelled using the Protégé tool. Properties available in the OWL language for recording the origin and source of the ontology elements were used to make reference to the standard where the item is identified and record the definitions provided in the ECSS standards. A total of 98 classes have been identified with their corresponding properties and relationships.

To illustrate the design procedure, this paragraph summarizes the representation of the review concept. It was added as a new class, and the definition provided in the standard: "activity undertaken to determine the suitability, adequacy and effectiveness of the subject matter to achieve established objectives..." was incorporated into the ontology as part of the class documentation. The definitions in the standard were also used to derive the types for the individuals in the classes: in the case of the reviews, ECSS-E-ST-40C defines six types of reviews: Software Requirement Reviews (SRR), Preliminary Design Reviews (PDR), Critical Design Review (CDR), Qualification Review (QR), Acceptance Review (AR) and Operational Readiness Review (ORR), each one having different objectives and related to a different set of deliverables provided as inputs to the review process. In addition, these deliverables may be successfully baselined[1] as part of the review conclusions, or discrepancies can be created requesting additional modifications. On the opposite, if the document needs further rework or modification, ECSS establish a new type of information item called Review Item Discrepancy (RID), whose proposed content is described in another ECSS standard: ECSS-M-ST-10-01 "Organization and conduct of reviews". From this information, additional classes are incorporated into the ontology: SoftwareReviews, TypesOfSofwareReviews, DataPackages, Documents and ReviewItemDiscrepancies. To represent the relationships between these items, new object properties are added, among others: (a) isDeliveredAsPartOf, between Documents and DataPackages, (b) isDelivered AsInput between Documents and SoftwareReviews, (c) isBaselinedAtTheEndOf between Documents and SoftwareReviews, (d) isRaisedAgainst between ReviewItemDiscrepancies and Documents and (e) isRaisedDuring between ReviewItemDiscrepancies and SoftwareReviews. Reverse properties for these relationships were also incorporated.

The RID management process also led to the consideration of additional relationships between RIDs and Reviews, as the RID may be dispositioned during the

---

[1] Baselining one document means that the document has been reviewed accepted and that from now on, any change on the document content shall be done following a formal change management procedure.

review (dispositioning means reaching an agreement on its implementation and the changes that are needed in the document), postponed to a future review or closed after verifying that the RID has been properly implemented. These relationships translate into additional object properties having the RID and the TechnicalReviews as domains and ranges.

The final ontology contains 48 classes and 215 properties.

5   The Planned and the Actuals

Annex A of ECSS-E-ST-40C includes a table showing the expected deliveries of document at each review: some documents need to be delivered only at particular reviews. In the proposed model, as types of reviews and types of document are recorded as individuals, the ontology need to incorporate a set of individuals or instances representing the proposed or expected requirements and knowledge embedded in the standards. If a particular document – the Software Design Document, for example, need to be delivered for the Preliminary Design Review, these information shall be recorded in the properties of the corresponding individuals. Of course, these relationships cover what is prescribed by the standard. But actual project delivery plans may differ. To avoid potential conflicts between the behavior requested in the standard and real projects' data, and it is necessary to establish a clear distinction between these two cases. To do that, additional object relationships were incorporated: isPrescribedToBeDeliveredAsPartOf, isPlannedToBeDeliveredAsPartOf and isActually DeliveredAsPartOf.

6   Versioning Information Items and Compound Items

When executing a project, information items and documents evolve and new versions or editions are created as a result of work progress. An ontology aimed to support information management needs to provide support to this characteristic of information items. One possibility is considering versions, issues or editions as different instances of the same information item, sharing a common identifier. Another choice is to have separate classes for the information items, and separate classes for their versions. Information common to all the versions are encoded in the upper-level class, and information specific for each version is encoded at the lower level. Additional, reflexive object properties between the versions are incorporated to distinguish between previous and next versions, branches or obsolete versions, all of them being subproperties of the isVersionOf general property.

Another issue regarding information items management is the fact that some of them may be made up of several files. For example, one specification may be split into one file created with a word processor, and an Excel sheet containing some additional data. To accommodate this situation, the ontology must incorporate the concept of Files, and a reflexive object property to encode relationships between them (e.g., for files that have the same content in different file formats).

7 Functional vs Physical Configuration of the Software

As previously stated, the purpose of the ontology is not to manage document information, but data about information types and their quality characteristics. Documents are just aggregates used to deliver sets of related information items in an easy-to-read way. For example, the Software Design Document (SDD) type includes the description of the software components, specifying for each component its identifier, the package, library or class the component belongs to, its type (task, subroutine, subprogram, package, file, etc.), executable type (computer instructions or non-executable data), purpose, function, subordinate units (units that are called by the component), dependencies, interfaces, resources, etc. For these components, forward and backward traceability must be provided from software requirements to design components and vice versa.

The elements or information items are usually considered from two different but related perspectives: the functional configuration and the physical configuration. The first one refers to the functionality of the system, as is based on the requirements and their verification or demonstration. The physical configuration refers to the physical items that constitute the software system: components, files, functions, classes, design elements, etc. The relationship between both configurations needs to be managed through traceability links: requirements are implemented in design elements that translate into software unit that are coded within source code files. This approach for software modeling is relevant: traceability between items in software is usually considered as a chain from requirements to test cases that moves through design and code. In this approach, the traceability at the functional configuration view moves from requirements to design elements to tests, and traceability at the physical configuration view moves from the initial, high level product decomposition (product tree) to source code files. Relationships between code files and design elements just represent the fact that the first ones serve as containers of the second ones. This approach avoids the artificial distinction between software design and implementation, as following Model Driven Design (MDD) philosophy, they constitute views of the software model, the latter having a higher degree of adaptation to the specific platform where the software will be executed.

Quality characteristics of software refer to those features that need to be incorporated and maintained to ensure its functionality, reliability, usability, maintainability, etc. These characteristics are usually assessed by means of software metrics like the size, complexity, percentage of comments, coupling between classes, etc. The ontology incorporates different properties to cover these metrics and the evolution of its values at different phases.

8 Conclusions

Standards represent the agreement reached by different, representative actors of the industry on the recommended way to do a work. Most of the European aerospace projects and industries use the ECSS standards as the guide for their activities. The generation of final and intermediate information products is one of the aspects regulated by these standards. The design of an ontology based on the ECSS standards constitute the basis for building an information system capable of supporting the data exchange requirements of the partners working together on distributed projects. The ontology also provides a common vocabulary that may be used to aggregate data generated by heterogeneous tools. During the design process, most relevant decisions were those related to the management of versions of information items, management of different baselines (versions of groups of items) and the relationship between the functional and the physical configuration of the software systems.

The ontology is not a document-oriented ontology, as the different artifacts, their characteristics and the relationships among artifacts have been identified independently of the methods used to publish them in different document deliverables. By doing that, the proposed solution incorporates the capability of managing information for different artifacts like requirements, problem reports, configuration items, design elements, etc., and for the different document and reports used to publish and exchange these documents. Preliminary assessments have demonstrated that the proposed ontology constitutes a useful tool to build an information system of aggregated data that give answers to common information access problems faced by engineers regarding impact of changes.

References

1. Corcho, Ó., Fernández-López, M., Gómez-Pérez, A., López-Cima, A.: building legal ontologies with METHONTOLOGY and WebODE. In: Benjamins, V., Casanovas, P., Breuker, J., Gangemi, A. (eds.) Law and the Semantic Web. LNCS (LNAI), vol. 3369, pp. 142–157. Springer, Heidelberg (2005)
2. Iqbal, R.: An analysis of ontology engineering methodologies: a literature review. Res. J. Appl. Sci. Eng. Technol. **6**(16), 2993–3000 (2013)
3. Park, J., Sung, K., Moon, S.: Developing graduation screen ontology based on the METHONTOLOGY approach. In: Fourth International Conference on Networked Computing and Advanced Information Management NCM 2008 (2008)
4. Prestes, P., et al.: Towards a core ontology for robotics and automation. Robot. Auton. Syst. **61**(11), 1193–1204 (2013)
5. ECSS-E-ST-40C. Space Engineering. Software. ESA-ESTEC. ECSS Secretariat (2009)
6. ECSS-M-ST-40C Rev. 1. Space Project Management. Configuration and Information Management. ESA-ESTEC. ECSS Secretariat (2009)
7. ECSS-Q-ST-80C. Space Product Assurance. Software Product Assurance. ESA-ESTEC. ECSS Secretariat.A.N. (2009)
8. Vyatkin, V.: Software engineering in industrial automation: state-of-the-art review. IEEE Trans. Ind. Inf. **9**(3), 1234–1249 (2013)

Information and Knowledge Extraction

Family Matters:
Company Relations Extraction from Wikipedia

Artem Kuznetsov[1], Pavel Braslavski[1], and Vladimir Ivanov[2](✉)

[1] Ural Federal University, Yekaterinburg, Russia
artkuznetsov.m@gmail.com, pbras@yandex.ru
[2] Innopolis University, Innopolis, Russia
v.ivanov@innopolis.ru

Abstract. The study described in the paper deals with the extraction of relations between organizations from the Russian Wikipedia. We experiment with two data sources for supervised methods – manual annotations made from scratch and relations from infoboxes with subsequent sentence matching, as well as different feature sets and learning methods – SVM, CRF, and UIMA Ruta. Results show that the automatically obtained training data delivers worse results than manually annotated data, but the former approach is promising due to its scalability. Evaluation of relations extracted from a subset of Wikipedia pages that are mapped to the Russian state company registry proves that external sources can enrich and complement official databases.

1 Introduction

Relation extraction (RE) between objects mentioned in text documents is an important area of information extraction. The task is not as well developed as named entity recognition (NER), which has independent significance, but is also a necessary preliminary step for RE.

RE research has made a significant progress since its advent in the 1990s; the development of the area during almost two decades can be tracked on the materials of two evaluation initiatives: MUC (1991–1997)[1] and ACE (2000–2008)[2].

The vast majority of RE research has been conducted on English data (see Sect. 2); there are only few studies on relation extraction for Russian. A pilot track on NER and fact extraction was organized by ROMIP in 2005[3]; however, participation was low. There has been no standard publicly available dataset suited for relation extraction task until recently. Open FactRuEval challenge[4] that has been conducted in spring 2016, partially solves this problem – organizers prepared and published a news corpus with labeled named entities (*persons* and *organizations*) and relations of four types (*commercial deal, meeting, person owning a company*, and *person employed in a company*).

---

[1] http://www.itl.nist.gov/iaui/894.02/related_projects/muc/.
[2] https://www.ldc.upenn.edu/collaborations/past-projects/ace.
[3] http://romip.ru/ru/2005/tracks/qa.html (in Russian).
[4] https://github.com/dialogue-evaluation/factRuEval-2016.

© Springer International Publishing Switzerland 2016
A.-C. Ngonga Ngomo and P. Křemen (Eds.): KESW 2016, CCIS 649, pp. 81–92, 2016.
DOI: 10.1007/978-3-319-45880-9_7

Our study deals with extraction of binary hierarchical relations (parent/daughter company, ownership, founding, governance, etc.) between organizations of various kinds from the Russian Wikipedia. Wikipedia data allowed us, on the one hand, to skip the NER step, on the other – to experiment with automatically gathered data for training.

The goal of our study is twofold:

- to compare several widely used supervised approaches and different shallow features in the task of RE from Russian documents and
- to explore the potential of automatically collected data for training.

We have manually annotated 7,059 contexts with company mentions from 4,662 Wikipedia pages. We used this data for training and testing. Moreover, we collected 2,799 relations between 3,025 companies from Wikipedia infoboxes (either directly from Wikipedia dump or through DBpedia), then identified 6,962 sentences mentioning these companies. Hypothesizing that these text fragments represent relations encoded in the infoboxes, we used the data for training (obviously, this assumption does not always hold and the resulting data is essentially noisy, see discussion in Sect. 3). The manually annotated data created within the study is freely available for research purposes.[5]

We compared three methods of building RE classifiers: Support Vector Machines (SVM, a universal classification method used in many applications), Conditional Random Fields (CRF, a sequence classification method, a *de facto* standard for NER and RE tasks), as well as an automatic rules induction algorithm. We used a set of shallow classification features – mostly lexical and part-of-speech features – and their combinations within a window of variable size. Since we aimed at creating a baseline, we did not employed syntactic features and left this option for our future work.

Based on the evaluation results, we can conclude that a straightforward use of Wikipedia data for RE learning produces useful results ($macro\ F_1 = 57.4\%$ for two relations) at virtually zero annotation costs, but manually annotated data of higher quality provides about 20% gain in terms of F-score ($macro\ F_1 = 69.1\%$). Using a large set of shallow features does not affect the extraction quality significantly – almost identical results can be obtained using tokens and lemmata only. The quality of relation extraction increases with context length for feature calculation and reaches a plateau at window size of nine words.

At the final stage, we estimated how the relations automatically mined from Wikipedia can supplement the existing official databases. To do this, we automatically extracted ownership relations between companies from about 6K Wikipedia pages mapped to the Russian registry of legal entities[6]. A comparison of the extracted relations and those from the registry shows that the proposed method can complement and enrich existing structured data sources.

---

[5] https://github.com/kriskk/OrganizationRelationRecognition.

[6] https://egrul.nalog.ru/ (in Russian).

2   Related Work

Relation extraction tasks and methods differ from each other in terms of the type of information to be extracted. Some of the recent works [10, 16] are aimed at extraction of a particular relation type between two classes of an ontology. In this case, no instances of the relation are extracted. Other works [15, 18] focus on extraction of instances of a particular relation type. In this case RE learning requires significant manual efforts, which leads to low scalability. Other approaches [1] propose methods that extract related pairs of concepts taking into account only strength of the relation, without considering its type.

Existing English corpora for relation extraction have been manually created during a series of shared tasks and evaluation initiatives [3]. Such text corpora are crucial for evaluation of extraction methods, however they will never be sufficient for all application domains. Thus, an important part of most relation extraction methods is the approach to training data acquisition and construction. There are four directions: supervised approaches, unsupervised approaches [4, 21], semi-supervised methods [5], and distant supervision [12], or self-supervised learning. In the past decade Wikipedia was intensively used in RE studies. Semi-supervised and distant supervision approaches are most relevant in the context of our work.

In [19] a bootstrapping semi-supervised method was proposed for "Semantifying Wikipedia" and identified Wikipedia link structure, taxonomies, infoboxes, etc. as useful data for self-supervised semantic enrichment. Their system KYLIN is based on a CRF extractor trained on a set of lexical features. The system uses concepts' mentions represented in a Wikipedia page as hyperlinks. The main purpose of the method was to fill infobox fields. A very similar approach is proposed in [9]. It also uses CRF and achieves precision of 91 % for the task of infobox attributes population. However, the performance was measured on all types of attributes, not just on relations between two entities.

A distant supervision approach to relation extraction was proposed by Mintz et al. [13] and provides a powerful idea to build a training set for relation extraction. The authors claimed that syntax-level features are important for relation extraction. Authors constructed a training set consisting of 800,000 pages and 900,000 relation instances from Freebase. The distant supervision means that any sentence with a pair of entities that participate in some known relation is likely to express that relation. The idea is very similar to our approach, but there are differences. First, we extract relations between page title entity and an entity mentioned in the page body. Second, our approach works with predefined types or classes of relations, and does not consider particular instances of relations. One implicit assumption of distant supervision is that the reference database is complete. Apparently, it cannot be true in practice and leads to a high number of false negative training examples. Min et al. [12] extended the idea and proposed the Multiple-Instance Multiple Label algorithm that learns (from positive and unlabeled data) and tested the algorithm on Wikipedia.

Recent works on distant supervision usually consider web-scale relation extraction and use the Linked Open Data cloud as a source of relations instances.

[2] describes an approach to an improved distant supervision approach, where statistical techniques help to strategically select training seeds with lesser lexical ambiguity. Authors propose the following relaxation to the "one sentence – one relation" assumption: "If two entities participate in a relation, any paragraph that contains those two entities might express that relation, even if not in the same sentence, provided that another sentence in the paragraph in itself contains a relationship for the same subject" [2].

When a training dataset is provided, one should employ an appropriate machine learning method for relation extraction. Conditional Random Fields [8] and SVM [6,11] are widely used for relation extraction tasks. A comprehensive survey of relation extraction methods can be found in [7,14].

3   Data

In our work we use articles about organizations and companies from the Russian Wikipedia[7]. Using Wikipedia data for relation extraction allows us to skip the NER step – we consider only relations between the title company (the company the article is about) and companies mentions in the page body that are marked as anchor text of outlinks to other companies' pages. To label relations in the text of the page, we employed two approaches: (1) manual annotation and (2) automatic extraction based on information presented in Wikipedia infoboxes. Automatically labeled data and a subset of manually annotated data are used for training; both approaches are tested on the held-out 'manual' data. Figure 1 shows an example of a Wikipedia page and relations labeled on the data preparation stage. In addition, we conducted a small experiment to find out how the relations extracted from Wikipedia pages correspond to the information presented in the official databases. We employed the JWPL library[8] for Wikipedia data processing.

3.1   Manual Annotation

To select Wikipedia pages for manual labeling, we compiled a wordlist of different organization types – *company, organization, holding, bank, factory*, etc. After that, we mined a list of Wikipedia categories containing these words and collected all the pages in these categories. Then, we selected only those pages that have links to other pages in the set; the final collection contained 10,512 Wikipedia pages.

The basic unit for annotation was a sentence containing inter-company links. The annotator was presented with the sentence and its context (±300 characters around the link) within a section of the wiki-page. The majority of sentences came from summary sections or from sections about companies' history.

---

[7] https://ru.wikipedia.org/.

[8] https://dkpro.github.io/dkpro-jwpl/.

We used *brat* tool[9] for manual annotation. The annotator's task was to link the highlighted organization to the organization in the title by one of the three relation types – *Holder*, *Subsidiary*, or *Other*. Since the annotators labeled only relations between the 'main' company and already highlighted other companies' mentions, it greatly simplified and speeded up the annotation process. The instruction required that the relation was expressed within a single sentence. For example, if a context contained a relation that required anaphora resolution, annotator was not supposed to set a link. Due to limited resources the whole annotation was performed by two annotators without overlap. This resulted in 7,154 annotated contexts in total, in particular 2,150 *Holder* relations, 992 – *Subsidiary*, and 4,012 *Other*.

3.2   Automatic Labeling

The second data source about company relations is infoboxes that represent important facts about the page subject in a structured way. As Wikipedia editor's guide states, infoboxes "are not 'statistics' tables in that they ... only summarize material from an article – the information should still be present in the main text".[10]

Similarly to the approach described in [20], we extracted ownership relations from infoboxes and then searched textual representations of them in the article body. We compiled a list of infobox fields that reflect ownership or governance relations and extracted 1,922 company pairs. Moreover, we extracted standardized company relations (*rel-parentCompany-ru, rel-owningCompany-ru, rel-parentOrganisation-ru*) from DBpedia[11], which resulted in 1,780 additional relations. The surplus is mainly due to relations presented in English pages' infoboxes that can be 'transferred' to their parallel Russian pages. After duplicates removal and normalization (inverting *Subsidiary* relations to *Holder*) we obtained 2,799 relations.

In the next step we extracted textual contexts presumably reflecting the infobox relationships. For each company in the relation we searched for an exact match of its counterpart on the corresponding page. For example, for the relation $X is\_Holder\_of Y$ we searched for mentions of Y on page X (and considered that the sought sentence expressed the *Subsidiary* relation) and vice versa – X's mentions on Y's page (assuming that these mentions expressed the *Holder* relation).

In addition, we required that the sentence was at least 30 characters long as a simple criterion for natural language sentences. We also sampled sentences with company mentions that were not members of infobox relations and regarded them as manifestations of *Other* relations (we needed them as negative class instances when training classifiers). After duplicates removal we got 6,471 contexts: 3,840 – *Holder*, 979 – *Subsidiary*, and 1,652 – *Other*.

---

[9] http://brat.nlplab.org/.
[10] https://en.wikipedia.org/wiki/Help:Infobox.
[11] http://wiki.dbpedia.org/.

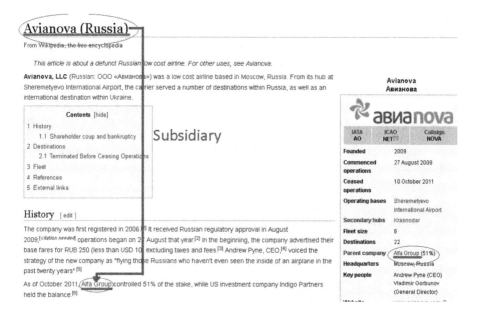

Fig. 1. Example: relation representation in the page body and infobox.

For example, infobox of the *TNK-BP* page indicates *Rosneft* as *Holder* and the page itself contains the following sentence:

> *At the end of October 2012, Rosneft has announced the acquisition of its competitor — TNK-BP oil company.*

Obviously, such automatic approach produces noisy data, for example infobox on the *Beltelecom* page mentions *Government of Belarus* as a holder, but the extracted sentence does not reflect this relation:

> *Sergei Popkov, the ex-head of Beltelecom, was appointed as Minister of Communications and Information Technology instead of Nikolai Pantelei.*

The 'manual' dataset (4,327 organizations) and 'automatic' one (3,004) have 970 entries in common. Out of 2,799 relations in 'automatic' and 2,383 in 'manual' datasets, 477 relations are presented in both. This comparison illustrates that the two approaches to data acquisition complement each other.

3.3   State Registry of Legal Entities

One of the goals of the study was to figure out to which extent the information from Wikipedia can enrich existing official databases. To this end, we used a set of 6,206 Wikipedia pages about companies that were automatically matched with records in the Russian registry of legal entities. The registry contains basic information about companies and organizations, including data about founders and owners.

4 Relation Extraction Learning

As we stated earlier, Wikipedia data allows us to simplify the task of relations extraction and skip the NER step. We cast the relation extraction problem as classification into three classes: *Holder/Subsidiary/Other*.

4.1 Classification Methods

Linear SVM. Support Vector Machines (SVM) showed their utility in a wide variety of tasks [6]. We treat linear SVM with bag-of-words features as a baseline in our experiments. Binary feature vectors are obtained based on the 12-word-long context around the company mention.[12] We used *scikit-learn* implementation of linear SVM.[13]

Conditional Random Fields. Sequence classifiers that take into account linear sentence structure proved to be very efficient in natural language processing, in particular – in information extraction tasks. Conditional random fields (CRF) is a sequential algorithm that became a *de facto* standard for NER and RE tasks. It treats a sentence as a sequence of chunks and marks each chunk with a class label (with additional 'None' label). We used the CRFSharp implementation[14] in our experiments. CRF allows accounting both for the left and right contexts of the current token and thus introduces window size as an additional parameter. We consider symmetric windows of size $2 \cdot x + 1$, i.e. x tokens on each side of the current token. We used a much richer feature set in case of CRF (see Sect. 4.2).

Rule Induction. An alternative approach to relation extraction is example-based rule induction. We took advantage of implementation of the WHISK algorithm [17] in Apache UIMA Ruta[15]. Rule generation algorithm is implemented as "TextRuler" plug-in for the Eclipse environment[16]. Unfortunately, we encountered performance issues when applying this algorithm to our data. As a workaround we splitted the training set into chunks of approximately 600 contexts each; the runtime for each chunk constituted about two hours.[17]

---

[12] We also performed experiments with lemmatized contexts, however, it did not affected classification accuracy.

[13] http://scikit-learn.org/stable/modules/generated/sklearn.svm.LinearSVC.html. We also experimented with other classifiers from the same library – MultinomialNB, BernoulliNB, RidgeClassifier, Perceptron, PassiveAggressiveClassifier, KNeighborsClassifier, SGDClassifier, NearestCentroid. They produced very similar results to those by SVM.

[14] https://crfsharp.codeplex.com/.

[15] https://uima.apache.org/ruta.html.

[16] https://eclipse.org/.

[17] It took about a week to process the complete dataset on a commodity desktop machine. However, it resuled in much lower quality in comparison to the *divide and conquer* approach – macro $F_1 = 37.9$ vs. 50.2.

We used morphological tags (see Sect. 4.2) as features on par with TextRuler internal labels.

UIMA Ruta produces rules of the following form (p and n indicate the number of positive and negatives outputs when applied to training set, respectively):

> *Org{→ MARKONCE(Holder)} Sush # SW; // p=6; n=0*
> *Org{→MARKONCE(Subsidiary)} SPECIAL COMMA # TokenAnnotation SPECIAL; // p=7; n=0*

4.2 CRF Classification Features

Each token in CRF method is described with the following features:

- *Token:* any alphanumeric sequence;
- *Lemma:* output of *mystem* morphological analyzer[18] in contextual disambiguation mode;
- *Script:* marks Cyrillic/Latin/Special symbols (punctuation marks and digits);
- *Part-of-speech (POS):* Due to rich Russian morphology, the standard approach is to encode part-of-speech tags (*noun, verb, adjective, adverb*, etc.) separately from the grammar tags (*number, case, animacy, person, tense, aspect*, etc.). This feature corresponds to the former notion, i.e. the very POS tags.
- *Grammar tags: mystem* output – 52 tags (*gender, case, tense*, etc.), each is a binary feature;
- *IsOrganization:* this is a binary feature that marks the organization mention – the potential relation member based on Wikipedia markup;
- *Feature bigrams:* combinations such as (*token, lemma*) and (*lemma, POS*);
- *Dictionary features:* based on a list of words that occurred frequently near company names in the training set plus synonyms and different organization names (e.g. *factory, corporation, bank*, etc.).

5   Results and Discussion

We splitted the manually labeled dataset into train (70 %) and test (30 %) sets, the latter was used for evaluation of all approaches. Table 1 shows the evaluation results (the cited CRF results correspond to all features and window of size 13). The table indicates quality measures for target classes only (i.e. evaluation results for *Other* class are not shown). It can be seen from the table that rule induction can deliver perfect precision, but a very low recall in case of *Holder* class. At the same time, rule induction is quite robust to the noise in the data and performs equally both with manual and automatic training data. Rule induction also delivers best recall for *Subsidiary* class and is on par with CRF in terms of general performance ($F1$) when the methods are trained on automatically gathered data. As expected, CRF outperforms other approaches, when trained on manual data.

---

[18] https://tech.yandex.ru/mystem/.

Table 1. Evaluation results for different learning methods and training sets (in percents).

Method	Training set	Holder			Subsidiary			Macro F1
		P	R	F1	P	R	F1	
Linear SVM	Manual	66.8	63.4	65.0	55.0	28.4	37.4	51.2
CRF	Manual	82.3	**75.7**	**78.8**	72.9	**50.1**	59.4	**69.1**
UIMA Ruta	Manual	**100.0**	25.7	40.8	**100.0**	42.3	59.5	50.2
Linear SVM	Automatic	47.0	41.4	44.0	26.0	18.1	21.4	32.7
CRF	Automatic	56.2	65.6	60.5	41.9	37.9	39.8	50.2
UIMA Ruta	Automatic	**100.0**	25.4	40.5	**100.0**	42.6	**59.8**	50.2

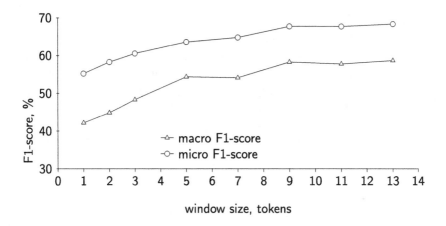

Fig. 2. Impact of context size on relation extraction quality.

Figure 2 illustrates that window size positively impacts F1-score that reaches a plateau at context length of nine words (the shown results are obtained without grammar features due to efficiency reasons).

Contribution of different features can be estimated based on Table 2. The table shows extraction results for CRF trained on manually obtained data with the window of nine words. The results allow us to conclude that the same extraction quality can be achieved with tokens and lemmata as features only; richer linguistic features such as POS and grammar features do not influence the resulting quality significantly.

At the final stage of our experiment we addressed the question, to what extent the automatically extracted relations can enrich the existing databases. To this end, we extracted relations from 6,206 Wikipedia pages that were automatically matched to the records of the Russian state registry of legal entities.

Table 2. Contribution of different features to overall relation extraction quality.

Feature set	F1, %
All features	67.8
w/o grammar tags	67.7
w/o dictionary	68.2
w/o bigrams	68.0
w/o script	67.3
w/o POS	67.7
Tokens only	67.4
Tokens and lemmata	67.5

We juxtapose the following three sets of relations: (1) Wikipedia + DBpedia — relations from infoboxes and DBpedia; (2) automatically extracted relations from Wikipeda articles; (3) relations from the registry. Figure 3 illustrates the overlap between these three sets. The results show that Wikipedia can enrich and complement existing official databases. News can be potentially even more valuable and dynamic source for relation extraction between companies.

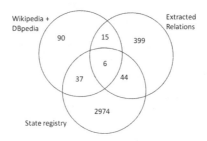

Fig. 3. Intersection of relations between 6,206 organizations from three different sources.

6   Conclusion

We conducted a pilot study aimed at extracting relations between companies from the Russian Wikipedia. We manually labeled a sizable dataset of sentences with *Holder/Subsidiary* relations and made it freely available for research purposes. We hope that these efforts will promote RE research on Russian language data.

We compared several supervised approaches to relations extraction – SVM and CRF with shallow features, as well as automatic rule generation. We also investigated automatic mining of labeled examples from Wikipedia. Rule induction, though computationally less effective, showed high precision results even when trained on noisy data. Although the automatically gathered training set was able to deliver decent results, the more elaborated manual dataset allowed for a better quality. Sequential method (CRF) outperformed SVM with bag-of-words features as expected. A wide variety of shallow features did not lead to improved results – the same quality was achieved with tokens and lemmata only as features. Rule induction, though computationally less effective, delivered high precision results even when trained on noisy data. Thus, we established several baselines for relation extraction methods from Russian documents.

Despite its size (more than 1.3 million articles), Russian Wikipedia contains relatively little information about companies and organizations, especially when compared to news stream and focusing on lesser-known organizations. In our future work we plan to transfer our methods to news data – it will include the NER step and switching from inherently unary relations to actual binary ones. We also plan to investigate the contribution of syntactic features to relation extraction for Russian.

Acknowledgements. The study was supported by RFBR grants #14-37-50950 "Research and development of algorithms for relation extraction from Wikipedia texts" and #15-29-01173 "Computational models and mathematical methods for big data analysis of trends and correlations in society". We thank Ksenia Zhagorina for providing us with Wikipedia pages matched to the state registry of legal entities, as well as Kontur.Focus (https://focus.kontur.ru/) for the very registry data. Last, but not least we thank Olga Rogacheva and Maria Belkova for annotating the data.

References

1. Astrakhantsev, N., Fedorenko, D., Turdakov, D.: Automatic enrichment of informal ontology by analyzing a domain-specific text collection. In: Proceedings of Dialog, pp. 29–42 (2014)
2. Augenstein, I., Maynard, D., Ciravegna, F.: Relation extraction from the web using distant supervision. In: Janowicz, K., Schlobach, S., Lambrix, P., Hyvönen, E. (eds.) EKAW 2014. LNCS, vol. 8876, pp. 26–41. Springer, Heidelberg (2014)
3. Doddington, G.R., Mitchell, A., Przybocki, M.A., Ramshaw, L.A., Strassel, S., Weischedel, R.M.: The automatic content extraction (ACE) program - tasks, data, and evaluation. In: Proceedings of LREC (2004)
4. Etzioni, O., Banko, M., Soderland, S., Weld, D.S.: Open information extraction from the web. CACM **51**(12), 68–74 (2008)
5. Etzioni, O., et al.: Web-scale information extraction in knowitall: (preliminary results). In: Proceedings of WWW, pp. 100–110 (2004)
6. Joachims, T.: Text categorization with suport vector machines: learning with many relevant features. In: Proceedings of ECML, pp. 137–142 (1998)
7. Konstantinova, N.: Review of relation extraction methods: what is new out there? In: Proceedings of AIST, pp. 15–28 (2014)
8. Lafferty, J.D., McCallum, A., Pereira, F.C.N.: Conditional random fields: probabilistic models for segmenting and labeling sequence data. In: Proceedings of ICML, pp. 282–289 (2001)

9. Lange, D., Böhm, C., Naumann, F.: Extracting structured information from Wikipedia articles to populate infoboxes. In: Proceedings of CIKM, pp. 1661–1664 (2010)
10. Maedche, A., Staab, S.: Discovering conceptual relations from text. In: Proceedings of ECAI, pp. 321–325 (2000)
11. McNamee, P., Mayfield, J.: Entity extraction without language-specific resources. In: Proceedings of CoNLL (2002)
12. Min, B., Grishman, R., Wan, L., Wang, C., Gondek, D.: Distant supervision for relation extraction with an incomplete knowledge base. In: Proceedings of HLT-NAACL, pp. 777–782 (2013)
13. Mintz, M., Bills, S., Snow, R., Jurafsky, D.: Distant supervision for relation extraction without labeled data. In: Proceedings of ACL, pp. 1003–1011 (2009)
14. Nastase, V., Nakov, P., Seaghdha, D.O., Szpakowicz, S.: Semantic relations between nominals. Synth. Lect. Hum. Lang. Technol. 6(1), 1–119 (2013)
15. Sarawagi, S.: Information extraction. Found. Trends Databases 1(3), 261–377 (2008)
16. Schutz, A., Buitelaar, P.: RelExt: a tool for relation extraction from text in ontology extension. In: Gil, Y., Motta, E., Benjamins, V.R., Musen, M.A. (eds.) ISWC 2005. LNCS, vol. 3729, pp. 593–606. Springer, Heidelberg (2005)
17. Soderland, S.: Learning information extraction rules for semi-structured and free text. Mach. Learn. 34(1–3), 233–272 (1999)
18. Weikum, G., Theobald, M.: From information to knowledge: harvesting entities and relationships from web sources. In: Proceedings of PODS, pp. 65–76 (2010)
19. Wu, F., Weld, D.S.: Autonomously semantifying Wikipedia. In: Proceedings of CIKM, pp. 41–50 (2007)
20. Wu, F., Weld, D.S.: Open information extraction using Wikipedia. In: Proceedings of ACL, pp. 118–127 (2010)
21. Yao, L., Haghighi, A., Riedel, S., McCallum, A.: Structured relation discovery using generative models. In: Proceedings of EMNLP, pp. 1456–1466 (2011)

A Bank Information Extraction System Based on Named Entity Recognition with CRFs from Noisy Customer Order Texts in Turkish

Erdem Emekligil[(⊠)], Secil Arslan, and Onur Agin

R&D and Special Projects Department Yapi Kredi Technology, Istanbul, Turkey
{erdem.emekligil,secil.arslan,onur.agin}@ykteknoloji.com.tr

Abstract. Each day hundred thousands of customer transactions arrive at banks operation center via fax channel. The information required to complete each transaction (money transfer, salary payment, tax payment etc.) is extracted manually by operators from the image of customer orders. Our information extraction system uses CRFs (Conditional Random Fields) for obtaining the required named entities for each transaction type from noisy text of customer orders. The difficulty of the problem arouses from the fact that every customer order has different formats, image resolution of orders are so low that OCR-ed (Optical Character Recognition) texts are highly noisy and Turkish is still challenging for the natural language processing techniques due to structure of the language. This paper mentions the difficulties of our problem domain and provides details of the methodology developed for extracting entities such as client name, organization name, bank account number, IBAN number, amount, currency and explanation.

Keywords: Named entity recognition · Turkish · Conditional random fields · Noisy Text · Banking applications

1 Introduction

Natural Language Processing (NLP) systems have been employed for many different tasks on Information Retrieval (IR). Specific task of extracting predefined types of data from documents is referred to as Named Entity Recognition (NER). Each banking process such as money transfers, tax payments etc. is performed in a maker-checker basis. Maker is responsible for entering the required data for initiating the customer orders, whereas checker has the last approval responsibility to prevent frauds at data entrance step. Customer orders are mostly received via fax channel that means each order document is in image format. Due to low resolution settings of the fax machines the OCR from customer order images generates noisy text and challenges even state-of-the-art NER systems. In this paper, we present a fast and robust information extraction system that harnesses NER to automatically extract necessary entities for banking processes from OCR-ed Turkish customer order documents; hence minimizes the effort and time consumed for data entrance.

© Springer International Publishing Switzerland 2016
A.-C. Ngonga Ngomo and P. Křemen (Eds.): KESW 2016, CCIS 649, pp. 93–102, 2016.
DOI: 10.1007/978-3-319-45880-9_8

There have been many applications of Turkish NER in the literature. Tatar and Cicekli [7] extracted different features from text by doing an automatic rule learning. Their method resulted with 91.08 % F-score on their TurkIE Turkish news data[1]. Yeniterzi [5] exploits morphology by capturing syntactic and contextual properties of tokens and she reports 88.94 % F-score on Tur et al's Turkish general news data [11] by using CRFs. Seker and Eryigit [4] also used CRFs along with morphological and lexical features. Moreover, they also made use of large scale person and location gazetteers. Their state of the art framework reports 91.94 % F-score on Tur et al's data.

Turkish newspaper domain is well studied and results are close to English. However, in both English and Turkish, results of unstructured, noisy data like Twitter are relevantly poor. Since what makes NER on customer orders difficult is noisy unstructured texts, we can state that to some degree, our problem is similar to NER on twitter data. Yamada et al. [9] report the best performing method of ALC 2015 Twitter NER shared task[2] with 56.41 % F-score on English Twitter data by exploiting open knowledge bases such as DBpedia and Freebase. Celikkaya et al. [12] trained a CRF model with news data and achieved 19 % F-score on their Turkish Tweet test data[3]. Kucuk and Steinberger [8] reported a diacritics-based expansion of their base NER system lexical resources that results with 38.01 % PLO (person, location, organization) overall F-score on Celikkaya et al's data and 48.13 % PLO overall F-score on their Turkish Twitter data. Eken and Tantug [10] exploited gazetteers along with some basic features. They used Celikkaya el al's Turkish Twitter data in training and their own annotated Turkish Twitter data while testing which they reported 64.03 % F-score.

The remaining of this paper is structured as follows: in Sect. 2, we explain the problems that differentiate our system from other NER systems. We provide details of our approach in Sect. 3 and share our experimental test results in Sect. 4. Finally, we conclude by giving our remarks in Sect. 5.

2   Problem Definition

In this section, we give brief information on three main challenges:

– Turkish as a language with complex morphology and low resources
– Noisy text generated by OCR systems
– Lack of consensus on document types in banking systems

2.1   Turkish

Turkish is one of the highly agglutinative and morphologically rich languages. In agglutinative languages each affix gives a different meaning to the word.

---

[1] TurkIE dataset contains approximately 55K tokens from news articles on terrorism from both online and print news sources.

[2] ACL 2015 Workshop on Noisy User-generated Text (W-NUT).

[3] Celikkaya et al's data contains approximately 5K tweets with about 50K tokens.

For instance, after adding plural suffix "-lar" to word "kitap" (book), it becomes "kitaplar" (books). Furthermore, by appending possessive suffix "-ım", the word "kitaplarım" (my books) gains a possessive meaning. Common NLP techniques that have been developed for English such as stemming and tokenization do not perform that well for Turkish. For instance, since Turkish is a agglutinative language, known stemming techniques cause us to lose important information which are embedded in morphology. Word "hesabımıza" (*to our account*) loses its possessive information after it is converted to "hesap" (*account*) by stemming. On the contrary to English which has the constituent order SVO (Subject-Verb-Object), Turkish has the order of SOV. Moreover, inverted sentences that changes SOV order, are often used in Turkish, which makes NER task on Turkish even harder. For example, sentences "Ali kitap oku" and "Kitap oku Ali" have the same meaning "Ali read the book".

2.2   Noisy Text

To convert fax documents to text, Optical Character Recognition (OCR) tool is used. However, the customer order document images often have low resolution, moreover the OCR is error-prone, thus may produce text deviated from its original form. For instance, as shown in Fig. 1, the word "nezdindeki" (*in care of*) is recognized as "nezrimdeki", which is not a valid Turkish word. Statistically speaking, this word is often used before account numbers, therefore holds quality information regarding whereabouts of account numbers. As we will talk about it in forthcoming sections, "nezdindeki" is used as a feature with many other words, and losing these words in OCR makes huge impact on system performance.

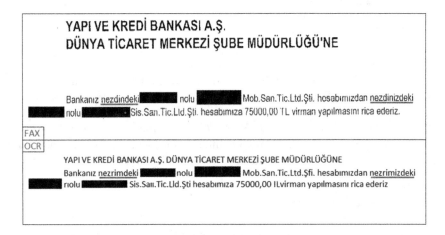

Fig. 1. Sample fax document and its OCR result. Words "nezdindeki" and its OCR result are underlined. Private information of client is masked for security reasons

2.3 Document Types

In banking domain, the customer transaction order documents that come via fax channel are created by customers. Since customers can create their own documents and no template is provided by the bank, resulting documents can be very diverse. Therefore, the documents might contain multiple transactions that increase entity counts on the page significantly. Also, spelling and capitalization errors that are made by the customers affect the system, since capitalization information of words are used as a feature.

a   b   c

Fig. 2. Sample blurred transaction documents that are received by fax and their annotated entities. Each entity type is annotated by a different color. The documents made unreadable by blurring to protect private customer information, but structures of documents and relations between different entity types can be clearly seen. (a) Document that contains table like structure. (b) Document that contains a tuple structure. (c) Unstructured document (Color figure online)

Although the language used in documents is Turkish, it may not be well formatted like newspapers. While some of the documents contain parts that are similar to tables (Fig. 2a), others may contain header-entity tuple structure (Fig. 2b) such as "Alıcı adı: Mehmet" (*Receiver name: Mehmet*), "Alıcı IBAN: TR01 2345 6789 ..." (*Receiver IBAN: TR01 2345 6789 ...*). As can be seen in Fig. 2, frequencies of entity types might differ from document to document and no explicit relation can be found between these entity types.

3 Proposed Methodology

In this section we briefly explain Conditional Random Fields (CRFs), then describe the features we have used with CRF along with its output named entities.

3.1 Conditional Random Fields

Conditional Random Fields (CRFs) [1] are conditional undirected probabilistic graphical models that are heavily used in Named Entity Recognition [14,15], Part of Speech (POS) Tagging [16] and many other tasks. We are interested in Linear Chain CRFs which solve the feature restriction problems in Hidden Markov Model (HMM) and label bias problem in Maximum Entropy Markov Model (MEMM) by combining good sides of both models.

Conditional distribution p(y|x) of classes y, given inputs x is defined in Linear chain CRFs as:

$$p\left(y|x\right) = \frac{1}{Z(x)} \prod_{t=1}^{T} \Psi_t\left(y_t, y_{t-1}, x_t\right) \tag{1}$$

where normalization factor Z and local function Ψ_t are defined as:

$$Z(x) = \sum_y \prod_{t=1}^{T} \exp\left\{ \sum_{k=1}^{K} \theta_k f_k(y_t, y_{t-1}, x_t) \right\} \tag{2}$$

$$\Psi_t\left(y_t, y_{t-1}, x_t\right) = \exp\left\{ \sum_{k=1}^{K} \theta_k f_k(y_t, y_{t-1}, x_t) \right\}. \tag{3}$$

The parameter θ is learned via optimization algorithms like Broyden Fletcher Goldfarb Shanno (BFGS) or Stochastic Gradient Descent (SGD). The feature function f is a non-zero function that represents features of state-observation pairs and y_{t-1} to y_t state transitions [2].

3.2 Selected Features

We have extracted features from most frequent 1-skip-2-grams, 2-skip-2-grams and 3-skip-2-grams [13] along with most frequent 1-grams and 2-grams. While N-grams denote most frequent phrases, skip-grams catch boundary words of entities. In NER, some entities are usually wrapped between same/similar words and to extract features from these patterns, we have used the most frequent skip-grams. Known stemmers for Turkish[4] cannot handle noisy text and cause loss of important information. Therefore, we have decided to do stemming by taking the first five characters of words only while extracting N-gram and skip-gram features.

Turkish person names are not only useful for CLIENT entity type, but also ORG entities, since some of the Turkish organization names contain person names. Therefore, a person name gazetteer[5] with 12.828 Turkish person first names is used for extracting features.

In cases like in Fig. 2b, entities come after words (tuples) that end with a colon character. Moreover, amounts may be written in parentheses and may contain

---

[4] Zemberek https://github.com/ahmetaa/zemberek-nlp.
[5] The gazetteer is provided by the bank.

dot and comma characters. These are some of the many reasons why we have decided to extract features from punctuation characters. For example, if the word begins with a parentheses and ends with a colon, "BEGIN_PARANTHESES", "END_COLON" boolean features will be extracted.

The information whether a word begins/ends with a capital or contains numbers can be considered as valuable for named entity recognition. To obtain that information, "d" digit, "p" punctuation, "s" lowercase letter, "S" uppercase letter shape features are extracted for each character in each word. After this pre-processing step, consecutive shape features are concatenated to form one complete feature. For instance, "dpdpdpS" feature will be extracted from word "1,300.00-EUR".

As we have discussed so far, we have decided on a subset of features from various studied features for NER in the literature [6]:

- Most frequent N-grams
- Most frequent skip-grams
- Person names
- Punctuation information
- Word shape information

POS tagging accuracy degrades significantly due to the unstructured documents as mentioned in Sect. 2.3. Furthermore, due to noisy text, first or last n characters of the words are untrustworthy. For these reasons, fixed length suffixes, prefixes and also POS tags are not used as features.

3.3   Named Entity Types

In our study, named entities are more customized than ENAMEX (person, organization, location) and NUMEX (money and percentage). To understand the problem better, these entity types are explained in Table 1.

4   Experimental Results

We have created training data by annotating 365 real transaction order documents that have been received by the bank during a period of two months. Annotation of training data is done by two different trained annotators, their results are inspected and merged by an expert. Test data has been gathered on a completely different time interval and annotated by another trained annotator, but controlled by the same expert. Details of this manually annotated data is given in Table 2.

We have used the open source Mallet[6] framework for our experiments on Hidden Markov Model (HMM) and CRF. As the first experiment, we have compared HMM with CRFs without features and observed that even when all features are employed, CRFs have provided better results in terms of accuracy. Therefore, we have decided to use CRFs in further evaluations.

Table 1. Descriptions of labels

Label	Description
ACCOUNT NUMBER	Ranges between 6–9 characters, but mostly given with 7–8 characters. In some cases it is concatenated to bank branch code with a hyphen character or contain space characters, which makes it harder to extract ("332-12345678" or "123 45 678")
AMOUNT	Amount could be numeric, completely alphabetic or both as in "1.500,00 TL (binbeşyüztürklirası)" has its alphabetic parts written concatenated
CLIENT	Corresponds to person names. But, documents may also contain irrelevant person names (bank employees name etc.), therefore not all person names are considered as CLIENT
CURRENCY	Currency types of different countries, could also be written as abbreviations. "Euro" could be written as "Avro" (in Turkish), "EUR" or "€"
EXPLANATION	Composed of multiple alphanumeric words. Since it does not have any special form, this entity is the hardest one to be extracted among others
IBAN	15 to 34 alphanumeric characters depending on the country. Since it begins with two uppercase characters, it is relatively easy to detect with a proper feature set
ORGANIZATION	Organization names, just as in ENAMEX. It may end with some suffixes such as "a.ş", "ltd.", "şti.", which makes it easier to predict

Table 2. Counts of entities of our training and test data. "O" denotes words that are not entities and not added to the total

Label	Training	Test
ACC	457	199
AMOUNT	736	391
CLIENT	227	66
CURR	542	264
EXPL	428	349
IBAN	1517	1061
ORG	3302	2028
O	27037	15574
Total	7209	4358

When previous and next tokens (also their features if available) are included in [-3,3] window as features (+window) for each word, overall CRF performance is increased by 21 %. We take this resulting method as our baseline and gradually added the features described in Sect. 3.2. As expected, punctuation specific features (+punc) especially increased F1 score of AMOUNT entity. Moreover, including word shape features (+wordshape) has tremendously improved IBAN entity, since "Sd", "d", "d", "d" features will be extracted respectively from IBANs such as "TR01 2345 6789 1234". Person name features (+names) have been included to the feature list to increase CLIENT detection performance. As a result, F-score of CLIENT entity is increased to 40.85 %. Finally, by using n-gram and skip-gram features (+gram) CRF gains the ability to predict EXPL entities. However, since predicting EXPL entities are hard due to their free form, only 21 % of these entities are correctly extracted (Table 3).

Table 3. F1 scores of CRF and HMM. + means that the feature is added to the model above

Features	ACC	AMOUNT	CLIENT	CURR	EXPL	IBAN	ORG	AVG
HMM	0	21.88	0	27.41	0	8.21	28.55	19.76
+all	0	46.65	0	65.88	0	6.78	29.26	23.45
CRF	5.85	52.56	0	86.39	5.03	47.92	32.72	40.26
+window	71.06	58.4	0	83.95	4.42	75.14	56.86	61.3
+punc	72.88	68.17	0	84.68	3.68	73.91	59.16	62.91
+wordshape	87.74	70.92	0	83.67	5.9	91.73	60.78	69.51
+names	87.19	70.83	40.85	83.27	5.8	91.95	61.61	70.17
+gram	86.33	72.23	36.73	86.51	21.05	89.62	67.18	**72.77**

Table 4. F1 scores of our method with all features on 5-fold cross validation and test set

Label	Cross-validation		Test set	
	RAW	CoNLL	RAW	CoNLL
ACC	91.04	90.63	86.86	86.33
AMOUNT	83.45	81.67	79.14	72.23
CLIENT	61.06	61.06	42.18	36.73
CURR	87.87	87.66	86.51	86.51
EXPL	65.99	56.49	28.23	21.05
IBAN	95.16	90.68	96.82	89.62
ORG	82.05	73.23	74.32	67.18
AVG	84.72	78.89	79.11	**72.77**

---

[6] Available from http://mallet.cs.umass.edu/.

Upon this point, given F1 measures were in CoNLL (Conference on Computational Natural Language Learning) metric. In CoNLL metric, a prediction is considered true positive, if all words of the entity are predicted correctly [3]. However, we also measured F1 scores using word by word predictions (RAW) and can conclude that segmentation is responsible from 7 % of the drop in F1 measure as given in Table 4.

5   Conclusions and Future Work

In this work, we have described our methodology that tackles the NER problem in noisy Turkish banking documents. In our method, we have trained CRFs with n-gram, skip-gram, punctuation and shape features along with the ones from a person name gazetteer. We compared F-scores (CoNLL metric) of each entity type with the proposed features step by step in order to see the impact on detection of specific entity types. Our final model has achieved 72.77 % F-score which is significantly lower than the models trained with the news data. We have expected the result to be worse than news domain due to noisy text produced by OCR systems and unstructured sentences.

This study is implemented as an information extraction system that is integrated to core banking. There are a thousand of maker (data entrance) and checker (data validator) human operators currently working for completing a 100 thousand customer orders each day to the bank. Our system automatizes the maker operators' data entrance effort, minimizes the time required to enter data and eliminates possible typing errors. After the system integration, 20 % of the manual workforce is saved. This is accomplished by digitalizing 53 % of the process workflow by replacing operators' workforce with the system. As a result overall cycle time of target processes is reduced significantly. For instance; book-to-book money transfer cycle time is decreased from 45 min to 13 min and EFT cycle time is decreased from 31 min to 19 min. On a yearly basis 53 % digitalization will mean 3.2 millon transactions to be completed without human workforce, since there are roughly 6.5 million transactions arriving each year.

In the future, text normalization methods might be applied to reduce the negative effects of noise. Also, ensemble methods like bagging and boosting can be tested to improve classification performance. In addition to the person name gazetteer as discussed in Sect. 3.2, an organization name gazetteer might be used to increase ORG entity prediction performance.

Some of the documents contain multiple orders in single sentence (compound order) such as money transfer followed by a currency exchange. These kind of documents require further processing since entity extraction is not enough and relations between entities should also be extracted. As a future work, we aim to extract these kind of relations between entities.

References

1. Lafferty, J., McCallum, A., Pereira, F.C.: Conditional random fields: probabilistic models for segmenting and labeling sequence data (2001)

2. Sutton, C., McCallum, A.: An introduction to conditional random fields. Mach. Learn. **4**(4), 267–373 (2011)
3. Nadeau, D., Sekine, S.: A survey of named entity recognition and classification. Lingvisticae Investigationes **30**(1), 3–26 (2007)
4. Seker, G.A., Eryigit, G.: Initial explorations on using CRFs for Turkish named entity recognition. In: COLING, pp. 2459–2474 (2012)
5. Yeniterzi, R.: Exploiting morphology in Turkish named entity recognition system. In: Proceedings of the ACL 2011 Student Session. pp. 105–110. Association for Computational Linguistics (2011)
6. Tkachenko, M., Simanovsky, A.: Named entity recognition: exploring features. In: KONVENS, pp. 118–127 (2012)
7. Tatar, S., Cicekli, I.: Automatic rule learning exploiting morphological features for named entity recognition in Turkish. J. Inf. Sci. **37**(2), 137–151 (2011)
8. Kucuk, D., Steinberger, R.: Experiments to improve named entity recognition on Turkish tweets. In: Proceedings of 5th Workshop on Language Analysis for Social Media, pp. 71–78 (2014)
9. Yamada, I., Takeda, H., Takefuji, Y.: Enhancing named entity recognition in Twitter messages using entity linking. In: ACL-IJCNLP, p. 136 (2015)
10. Eken, B., Tantug, C.: Recognizing named entities in Turkish tweets. In: Proceedings of 4th International Conference on Software Engineering and Applications, Dubai (2015)
11. Tur, G., Hakkani-Tur, D., Oflazer, K.: A statistical information extraction system for Turkish. Nat. Lang. Eng. **9**(02), 181–210 (2003)
12. Celikkaya, G., Torunoglu, D., Eryigit, G.: Named entity recognition on real data: a preliminary investigation for Turkish. In: 7th International Conference on Application of Information and Communication Technologies (AICT), pp. 1–5 (2013)
13. Guthrie, D., Allison, B., Liu, W., Guthrie, L., Wilks, Y.: A closer look at skip-gram modelling. In: Proceedings of 5th international Conference on Language Resources and Evaluation (LREC-2006), pp. 1–4 (2006)
14. Settles, B.: Biomedical named entity recognition using conditional random fields and rich feature sets. In: Proceedings of International Joint Workshop on Natural Language Processing in Biomedicine and Its Applications, pp. 104–107 (2004)
15. Klinger, R., Friedrich, C.M., Fluck, J., Hofmann-Apitius, M.: Named entity recognition with combinations of conditional random fields. In: Proceedings of 2nd Biocreative Challenge Evaluation Workshop (2007)
16. Sha, F., Pereira, F.: Shallow parsing with conditional random fields. In: Proceedings of 2003 Conference of the North American Chapter of the Association for Computational Linguistics on Human Language Technology. vol. 1, pp. 134–141 (2003)

A New Operationalization of Contrastive Term Extraction Approach Based on Recognition of Both Representative and Specific Terms

Aliya Nugumanova[1](✉), Igor Bessmertny[2], Yerzhan Baiburin[4], and Madina Mansurova[3]

[1] D. Serikbayev East Kazakhstan State Technical University,
Ust-Kamenogorsk, Kazakhstan
yalisha@yandex.kz
[2] Saint Petersburg National Research University of ITMO,
Saint Petersburg, Russia
igor_bessmertny@gmail.com
[3] Al-Farabi Kazakh National University, Almaty, Kazakhstan
mansurova01@mail.ru
[4] East Kazakhstan State University, Ust-Kamenogorsk, Kazakhstan
ebaiburin@gmail.com

Abstract. A contrastive approach to term extraction is an extensive class of methods based on the assumption that the words frequently occurring within a domain and rarely beyond it are most likely terms. The disadvantage of this approach is a great number of type II errors – false negatives. The cause of these errors is in the idea of contrastive selection when the most representative high frequent terms are extracted from the texts and rare terms are discarded. In this work, we propose a new operationalization of the contrastive approach, which supports the capture of both high frequent and low frequent domain terms. Proposed operationalization reduces the number of false negatives. The experiments performed on the texts of the subject domain "Geology" show promising of proposed approach.

Keywords: Contrastive term extraction · Termhood · Mutual information · LSA

1 Introduction

At present, in the field of information extraction there are numerous methods aimed at automated extraction of knowledge structures from natural language texts [1–4]. Among the knowledge structures being extracted, the simplest ones are lists of terms and the most complex ones are domain thesauri and ontologies. All these structure are designed for setting explicit specifications of subject domain to eliminate uncertainty and ambiguity in the knowledge exchange between humans and applications.

In this work, we focus on extraction of simple but valuable knowledge structures – lists of single word terms. Like the authors of [5], we consider domain terms to be words used by experts to describe conceptual apparatus of the domain. Lists of terms are used everywhere when it is necessary to convey in a structured compressed form

© Springer International Publishing Switzerland 2016
A.-C. Ngonga Ngomo and P. Křemen (Eds.): KESW 2016, CCIS 649, pp. 103–118, 2016.
DOI: 10.1007/978-3-319-45880-9_9

the semantics of a text, topic or domain. Apart from its original value, lists of terms are used as building material for more complex structures, for example, ontologies. Therefore, the significance of automated term extraction can hardly be overestimated.

The process of automated term extraction is very often accompanied by the process of validation. Validation is usually carried out by domain experts in manual or semi-automatic mode and allows confirming or disproving the terminological status of the words being extracted. Therefore, the authors of [6] think it to be more correct to call the words being extracted not terms but terms-candidates from which, in the process of validation, true domain terms will be selected. As a rule, to support the process of validation, the output list of terms-candidates are ranked by the degree of their termhood – an index defining how much a term-candidate corresponds to the conceptual apparatus of the domain [6, 7]. Sometimes, the process of ranking replaces the process of validation, in this case the candidates which termhood values exceed the specified threshold are deemed as terms.

The problem of appropriate measuring of termhood is central in the task of term extraction. In [8], the authors present a detailed review of the corresponding methods and conclude that most of them measure the termhood heuristically: on the basis of assumptions about the character of distribution of terms in texts. The popularity of heuristic approaches can be accounted for the conditions of uncertainty, which accompany the process of terms search. Both the terms themselves and features allowing to recognize terms are unknown. In work "Mathematical discovery", G. Pólya noted that a clearly formulated problem must precisely specify the condition, which has to be satisfied by the unknown [9]. If such conditions cannot be specified, it is expedient to use heuristic methods.

In this work, we study a popular heuristic approach based on contrastive term extraction [8, 10]. This approach uses an idea that the words, which often occur in the domain texts and seldom in the texts of other domains are most likely terms [8]. Like all heuristics, the contrastive method suffers from a grave shortcoming, such as the loss of recall. It makes a great number of the second type errors, i.e. false negatives. The cause of errors is in the idea of contrastive selection when the most representative (high frequent) words are extracted from texts and rare words are discarded. According to the opinion of the authors [11], the practice of exclusion of rare words is usual for the information retrieval tasks but it is not always useful for term extraction. As an example, the authors refer to the collection of medicinal abstracts in which rare terms denoting side effects due to administration of medicinal preparations make up 68 %. Exclusion of these rare terms from consideration would result in great losses, when specifying a subject domain.

The aim of this work is to overcome the above-mentioned shortcoming, i.e. to enhance the recall of contrastive term extraction based on the separated capture of rare and representative terms. For this, we use mutual information criterion described in [12] in detail. For representative terms extraction we use average weighted mutual information, and for rare terms extraction we use pointwise mutual information. However, we recognize that mutual information has own drawbacks. It makes errors of the first type, i.e. false positives. The cause of the errors is in the presence of the so-called conjugate words, i.e. words that "accompany" the domain, but not related to it directly. In [13] as an example quoted the Gulf and Kuwait words, which are closely related to the domain "Oil" due to their frequent occurrence in texts associated with this domain. However, in fact they are not terms.

Therefore, in this work we have to adjust the estimates obtained with the help-means of mutual information involving a special technique based on the analysis of the causes of false positives. The structure of the work is as follows. Section 2 presents the review of related work dealing with the method of contrastive term extraction. Section 3 contains the necessary theoretical information advancing our approach. Section 4 describes the method being proposed by us allowing to away with the existing drawbacks of the contrastive approach. Section 5 presents the case-study on term extraction from the texts of the subject domain "Geology". In Sect. 5, we formulate conclusions and present a plan for further investigations.

2   Related Work

A contrastive approach is a general name of the methods revealing terms from the point of their different occurrence within and beyond a subject domain. All these methods are governed by general idea of defining the termhood of words based on comparison of their distribution in two collections: target domain and alternative. As an alternative collection, we may use either a contrastive collection, i.e. formed from the texts of different domains, or general collection, i.e. formed from the texts which do not refer to any domain [14]. The differences between contrastive methods are only in the ways of operationalization of the idea in their basis. In this case, we speak about operationalization of such a fuzzy concept as termhood by means of measurable indicators or procedures based on the contrastive approach.

One of the first works concerning contrastive term extraction is [15]. To evaluate the termhood, its authors introduce a new intuitively perceived measure called "weirdness". The weirdness is calculated for each term-candidate and is the ratio of the frequency of term occurrence in a target collection to the frequency of occurrence in a general collection. For usual words, the weirdness formula restores values close to 1, and for terms – values much more exceeding 1 as in this case the denominator of the formula is close to 0. In their later works, for example, in [16], the authors present a modified variant of the formula as the initial formula develops singularity when the denominator turns 0.

In [17], the idea of contrastive term evaluation is formed as not one but two statements: (1) the words more rarely used in the target collection must have a lower value; (2) the words more frequently used in the target collection must have a higher value, bit with reserve that they do not often occur in the contrastive collection or in a limited set of texts of the target collection. The authors operationalize these statements in the form of metrics called by them relevance:

$$Relevance = \frac{1}{log_2(2 + \frac{f_{SL} * N_{SL}^t}{f_{GL}})} \tag{1}$$

where, f_{GL} and f_{SL} are relative frequencies of the word in the target and contrastive collections, respectively; N_{SL}^t is a relative number of texts of the target collection in which this word occurs. The reduced metrics copes well with extraction of representative terms but artificially decreases the value of rare terms.

In [18], the authors propose a somewhat other method for evaluation of termhood on the basis of the contrastive approach. The method is based on the well-known TF-IDF formula according to which the weight of the word in the document is the higher, the higher the frequency of its occurrence in this document and the lower its dispersion in the whole collection. In the new formula, which the authors call "term frequency-inverse domain frequency" they evaluate the weight of the word not in the document but in the target collection. According to the new formula, the weight of the word is the higher, the higher the relative frequency of its usage in target collection and the lower its relative dispersion in all collections:

$$TF \cdot IDF = TF(t, D) \cdot IDF(t) = \frac{n_{t,D}}{\sum_k n_{k,D}} \cdot log\left(\frac{|TS|}{|\{d : t \in d\}|}\right) \qquad (2)$$

where, $n_{t,D}$ is the number of the word t entry into the target collection D, $\sum_k n_{k,D}$ is the sum of all words entries into the target collection D, $|TS|$ is the number of documents in all used collections, $|\{d : t \in d\}|$ is the number of all documents into which the word t enters at least once. Thus, the authors consider all words with high concentration within a narrow subset of documents to be terms. For a definite part of terms it is undoubtedly a correct approach but for rare terms it is of title use.

The authors of [19] evaluate the termhood on the basis of formula TF-IDF, too. They call their variant of this formula a contrastive weight and define it as a measure, which is the higher, the higher the frequency of the word usage in the target collection and the lower the relative frequency of its usage in contrastive collections:

$$Contrastive\ Weight = TF(t, D) \cdot IDF(t) = log\left(f_t^D\right) \cdot log\left(\frac{F_{TC}}{\sum_j f_t^j}\right) \qquad (3)$$

where, f_t^D is the frequency of the word usage in the target collection, $\sum_j f_t^j$ is the sum of frequencies of usages of all words in all collections including the target one. As the authors themselves note, the contrastive weight evaluates the termhood of words significantly better than pure frequencies, however, the efficiency of the method determined with the help of F-measure, according to their words, is not evident.

In [20], the formula of contrastive weight undergoes a critical evaluation. As the authors note, the contrastive weight and similar metrics evaluate in fact not the reference of terms to a domain but their prevalence. To rectify the mentioned drawback, the authors evaluate the termhood on the basis of two indices: DP (domain prevalence) and DT (domain tendency). The formula for calculation of DP is in fact a subdued variant of the contrastive weight (3). Its high value indicates the prevalent distribution of the word in the target collection compared to distribution of the word in the target collection compared to distribution of other words in this collection:

$$DP = log_{10}(f_t^D + 10) \cdot log_{10}\left(\frac{F_{TC}}{f_t^D + f_t^{\bar{D}}} + 10\right) \qquad (4)$$

where, f_t^D and $f_t^{D^-}$ are frequencies of the word in the target and contrastive collections, respectively, $F_{TC} = \sum_j f_j^D + \sum_j f_j^{D^-}$ are sums of frequencies of all terms-candidates usage in the target and contrastive collections, respectively.

The formula for calculation of DT is a subdued variant of the weirdness formula, i.e. it penalizes the words, which often occur in the contrastive collections. Its high value indicates the prevalent distribution of the word in the target collection compared to its distribution in the contrastive collections:

$$DT = log_2\left(\frac{f_t^D + 1}{f_t^{\bar{D}} + 1} + 1\right) \qquad (5)$$

The measures of DP and DT are combined into one general index called discriminative weight DW. It is the product of DP and DT measures. According to the authors' opinion, the discriminative weight possesses a differentiating ability. It pushes up the terms which often occur in the target collection and rarely in the contrastive collections. However, the indices DT and DP correlate with each other quite well. For example, in our experiments, the correlation values of these indices made up 0.71–0.82. To reveal the nature of this correlation, we divided all terms-candidates into 4 not intersecting groups depending on DT and DP: (1) the values of DT and DP are lower than average; (2) the value of DT is lower than average and the value of DP is not lower than average; (3) the values of DT is not lower than average and the values of DP is lower than average; (4) the values of DT and DP are not lower than average. Both expert evaluations and evaluations based on formula (5) showed the same result: with few exceptions, only candidates from the third and fourth groups can be recognized as terms, this corresponding to high values of DT index. Such result indicates high informational content of DT index and redundancy of DP index.

The reservation "with few exceptions" is not coincidental. Validation of the proposed method demonstrates the fact that in the area of low values of termhood, among usual words, there occur terms which we call rare terms. These are terms, which occur 1–2 times in the target collection and not a single time in the contrastive collection. Extraction of such terms requires the use of more distinctive instruments of differentiation.

Not only is the work [20] distinguished by the use of at once several indices for evaluation of termhood. In [21] for this purpose, at once 3 indices are used: domain pertinence DP, domain consensus DC and lexical cohesion LC, assigned for evaluation of cohesion of verbose terms. As a result, the total evaluation of the word termhood in the target collection Di is formed from the linear combination of the three enumerated measures. The measure of pertinence DR is the measure of weirdness generalized for the case of a contrastive collections set:

$$DR(t, Di) = \frac{freq(t, Di)}{\underset{j}{\max}(freq(t, Dj))} \tag{6}$$

The measure of consensus DC allows to take into account the distribution of words in separate documents. It is defined via normalized frequencies φ_k of the word occurrence in the documents of target collection Di and is the higher, the more uniformly the word is distributed in these collections:

$$DC(t, Di) = -\sum_{d_k \in Di} \phi_k log \phi_k \tag{7}$$

Introducing the measure of consensus, the authors emphasize its high importance. According to their opinion, the terms which frequently occur in a great number of documents of target collection must be evaluated higher than the terms which frequently occur in a restricted number of documents. It should be noted that this statement is completely antagonistic to heuristic used in [18]. The authors of [22] also note this interesting fact illustrating a wide interpretation of the notion 'termhood' and high uncertainty in the choice of criteria of termhood.

In [23], termhood is defined not via the ratio of the word usage frequency in the target and contrastive collections but via the difference of their ranks. The rank of the word in the collection is its position in the dictionary compiled from all words of the collection sorted out according to the increase in the word usage frequencies. The index of the termhood expressed via the difference of word ranks in the target D and contrastive G collections has the form:

$$thd(t, D) = \frac{rank(t, D)}{|V(D)|} - \frac{rank(t, G)}{|V(G)|} \tag{8}$$

where $V(D)$ and $V(G)$ are vocabularies of the corresponding collections. The index has values from -1 to 1, the value 1 corresponds to the case when the word has the highest rank in the target collection and a zero rank in the contrastive collection. Ranking of words allows to draw up the most representative words. However, as the authors note, their approach does not allow to extract only terms.

The last work we would like to note in this number is the [24]. It also develops the idea of penalties and rewards which was introduced into the basic construction of TF-IDF formula and proposes a new variant of this formula called "term frequency – disjoint corpora frequency":

$$TF \cdot DCF = \frac{f_t^D}{\prod_{g \in G} 1 + log(1 + f_t^g)} \tag{9}$$

where f_t^D and f_t^g are frequencies of this word occurrences in target and contrastive collections respectively, G is a set of all contrastive collections.

The authors prove on a series of experiments that their formula is the best by precision of terms extraction compared to the values proposed in works [18, 23] and in a number of other works. They justify the use of the product in the denominator of the formula by the fact that penalty much increase in a geometrical progression for each use of the word in the next in turn contrastive collection. According to the opinion of the authors, the termhood of words which are used few times in a great number of contrastive collections must be evaluated lower than that of the words which are used many times but a small number of contrastive collections.

Thus, in this review we considered 8 most interesting contrastive methods of the termhood value operationalization. All these methods are heuristic, i.e. based on assumptions concerning the character of terms distribution in target and contrastive collections. A comparative analysis of these statements show that there take place both coincidences of positions of different authors and grave divergences, this indicating the presence of unsolved problems in this field.

3 Theoretical Background

The authors of [25] note that the term extraction methods based on heuristic assumptions are often criticized for the absence of theoretical strictness. The authors say that such criticism becomes evident when simple but important questions on the methods of operationalization of these or those heuristics are put, for example "Why are different bases of logarithms used in metrics?" or "Why combination of two weights in metric is based on their product but not their sum?"

However, popularity of using similar ad-hoc metrics can be easily explained. Termhood is a notion which is rather difficult to be formalized and it is easier to express it with the help of heuristics captured of a set of termhood aspects only those which are the most evident and available for formalization. Meanwhile, in the field of automated extraction of terms there have been developed strict statistical criteria based on mathematical grounds of information theory and probability theory.

Mutual information refer to the number of such criteria [12]. The notion of mutual information goes back to the more fundamental and more general notion of information theory – informational divergence, also known as the distance of Kullback–Leibler or relative entropy. Informational divergence is an asymmetric measure of the distance between two discrete distributions $P = \{p_i\}$ and $Q = \{q_i\}$:

$$D(P\|Q) = \sum_i p_i log\left(\frac{p_i}{q_i}\right) \qquad (10)$$

here and further, the base of logarithms is taken as standard value 2. As a rule, one of the distributions being compared is 'true' (observable) and the second one is expected (under test). Therefore, informational divergence may be treated as the measure of how much the "true" distribution diverges with the expected, approximate one. Mutual information is a particular case of informational divergence when distribution P is a joint distribution of two random discrete values X and Y and distribution Q is the product of marginal distributions of these random values [26]:

$$MI(X, Y) = D(P(X, Y)||P(X) \times P(Y)\} = \sum_{i,j} p(x_i, y_j) \cdot log\left(\frac{p(x_i, y_j)}{p(x_i)p(y_j)}\right) \quad (11)$$

If random values X and Y are independent, the probability of their joint distribution is equal to the product of probabilities of their marginal distributions $p(x_i, y_j) = p(x_i)p(y_j)$, then $log\left(\frac{p(x_i, y_j)}{p(x_i)p(y_j)}\right) = log 1 = 0$. Hence, mutual information of these values is equal to 0. Intuitively, this can be explained as follows: if two random values are independent, the appearance of one of them does not give any information in regard to the appearance of the other. Correspondingly, mutual information can be treated as the measure of correlation of these values. The notion of mutual information determined for two random values X, Y is closely related to the pointwise mutual information, defined for a concrete pair of outcomes (x, y) of these random values:

$$PMI(x, y) = log\frac{p(x, y)}{p(x) \cdot p(y)} \quad (12)$$

Having compared formulas (11) and (12), one may note that mutual information is an average weighted estimation of pointwise mutual information values on all pairs of random values X and Y outcomes:

$$MI(X, Y) = \sum_{i,j} p(x_i, y_j) \cdot PMI(x_i, y_j) \quad (13)$$

The obtained formulas (12), (13) fit well to the problem of contrastive term extraction. To derive the fit formulas, the authors of [12] introduce designations as is shown in Table 1, and on the basis of this table they evaluate distribution probabilities of two random values: X is the presence of the word in the document; Y is the ratio of the document to the subject domain (see Tables 2, 3 and 4).

For each of the four possible outcomes presented in Table 4, its own formula of pointwise mutual information is derived on the basis of formula (12). Of basic interest is the formula for the outcome $t \wedge d$:

$$
\begin{aligned}
PMI\left(t \wedge d\right) &= log\frac{p(t \wedge d)}{p(t)p(d)} = log\frac{A/N}{((A+B)/N) \times ((A+C)/N)} \\
&= log\frac{A \times N}{(A+B) \times (A+C)}
\end{aligned} \quad (14)
$$

Table 1. The contingency table describing the distribution of words in collections

Number of documents	In target collection	In contrastive collection	Total
Containing this word	A	B	A+B
Not containing this word	C	D	C+D
Total	A+C	B+D	N = A+B+C+D

Table 2. Marginal distribution of random variable X

Outcomes	Probability
t : the document contains the word	$p(t) = (A+B)/N$
$\bar{t}$: the document does not contain the word	$p(\bar{t}) = (C+D)/N$
Σ:	$p(t) + p(\bar{t}) = 1$

Table 3. Marginal distribution of random variable Y

Outcomes	Probability
d : the document refers to the subject domain	$p(d) = (A+C)/N$
$\bar{d}$: the document does not refer to the subject domain	$p(\bar{d}) = (B+D)/N$
Σ:	$p(d) + p(\bar{d}) = 1$

Table 4. Joint distribution of random variables X, Y

Outcomes	Probability
$t \wedge d$: the document contain the word and refers to the subject domain	$p(t \wedge d) = A/N$
$\bar{t} \wedge d$: the document does not contain the word and refers to the subject domain	$p(\bar{t} \wedge d) = C/N$
$t \wedge \bar{d}$: the document contain the word and does not refer to the subject domain	$p(t \wedge \bar{d}) = B/N$
$\bar{t} \wedge \bar{d}$: the document does not contain the word and does not refer to the subject domain	$p(\bar{t} \wedge \bar{d}) = D/N$
Σ:	1

The given formula allows to evaluate the amount of information carried by the fact of the presence of this word in the document of the subject domain [27]. The formulas of point wise mutual information for the other three outcomes have the following form:

$$PMI\left(\bar{t} \wedge d\right) = log \frac{p(\bar{t} \wedge d)}{p(\bar{t})p(d)} = log \frac{C \times N}{(C+D) \times (A+C)} \tag{15}$$

$$PMI\left(t \wedge \bar{d}\right) = log \frac{p(t \wedge \bar{d})}{p(t)p(\bar{d})} = log \frac{B \times N}{(A+B) \times (B+D)} \tag{16}$$

$$PMI\left(\bar{t} \wedge \bar{d}\right) = log \frac{p(\bar{t} \wedge \bar{d})}{p(\bar{t})p(\bar{d})} = log \frac{D \times N}{(C+D) \times (B+D)} \tag{17}$$

The formula of average weighted mutual information is derived on the basis of formula (13) and obtained values of PMI similarly:

$$MI = \frac{A}{N} log \frac{A \times N}{(A+B)(A+C)} + \frac{C}{N} log \frac{C \times N}{(C+D)(A+C)} + \frac{B}{N} log \frac{B \times N}{(A+B)(B+D)} + \frac{D}{N} log \frac{D \times N}{(C+D)(B+D)} \tag{18}$$

Despite the fact that both formulas of mutual information (14) and (18) evaluate the amount of information, which is carried by the word on the subject domain, there is a principal difference between them, which is best illustrated by the known expression "briller par son absence".

In other words, average weighted mutual information evaluates the relation between the word and subject domain taking into account not only the probability of its presence in target collection (outcome $t \bigwedge d$) but also absence (outcome $\bar{t} \bigwedge d$) as well as probabilities of it presence and absence in contrastive collection (outcomes $t \bigwedge \bar{d}$ and $\bar{t} \bigwedge \bar{d}$, respectively). The work [12] states the following important fact which is the effect of the mentioned difference: pointwise mutual information is biased to the side of rare terms, while average – weighted mutual information normalizes bias on account of using weights.

Let us consider on account of what there takes place the bias of estimations obtained with the help of pointwise mutual information, i.e. let us determine under what conditions formula (18) has a maximum. In the reduced formula, values N and $A + C$ (the total number of documents and the number of documents of target collection, respectively) are constants as they do not depend on distribution of words. Therefore, they can be ignored and function $log\left(\frac{A}{A+B}\right) = log\left(\frac{1}{1+B/A}\right)$ must be considered. As we speak about logarithms by base 2 (increasing function), the function aspire to maximum when B/A aspire to minimum. The expression B/A reaches minimum only at $B = 0$, i.e. when the word never occurs in contrastive collection. It does not matter what A equals to, i.e. to it is not important whether the word occurs in target collection 50 or 10 times, evaluation of these words will be the same. In all other cases, when B is not equal to 0, the value of A does not effect the termhood evaluation, only the ratio B/A matters. For example, if the first word occurs 50 times in the target collection and 10 times in contrastive collection and the second word occurs 10 times in target and once in contrastive collection, then the termhood of the second word will be evaluated higher, since $B_1/A_1 = 10/50 > B_2/A_2 = 1/10$.

4   Proposed Approach

We conditionally divide all the words related to the domain into 2 classes. We refer the frequently used, representative words of the domain to the first class and rarely used, specific words of the domain – to the second class. For example, in the field of geological sciences the words "geology", "rock", "crust" refer to the first class, and the words "breccia", "feldspar", "columbite" – to the second class.

We consider that, since the words referring to these two classes are quite differently distributed in the domain, it is necessary to use not one but two criteria to extract them. In particular, as we showed in the previous section, to extract words of the first class, the average weighted mutual information fits well and for extraction of the second class words – pointwise mutual information (see Table 5).

To our mind, the main drawback of the methods considered in Section "Related work" is that they aim to combine several differently oriented metrics in one criterion uniting them by multiplication or addition. Thereby, the mentioned methods try to find

Table 5. The criteria used for term extraction

Class	Representative words	Specific words
Used criteria	The average weighted mutual information MI	Pointwise mutual information PMI
Condition of selection into class	$A > 0$, $MI > MI_{crit}$	$MI \leq MI_{crit}$, $PMI > PMI_{crit}$
Notation	MI_{crit} – is a critical value of criterion MI determined on the basis of Student's test at the level of significance $\alpha = 0.01$ and the number of degrees of freedom equal to the dimensions of the subdictionary.	PMI_{crit} – is a critical value of criterion PMI determined on the basis of Student's test at the level of significance $\alpha = 0.01$ and the number of degrees of freedom equal to the dimensions of the subdictionary.

a compromise where it cannot be. Unlike the authors of these methods, we believe that a method of separate per class term capture will allow to enhance the term extraction recall though it will not solve the problem of the term extraction precision.

The problem of the term extraction precision is that among the words of both the first and the second class there can occur not only terms but also the words. These words are statistically distributed in the same way as terms but in fact they are not terms. They form a set of false positives. For example, in the target collection of texts on geology used by us the words "geyserite" and "resort" occur only in one document and not a single time – in contractive collection. Correspondingly, as we noted in the previous section, the both words are characterized by the maximum value of pointwise mutual information though one of them is a term and the second one is not.

This is caused by imperfection of target and contrastive collections: if these collections were sufficiently full and extensive, the term "geyserite" would occur more frequently in target collection, while the word "resort" would more frequently occur in contrastive collection. Thus, after formation of both classes we have to eliminate the words, which are not terms from the each class. For this, let us analyze the causes of false positives in each class under study and propose the methods for elimination of these false positives (see Table 6).

As follow from this table, to recognize false terms among specific words, it is not sufficient to have only information on indices of word occurrence. More extensive information is needed and we can gain it using, apart from the indices of occurrence of words, the indices of their co-occurrence. This idea is not new. It is noted in [28] that traditionally the importance of a term-candidate is determined on the basis of analysis of how it is related to the other words of the text or collection. A term-candidate is considered to have a large weight if it is related to either a large number of other terms-candidates or terms-candidates which themselves have a large weight.

In this work, we, like the authors of [29], are guided by this idea and consider a co-occurrence matrix as an instrument for measuring the relations between words. The contribution of our work is the method of construction and processing of this matrix. At first, we form a matrix "documents-terms" the rows of which are documents of target

Table 6. The reasons of false alarms and how to resolve them

Class	Representative words	Specific words
The cause of false alarms	The word is often found in the target collection $(A \gg 0)$, but it is quite common and contrasting collection $(A \sim B)$ Uncertainty type $\frac{\infty}{\infty}$	The word is rare in contrast collection $(B \sim 0)$, but it is rarely found in a target collection $(B \sim A)$. $(B \sim A)$. Uncertainty type $\frac{0}{0}$
Result	The ratio A/B did not limited to the top terms and limited to a certain threshold value for not R_{crit}	The ratio B/A is equal to 0 (or very close to 0) for both terms, and not for the terms.
Recognition method of false detection	It is to verify the conditions of $A/B < R_{crit}$, where R_{crit} – is a critical value of criterion A/B determined on the basis of Student's test at the level of significance $\alpha = 0.01$ and the number of degrees of freedom equal to the dimensions of the subdictionary.	Stepping outside of the contrast analysis.

collection and the columns are extracted specific and representative words. Then, we subject this matrix to the operation of singular decomposition to get rid of noise and rarity.

And only after that we go on to formation of the matrix "terms-terms". Since in the matrix "documents-terms" each term is a vector-column, the semantic relation between any two terms can be treated as closeness or distance between vectors corresponding to these terms using a cosine measure:

$$r_{ij} = cos\left(\bar{T}_i, \bar{T}_j\right) = \frac{\bar{T}_i \cdot \bar{T}_j}{|\bar{T}_i| \cdot |\bar{T}_j|} \tag{19}$$

where $\bar{T}_i, \bar{T}_j$ are vector-column of the matrix "documents-terms" corresponding to i-th and j-th terms, respectively, (i, j run over the whole list of terms), r_{ij} is the value of closeness, an element of the matrix of semantic relations. Determination of cosine in the first quadrant of Cartesian coordinates allows to state that maximum possible value of closeness between terms is equal to 1 and minimum possible one is equal to 0. As we are only interested in the strongest and stable relations, we will not take into account the value of closeness lower than a certain threshold. For each specific word we will evaluate the number of strong relations with other words and if this number is higher than a certain threshold value, we will consider this specific word to be term.

5   Experiments

To conduct the experiments, we used the textbook on general geology [30]. We divided all chapters of the textbook to documents and executed the necessary operations on preparation of target collection (i.e. tokenization and lemmatization). Table 7 shows representative words of the domain "Geology" among which there are both terms and non-terms.

Table 7. Table of representative terms in the domain "Geology"

№	Word	Value MI	Value A/B	Term
1	**rock**	**0,552**	**6,15**	**Yes**
2	chapter	0,524	2,60	No
3	**surface**	**0,495**	**5,13**	**Yes**
4	**process**	**0,436**	**3,34**	**Yes**
5	**geological**	**0,402**	**5,78**	**Yes**
6	**mountain**	**0,398**	**4,44**	**Yes**
7	**temperature**	**0,353**	**5,64**	**Yes**
8	**terrestrial**	**0,345**	**4,32**	**Yes**
9	**education**	**0,345**	**4,32**	**Yes**
10	**dynamics**	**0,344**	**5,73**	**Yes**
11	result	0,13	2,60	No
12	name	0,136	3,04	No
13	structure	0,136	3,04	No
14	material	0,126	2,27	No
15	condition	0,125	2,41	No
16	**lithosphere**	**0,103**	**4,17**	**Yes**

The decision on inclusion of the word into the class of representative words was made on the basis value MI, it had to be higher than the critical value 0.0192, and the decision on the termhood was made on this basis of value A/B, it must be higher than the critical value 3.1447. Table 8 presents specific words among which there are also both terms and non-terms.

The decision on inclusion of the word into the class of specific words was made on the basis of value PMI, it must be higher than the critical value 0.9449, and the words with the value MI > 0.0192 did not enter the class (they were already included into the class of representative words). The decision on the termhood of words was taken on the basis of the number of relations, it had to be higher than the threshold value 30. Unfortunately, we did not find the way to estimate this value not empirically and, most likely, our further works will deal with it. It is interesting that when our method selected the word "bomb" as a specific term we were puzzled. However, later, we found out that the word "bomb" is a specific geological term meaning a lump or a piece of lava blown out during eruption in a liquid or plastic state from the crater and acquiring a specific form in the course of flying and solidification in air.

Table 8. Table of specific terms in the subject area "Geology"

№	Word	Value PMI	Value B/A	Number of strong relation	Term
1	**bomb**	**1**	**0**	**136**	**Yes**
2	bombardment	1	0	4	No
3	**breccia**	**1**	**0**	**134**	**Yes**
4	storm	1	0	10	No
5	**bay**	**1**	**0**	**85**	**Yes**
6	vector	1	0	24	No
7	**hematite**	**1**	**0**	**45**	**Yes**
8	**hydrohematite**	**1**	**0**	**54**	**Yes**
9	abundance	1	0	2	No
10	**feldspar**	**1**	**0**	**66**	**Yes**
11	squid	1	0	9	No
12	stone	1	0	29	No
13	reed	1	0	13	No
14	resort	1	0	8	No
15	resort	1	0	27	No
16	**abrasion**	**1**	**0**	**82**	**yes**

Thus, from 1033 representative words were selected 617 terms, from 6489 specific words were selected 1266 terms. This once again confirms the assertion of authors of [11] that the conceptual apparatus of the domain is formed not so much by representative terms as by specific terms.

6   Conclusion

The main problem of the existing methods of automatic term extraction on the basis of contrasting approach is the need to find a compromise between the recall and the precision. Usually the choice of heuristics for the term selection is aimed at the most powerful assumption, resulting in not fully recovered rare terms or wrongly extracted words which are not terms. The proposed approach is based on the average weighted and pointwise mutual information. It provides separate extraction of both representative and specific terms that can simultaneously improve both the recall and the precision of term extraction. The results of our experiments on the "Geology" domain demonstrate that proposed method extracts much more valuable and deep information about the domain compared to the strong, but rough heuristic filters.

Our future work will be focus on comparing our method with other best heuristic filters. Especially we plan to compare our metrics with the TF-DCF metrics proposed in [24]. We conducted some opening experiments and were agree that TF-DCF is a very promising approach balancing extraction of representative and rare terms. For example, TF-DCF as well as PMI gives good estimates for rare terms such as Paleozoic, diopside, kyanite, granulite, cordierite, prominence etc. However, it gives low estimates for rare terms such as geomorphology, Paragneiss, zeolite, pseudomorphism, radioactivity whereas PMI give them high estimates.

References

1. Medelyan, O., Manion, S., Broekstra, J., Divoli, A., Huang, A.-L., Witten, I.H.: Constructing a focused taxonomy from a document collection. In: Cimiano, P., Corcho, O., Presutti, V., Hollink, L., Rudolph, S. (eds.) ESWC 2013. LNCS, vol. 7882, pp. 367–381. Springer, Heidelberg (2013)

2. Medelyan, O. et al.: Automatic construction of lexicons, taxonomies, ontologies, and other knowledge structures. In: Wiley Interdisciplinary Reviews: Data Mining and Knowledge Discover, vol. 3, no. 4, pp. 257–279 (2013)

3. Fan, J., et al.: Automatic knowledge extraction from documents. IBM J. Res. Dev. **56**(3.4), 5:1–5:10 (2012)

4. Aggarwal, C.C., Zhai, C.X.: Mining Text Data. Springer Science & Business Media, New York (2012)

5. Nenadi, G., Ananiadou, S., McNaught, J.: Enhancing automatic term recognition through recognition of variation. In: Proceedings of the 20th International Conference on Computational Linguistics, p. 604. Association for Computational Linguistics (2004)

6. Ahrenberg, L.: Term extraction: A Review Draft Version 091221 (2009)

7. Kageura, K., Umino, B.: Methods of automatic term recognition: a review. Terminology **3** (2), 259–289 (1996)

8. Wong, W., Liu, W., Bennamoun, M.: Determination of unithood and termhood for term recognition. In: Handbook of Research on Text and Web Mining Technologies. IGI Global (2008)

9. Polya, G.: Mathematical Discovery: On Understanding, Learning, and Teaching Problem Solving. Wiley, New York (1981)

10. Heylen, K., De Hertog, D.: Automatic term extraction. In: Handbook of Terminology, vol. 1 (2014)

11. Weeber, M., Baayen, R.H., Vos, R.: Extracting the lowest-frequency words: pitfalls and possibilities. Comput. Linguist. **26**(3), 301–317 (2000)

12. Yang, Y., Pedersen, J.O.: A comparative study on feature selection in text categorization. In: ICML, vol. 97, pp. 412–420 (1997)

13. Kim, S.N., Cavedon, L.: Classifying domain-specific terms using a dictionary. In: Australasian Language Technology Association Workshop 2011, p. 57 (2011)

14. da Silva Conrado, M., Pardo, T.A.S., Rezende, S.O.: A machine learning approach to automatic term extraction using a rich feature set. In: HLT-NAACL, pp. 16–23 (2013)

15. Ahmad, K., et al.: University of surrey participation in TREC8: weirdness Indexing for logical document extrapolation and retrieval (WILDER). In: TREC (1999)

16. Gillam, L., Tariq, M., Ahmad, K.: Terminology and the construction of ontology. Terminology **11**(1), 55–81 (2005)

17. Peñas, A., et al.: Corpus-based terminology extraction applied to information access. In: Proceedings of Corpus Linguistics, pp. 458–465 (2001)

18. Kim, S.N., Baldwin, T., Kan, M-Y.: An unsupervised approach to domain-specific term extraction. In: Australasian Language Technology Association Workshop 2009, pp. 94–98 (2009)

19. Basili, R.: A contrastive approach to term extraction. In: Proceedings of the 4th Terminological and Artificial Intelligence Conference (TIA 2001) (2001)

20. Wong, W., Liu, W., Bennamoun, M.: Determining termhood for learning domain ontologies using domain prevalence and tendency. In: Proceedings of the Sixth Australasian Conference on Data Mining and Analytics, vol. 70, pp. 47–54. Australian Computer Society, Inc. (2007)

21. Sclano, F., Velardi, P.: Termextractor: a web application to learn the shared terminology of emergent web communities. In: Gonçalves, R.J., Müller, J.P., Mertins, K., Zelm, M. (eds.) Enterprise Interoperability II, pp. 287–290. Springer, London (2007)

22. Astrakhantsev, N.A., Fedorenko, D.G., Turdakov, D.Y.: Methods for automatic term recognition in domain-specific text collections: a survey. Program. Comput. Softw. **41**(6), 336–349 (2015)

23. Kit, C., Liu, X.: Measuring mono-word termhood by rank difference via corpus comparison. Terminology **14**(2), 204–229 (2008)

24. Lopes, L., Fernandes, P., Vieira, R.: Estimating term domain relevance through term frequency, disjoint corpora frequency-tf-dcf. Knowl.-Based Syst. (2016)

25. Wong, W., Liu, W., Bennamoun, M.: Determining termhood for learning domain ontologies in a probabilistic framework. In: Proceedings of the Sixth Australasian Conference on Data Mining and Analytics, vol. 70, pp. 55–63. Australian Computer Society, Inc. (2007)

26. Prelov, V.: Mutual information of several random variables and its estimation via variation. Prob Inf Transm. **45**(4), 295–308 (2009)

27. Manning, C.D., et al.: Introduction to Information Retrieval, vol. 1, p. 496. Cambridge University Press, Cambridge (2008)

28. Hasan, K.S., Ng, V.: Automatic keyphrase extraction: a survey of the state of the art. In: ACL, vol. 1, pp. 1262–1273 (2014)

29. Matsuo, Y., Ishizuka, M.: Keyword extraction from a single document using word co-occurrence statistical information. Int. J. Artif. Intell. Tools **13**(01), 157–169 (2004)

30. Sokolovsky, A.K. (ed.): A Textbook of General geology: In 2 volumes, vol. 1, p. 448. KDU, George Town (2006)

Ontology-Based Information Extraction for Populating the Intelligent Scientific Internet Resources

Irina R. Akhmadeeva[1]([⊠]), Yury A. Zagorulko[1],
and Dmitry I. Mouromtsev[2]

[1] A.P. Ershov Institute of Informatics Systems, Siberian Branch of the Russian
Academy of Sciences, Novosibirsk, Russia
{i.r.akhmadeeva,zagor}@iis.nsk.su
[2] ITMO University, St. Petersburg, Russia
mouromtsev@mail.ifmo.ru

Abstract. The paper considers the problems of ontology-based collection of information from the Internet about scientific activity for the population of the Intelligent Scientific Internet Resource. An approach to automating this process is proposed, which combines metasearch and information extraction methods based on ontology, thesaurus and pattern technique. In accordance with the approach, specific methods of information extraction adjustable to the knowledge area and types of information resources are developed for every type of entities (ontology class). Each of these methods includes a set of query templates and a set of information extraction patterns. The query templates constructed on the basis of an ontology class description are used to generate queries to search engines in order to collect web documents containing information about the individuals of this class. Web documents gathered using metasearch methods are analyzed by applying the information extraction patterns. For every kind of information to be extracted, these patterns give text markers defining their position in a web document. The patterns are generated on the basis of an ontology taking into consideration the structure of web documents. Several patterns can be combined together to extract information about related entities. To improve the recall of information extraction, the patterns use alternative terms in different languages from the thesaurus (synonyms and hyponyms) to describe the markers. Experiments showed that the proposed approach allows us to achieve an acceptable recall of the extraction from the Internet of information about scientific activity.

Keywords: Scientific activity · Knowledge area · Ontology · Thesaurus · Information extraction · Metasearch

1 Introduction

Though humanity has accumulated a huge amount of information related to various areas of knowledge and a great bulk of this information is available directly on the Internet, the problem of supplying the scientific community with information on the subjects of interests has no satisfactory solution yet.

© Springer International Publishing Switzerland 2016
A.-C. Ngonga Ngomo and P. Křemen (Eds.): KESW 2016, CCIS 649, pp. 119–128, 2016.
DOI: 10.1007/978-3-319-45880-9_10

This situation is partly attributed to the specifics of scientific knowledge representation on the Internet, which is weakly formalized, insufficiently systematized and distributed among various Internet sites, electronic libraries, and archives.

Another reason is that modern information systems use a rather limited set of methods for information representation, search, and interpretation. As a rule, data and knowledge are represented in these systems as text documents or a set of information resources, though in our opinion the most human-friendly form of information representation is a network of interrelated facts built on an ontology. This mode of information interpretation facilitates its perception and allows content-based search and convenient navigation through it.

Convenient access to information processing means developed in various knowledge areas also remains a problem. Information processing methods, even those already implemented and presented on the Internet, are inaccessible for a wide range of users because of their poor systematization and the absence of semantic information about them.

To solve the problems discussed above, we have suggested a conception of an Intelligent Scientific Internet Resource (ISIR) providing content-based access to systematized scientific knowledge and information resources related to a certain knowledge area and to their intelligent processing facilities [1]. We call the ISIR an intelligent internet resource because not only information representation and systematization, but also all the functionalities of this resource are based on the formalisms of ontology [2] and ontology inference tools.

The ontology is the core of the ISIR knowledge system containing, along with a description of various aspects of the modeled knowledge area, a description of the structure and typology of information resources and methods of intelligent information processing facilities associated with this area. Information about the basic entities of a modeled knowledge area, relevant scientific information resources and the web-services implementing information processing methods is stored in the content of the ISIR, whose structure is based on the ontology.

To keep the ISIR knowledge system up to date, it is necessary to develop methods for populating the ISIR with relevant information from the Internet about scientific activities related to the ISIR knowledge area. This paper suggests an approach to automating this process.

Note that the process of information collection for the ISIR is guided by the ISIR ontology. The task of ontology-based information collection can be divided into two steps. The first step is information retrieval, and its aim is to find relevant resources in the Internet. The second step, information extraction, focuses on extracting certain kinds of information from web documents.

The rest of the paper is organized as follows. In the next section, we describe in more detail the proposed approach to ontology-based information collection for populating the ISIR. Subsection 2.1 dwells on a meta-search method of for retrieving web documents relevant to the ISIR knowledge area. Subsection 2.2 presents an information extraction method based on ontologies and thesauri. In Sect. 3, we study the performance of some parts of the subsystem intended to populate the ISIR. Section 4 presents the state of the art on ontology-based information extraction. Finally, Sect. 5 gives some conclusions and directions for future research.

2   The Approach Suggested

It is complicated to collect information for the ISIR because we need to extract information of a great many kinds presented on the Internet in a variety of modes. In particular, we have to collect information about organizations, persons, projects, conferences, publications and other entities. The information can be presented in the form of web-pages with various structures and text documents of different formats. Therefore, we considered it impractical to use information extraction methods based on learning by example (see, for example, [3]), which are very popular now, and applied an approach based on ontology: for every type of entities of the modeled knowledge area, we developed a specific method of information extraction adjustable to the knowledge area and kinds of Internet resources and documents.

To implement this approach, we developed a subsystem automating the information collection about the basic entities of the ISIR knowledge area and relevant Internet resources (see Fig. 1).

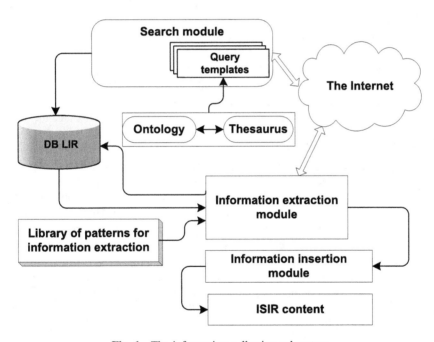

Fig. 1. The information collection subsystem

The information collection subsystem consists of the following modules: a search module, an information extraction module, an information insertion module, a library of information extraction patterns, and a data base for the storage of links to the relevant Internet resources (DB LIR).

2.1   Collection of Links to Relevant Internet Resources

The search module retrieves links to relevant Internet resources using the search queries generated on the basis of the ontology concepts and thesaurus terms presenting the concepts of the modeled knowledge area.

For every basic concept (ontology class), a set of templates of search queries are generated in all the languages used in the ISIR. The queries are generated in order to get a list of links to the resources containing information not presented in the ISIR content. There are two cases of information absence in the ISIR content. First, when there is no information at all about an individual of an ontology class. Second, when an individual is present in the content but information about some of its attributes and relationships is absent. Each of these cases requires a specific set of query templates.

In the first case, query templates intended for searching new individuals of an ontology class are generated. These templates include names of classes, attributes and relationships. In the second case, query templates are designed to retrieve additional information about the existing individuals of an ontology class. These templates can include, apart from the names of classes, attributes and relationships as well as the known values of attributes.

The search module is invoked with a set periodicity. This module uses Google, Yandex and Bing search systems via their application programming interfaces. It applies the meta-search method and filtering duplicates and non-relevant links to the Internet resources [4]. Then it adds relevant links in the DB LIR.

During searching, the efficiency of every search template is evaluated on the basis of the number of retrieved relevant links. The more efficient a template, the more frequently it is used.

Since the search module makes use of general-purpose search engines, it is necessary to filter web documents non-relevant to the knowledge area. Relevance of a link is evaluated using characteristic vectors built both for the search query and the Web page downloaded by its link. These vectors include the absolute frequencies of the occurrence of terms (words) in the texts of a query or a Web page. These vectors do not contain stop-words, i.e. words having no information content, such as prepositions and generally used words. To take into account the occurrence of terms in different morphological forms, a word stemmer is used in vector building.

To solve the problems with words ambiguity, names of ontology concepts close to the concepts contained in a query are included in the query vector. The relevance of a Web page with respect to a query is calculated as the value of cosine similarity measure between the vector of the query and the vector of the Web page [5].

2.2   Information Extraction

At the second stage, we extract from relevant web documents information about the entities of the ISIR knowledge area presented by the individuals of ontology classes.

The web documents can be presented in various formats (HTML, DOC, PDF, TXT, etc.). However, since HTML is the main format for information presentation on the Internet, the proposed methods of information extraction were designed for HTML pages.

As a rule, information extraction methods dealing with HTML pages extract information using the tags of these pages. The templates used in these methods are usually defined by a set of xpaths[1] on a page, which allows extracting similar data from one site [6]. The disadvantage of this approach is that these templates are strongly dependent on the structure of the page, which is specific for a site. A slight change in the page structure can disable the template so that it cannot extract information from this page. Moreover, in order to extract information from a new kind of a site, it is necessary to build (manually or automatically) a new site-specific template.

The main idea of our approach is to eliminate the use of tags in the templates. Instead, we define common elements for all HTML pages and extract information on their basis. These elements are described in the ontology of the web document structure used for constructing patterns for information extraction and for analyzing web documents.

The concepts of this ontology are structural elements that can be found virtually on any web page. They are "Site menu", "Menu item", "Main content", "Title", "Block", "List", "Table", "Paragraph", "Link", "Page footer", "Page header", etc. The ontology of the web document structure describes possible relationships between these elements: is-a, part-of, refers-to.

Information extraction pattern is an XML document that contains markers defining in it the positions of individuals, relationships and attributes, as well as the engines implementing the algorithm analyzing the corresponding fragments of HTML pages and extracting from them the information required. Importantly, information about entities can be given in various modes. For each of these modes, a specific pattern is constructed using the ontology class and the HTML page structure. A fragment of a pattern for extracting information about conferences is shown in Fig. 2.

Each pattern is described by a **Class** block and contains attribute blocks (**Attr**), relation blocks (**Relation**) and arguments of relation blocks (**Object**). Each of these blocks can be described by an alternative marker or a group of alternative markers (**Marker**) that specify the properties of the text containing information to be extracted. Several patterns can be combined together to extract information about related entities.

Each marker block includes a term of the thesaurus, a page element that should contain this term, and a page element that should contain information of interest to this marker.

In order to apply these patterns, we should be able to extract elements from analyzed HTML pages (menu, main content, etc.). For this purpose, the methods based on the analysis of a DOM[2] tree of the page are developed. Note that there is a problem with dynamic content. On the pages dynamically generated on the user side, the content is loaded during the execution of scripts (e.g. JavaScript). This complicates the analysis of such pages because their DOM tree changes during script execution. To solve this problem, we apply the Selenium WebDriver tool[3], which allows us to interact with a browser and analyze a DOM tree in its final form.

---

[1] http://www.w3.org/TR/xpath.

[2] http://www.w3.org/TR/2003/REC-DOM-Level-2-HTML-20030109/.

[3] http://docs.seleniumhq.org/projects/webdriver/.

```
<Class Name= "Event" engine = "FragmentSearch" >
    <Marker Term = "about conference" PType="Menu" FragType="Page" />
    <Marker Term = "conference" PType="Menu" FragType = "Page" />
    <Marker Term = "symposium " PType="Menu" FragType = "Page" />
    <Marker Term = "workshop" PType="Menu" FragType = "Page" />
    <Attr Name= "Title" type= "string">
        <Marker Term = "conference" PType="Head" FragType = "Head" />
        <Marker Term = "symposium " PType="Head" FragType = "Head" />
        <Marker Term = "workshop" PType="Head" FragType = "Head" />
    </Attr>
    <Attr Name= "Description" type= "text">
        <Marker Term = "about conference" PType= "Head" FragType= "Block" />
        <Marker Term = "about symposium" PType= "Head" FragType= "Block" />
        <Marker Term = "information" PType= "Head" FragType= "Block" />
    </Attr>

    <Relation Name= "personConferenceParticipant" >
        <Marker Term = "authors" PType="Menu" FragType="Page" />
        <Marker Term = "participants" PType="Menu" FragType="Page" />
        <Marker Term = "speakers" PType="Menu" FragType="Page" />
        <Marker Term = "conference program" PType="Menu" FragType="Page" />
        <Marker Term = "program" PType="Menu" FragType="Page" />
        <Marker Term = "committee" PType="Menu" FragType="Page" />
        <Marker Term = "agenda" PType="Menu" FragType="Page" />
        <Marker Term = "organization committee" PType="Menu" FragType="Page" />
        <Marker Term = "programm committee" PType="Menu" FragType="Page" />
        <Marker Term = "organization" PType="Menu" FragType="Page" />

        <Object Name= "Person" engine = "PersonsList" />
    </Relation>

    <Relation Name = "proceedings" >
        <Marker Term = "proceedings" PType= "Menu" FragType="Page" />
        <Marker Term = "papers" PType= "Menu" FragType="Page" />
        <Marker Term = "publications" PType= "Menu" FragType="Page" />

        <Object Name = "Publication" engine ="PublicationsList" />
    </Relation>
```

Fig. 2. A fragment of a pattern for the ontology class *Event*

To analyze the DOM tree of pages, two methods were combined: the method using heuristics and logistic regression. Logistic regression is a statistical model used to predict the likelihood of an event by fitting data to a logistic curve. Our approach uses logistic regression to classify the leaf nodes of a DOM tree as belonging or not belonging to the main content. Other elements of an HTML page are extracted with the help of heuristics.

To train the logistic regression classifier, the collection [7] consisting of 621 manually assessed news articles from 408 different web sites was used. For each node of a DOM tree, a number of features were calculated: the length of the text content, the number of sibling nodes, the number of images of the parent node, the number of links of the parent node, the depth of a node in the DOM tree. These features were used to classify the nodes of the DOM tree into two classes: the main content and not the main content.

After analyzing an HTML page and selecting its basic elements, information is extracted in accordance with an information extraction pattern. Thus, during the pattern recursive processing, corresponding elements of an HTML page are extracted and processed by special engines specified in the pattern. As a result, an individual of a given ontology class and its relationships with other individuals are generated.

3   Evaluation

In this section, we will describe the evaluation process of the approach proposed. To perform evaluation, we ran a whole pipeline including all the steps of ontology population so as to collect information about Persons, Organizations, Conferences and Projects. The resulting collection of relevant web resources was successfully used to extract instances of the mentioned ontology classes.

The experiments showed that the performance of the whole subsystem of information collection largely depends on the performance of the HTML page analysis algorithm. This follows from the fact that the markers in patterns are based on the page elements selected during the analysis.

Therefore, the performance of HTML page analysis was evaluated separately, and in particular the quality of the main page content retrieval. For this we used a collection of HTML pages from CleanEval [8]. CleanEval was an arbitrary web page cleaning competition, with the task of preparing web data to be used as a corpus designed for research and development in linguistics and language technologies. A test set of 684 web documents used in that competition is still available for performance evaluation of new algorithms. The corpus was collected from the URLs returned by making queries to Google. The queries consisted of four words of high frequency in the corpus language. In [9], an evaluation script was created calculating precision, recall and F-score for the given gold standard and the results of an estimated algorithm. Precision and recall were determined from a word-level sequence alignment of the automatically cleaned pages with the corresponding gold standard versions, using a fast heuristic algorithm implemented in the Python difflib package[4].

Table 1 shows the results of the evaluation of the two algorithms of HTML page analysis using different methods for determining the main content: logistic regression and the method using heuristics. The logistic regression classifier shows better recall with almost the same precision.

Table 1. Results of page analysis on the dataset from CleanEval

Method	Recall	Precision	F1
Heuristics	77.12 %	89.75 %	82.96 %
Logistic regression	94.57 %	88.46 %	91.41 %

---

[4] https://docs.python.org/2/library/difflib.html.

4 Related Work

Many researchers deal with the problem of collecting information from the Internet. However, as shown in survey [10], the majority of such studies are aimed at the extraction of information needed for the solution of the tasks of electronic commerce or analysis of social networks [11], and only a minor part of this research collects information needed for research [12].

Ontology-based information extraction (OBIE) has emerged as a subfield of Information Extraction. According to this approach, information extraction is based on ontologies and, generally, the output is also presented through an ontology. An ontology-based information extraction system can be defined as a process that analyzes unstructured or semi structured natural language text through a mechanism guided by ontologies to extract certain types of information and presents the output using ontologies [13].

In [14], the authors describe the application of ontology-based extraction in the context of a practical e-business application for the EU MUSING Project. Instead of analyzing the entire Web, they use a predefined set of information sources. They follow page links on the main pages of company sites that contain keywords such as "contact us", "about us", etc., and use a number of domain specific gazetteer lists and named entity recognition to target specific concepts.

Similarly to our approach, the OntoSyphon system [15] generates search phrases for each ontological concept to be learnt. This system can work with a local corpus and with the entire Web. This system uses search engines to retrieve relevant documents from the Web. The authors focus only on learning instances on the basis of Hearst templates to populate an ontology.

In [16], a method called PANKOW (Pattern-based Annotation through Knowledge On the Web) is proposed. Here, they generate search phrases in a similar fashion using linguistic patterns but instead of downloading links returned by search engines they just take the number of found web documents as an indicator of the pattern strength.

The SOBA OBIE system extracts information from HTML tables into a knowledge base with F-Logic [17]. The source is a corpus of web pages about soccer games. SOBA consists of a web crawler, linguistic annotation components and a component for the transformation of linguistic annotations into a knowledge base.

5 Conclusion and Future Work

The paper presents an approach to collecting information about scientific activities from the Internet in order to populate the Intelligent Scientific Internet Resources. An important merit of the ISIR is an appreciably shorter time required to access and analyze information, which is attributed to the accumulation, directly in the ISIR content, of the semantic descriptions of the basic entities of a modeled knowledge area, the Internet resources relevant to this area, and the information processing facilities used in it.

The proposed approach combines metasearch and ontology-based information extraction methods. In accordance with the approach, for every type of entities (ontology class), specific methods of information extraction adjustable to the knowledge area and types of information resources are developed.

Each of these methods includes a set of query templates and a set of information extraction patterns. The query templates constructed on the basis of an ontology class description are used to generate queries in order to get a list of links to the resources containing information not presented in the ISIR content. Web documents gathered using metasearch methods are analyzed with the help of information extraction patterns. These patterns are generated on the basis of an ontology and take into consideration the structure of web documents. Several patterns can be combined together to extract information about related entities. However, generating these patterns is quite a laborious task; therefore, a complete automation of this process requires further research.

Experiments have showed that the proposed approach is promising; it allows us to achieve an acceptable recall of extracting information about scientific activities from the Internet. It has been found that the performance of the whole subsystem of information collection largely depends on the performance of the HTML page analysis method; therefore, further efforts to develop this approach should focus on improving this part of the subsystem.

Currently, our approach deals only with HTML documents since HTML is the most popular Internet format. However, information about scientific activities can be presented in other formats as well: for example, papers are often presented in the PDF format. Therefore, we plan to develop methods for extracting information from the documents of various formats.

In addition, we are going to develop information extraction methods relying on manual annotations made by web masters on the basis of microdata vocabularies (for example, from schema.org).

Acknowledgments. The authors are grateful to the Russian Foundation for Basic Research (grant № 16-07-00569) for financial support of this work.

References

1. Zagorulko, Y., Zagorulko, G.: Ontology-based technology for development of intelligent scientific internet resources. In: Fujita, H., Guizzi, G. (eds.) SoMeT 2015. CCIS, vol. 532, pp. 227–241. Springer, Heidelberg (2015)
2. Guarino, N.: Formal ontology in information systems. In: Proceedings of FOIS 1998, Trento, Italy. IOS Press, Amsterdam, pp. 3–15 (1998)
3. Zhai, Y., Liu, B.: Extracting web data using instance-based learning. In: Ngu, A.H., Kitsuregawa, M., Neuhold, E.J., Chung, J.-Y., Sheng, Q.Z. (eds.) WISE 2005. LNCS, vol. 3806, pp. 318–331. Springer, Heidelberg (2005)
4. Meng, W., Yu, C., Liu, K.L.: Building efficient and effective metasearch engines. ACM Comput. Surv. (CSUR) **34**(1), 48–89 (2002)
5. Manning, C.D., Raghavan, P., Schutze, H.: An Introduction to Information Retrieval. Cambridge University Press, Cambridge (2008)
6. Gentile, A.L., et al.: Unsupervised wrapper induction using linked data. In: Proceedings of the Seventh International Conference on Knowledge Capture, pp. 41–48. ACM (2013)

7. Kohlschütter, C., Fankhauser, P., Nejdl, W.: Boilerplate detection using shallow text features. In: Proceedings of the Third ACM International Conference on Web Search and Data Mining, pp. 441–450. ACM (2010)
8. Baroni, M., et al.: Cleaneval: a competition for cleaning web pages. In: Proceedings of the Sixth International Conference on Language Resources and Evaluation (LREC 2008) (2008)
9. Evert, S.: A lightweight and efficient tool for cleaning web pages. In: Proceedings of the Sixth International Conference on Language Resources and Evaluation (LREC 2008) (2008)
10. Ferrara, E., De Meo, P., Fiumara, G., Baumgartner, R.: Web data extraction, applications and techniques: a survey. Knowl.-Based Syst. **70**, 301–323 (2014)
11. Bernabe-Moreno, J., Tejeda-Lorente, A., Porcel, C., Fujita, H., Herrera-Viedma, E.: CARESOME: a system to enrich marketing customers acquisition and retention campaigns using social media information. Knowl.-Based Syst. **80**, 163–179 (2015)
12. Cobo, M.J., Martinez, M.A., Gutierrez-Salcedo, M., Fujita, H., Herrera-Viedma, E.: 25 years at knowledge-based systems: a bibliometric analysis. Knowl.-Based Syst. **80**, 3–13 (2015)
13. Wimalasuriya, D.C., Dou, D.: Ontology-based information extraction: an introduction and a survey of current approaches. J. Inf. Sci. **36**(3), 306–323 (2010)
14. Saggion, H., Funk, A., Maynard, D., Bontcheva, K.: Ontology-based information extraction for business intelligence. In: Aberer, K., Choi, K.-S., Noy, N., Allemang, D., Lee, K.-I., Nixon, L.J., Golbeck, J., Mika, P., Maynard, D., Mizoguchi, R., Schreiber, G., Cudré-Mauroux, P. (eds.) ASWC 2007 and ISWC 2007. LNCS, vol. 4825, pp. 843–856. Springer, Heidelberg (2007)
15. McDowell, L.K., Cafarella, M.: Ontology-driven information extraction with OntoSyphon. In: Cruz, I., Decker, S., Allemang, D., Preist, C., Schwabe, D., Mika, P., Uschold, M., Aroyo, L.M. (eds.) ISWC 2006. LNCS, vol. 4273, pp. 428–444. Springer, Heidelberg (2006)
16. Cimiano, P., Handschuh, S., Staab, S.: Towards the self-annotating web. In: Proceedings of the 13th International Conference on World Wide Web, pp. 462–471. ACM (2004)
17. Buitelaar, P., et al.: Ontology-based information extraction with soba. In: Proceedings of the International Conference on Language Resources and Evaluation (LREC) (2006)

Thesaurus-Based Method of Increasing Text-via-Keyphrase Graph Connectivity During Keyphrase Extraction for e-Tourism Applications

Ilya Paramonov, Ksenia Lagutina, Eldar Mamedov[(✉)],
and Nadezhda Lagutina

P.G. Demidov Yaroslavl State University,
Sovetskaya Str. 14, 150003 Yaroslavl, Russia
`ilya.paramonov@fruct.org`, `lagutinakv@mail.ru`,
`eldar.mamedov@e-werest.org`, `lagutinans@gmail.com`

Abstract. The paper is devoted to solving the task of automatic extraction of keyphrases from a text corpus relating to a specific domain so that the texts linked by common keyphrases would form a well-connected graph. The authors developed a new method that uses a combination of a well-known keyphrase extraction algorithm (e.g., TextRank, Topical PageRank, KEA, Maui) with thesaurus-based procedure that improves the text-via-keyphrase graph connectivity and simultaneously raises the quality of the extracted keyphrases in terms of precision and recall. The effectiveness of the proposed method is demonstrated on the text corpus of the Open Karelia tourist information system.

Keywords: Keyphrase extraction · Graph connectivity · Thesaurus

1 Introduction

Automatic keyphrase extraction is a search of words and phrases that describe main topics of the corresponding text. It is widely used in many natural language processing and information retrieval tasks, such us text summarization, clustering, categorization, opinion mining, and document indexing. Nowadays there are several well-known methods that can rather effectively extract keyphrases from texts of various types [3].

As the extracted keyphrases correctly identify the main topics and significant terms of the text, they can be used for tagging or linking the texts from which these keyphrases were extracted. This can be useful for improving navigation in websites and information systems. A possible scenario can imply that the user explores some text and clicks on one of its keyphrases, the system shows a list of texts corresponding to this keyphrase, from which the user chooses a text for further reading. For success of such scenarios the extracted keyphrases should be not only relevant but also make up a well-connected graph that makes navigation

© Springer International Publishing Switzerland 2016
A.-C. Ngonga Ngomo and P. Křemen (Eds.): KESW 2016, CCIS 649, pp. 129–141, 2016.
DOI: 10.1007/978-3-319-45880-9_11

over keyphrases convenient for the user. However, the existing algorithms of keyphrase extraction often do not supply the desirable degree of connectivity.

Our research is targeted at e-Tourism applications, e.g., at websites and information systems containing information about landmarks and other tourist objects including their descriptions, photographs, geographic location, and so on. Unfortunately, commonly used methods of keyphrase extraction are targeted primarily at scientific articles [7,9] or news texts [6,14] and do not show good enough performance when applied to tourist object descriptions. The specificity of such descriptions consists in the fact that they often describe a specific object and do not mention its general category explicitly. For example, the text may provide a description of a church and does not contain the word "architecture", which is a good keyphrase that cannot be extracted by existing algorithms. The same is true for dates, geographical names and other proper names that rarely appear amongst the corpus.

Therefore, it becomes necessary to create a method to extract keyphrases taking into account specificity of the domain as well as requirement of connectivity.

In this paper we propose a method combining a well-known keyphrase extraction algorithm with thesaurus-based procedure that supplements the results of the keyphrase extraction with semantically similar terms taking into account specificity of the domain (particularly, the importance of geographic names) while filtering out rare keyphrases to improve the text-via-keyphrase graph connectivity.

The rest of the paper is organized as follows. In Sect. 2 we state the objective of the research and determine measures suitable to evaluate its achievement. Section 3 overviews related works including main algorithms of keyphrase extraction used in this research. In Sect. 4 we provide results of our experiments with the existing methods and deduce that their performance is far from being enough to solve the task in question. Section 5 describes the proposed thesaurus-based post-processing procedure, demonstrates results of its application and reveals its advantages and flaws. In Sect. 6 we suggest improvements for the procedure targeted to eliminate its weaknesses and show that the proposed method with the improved post-processing procedure has enough performance to reach the objective. Conclusion summarizes the main results of the work.

2 Objective

Let $V = \{v_1, \ldots, v_n\}$ be a corpus of texts describing objects from the e-Tourism domain. Our objective is to develop a method for extraction of a set of keyphrases K_i for each text v_i that satisfies the requirements unveiled below.

Let $G = (V, E)$ be a text-by-keyphrase connected graph (TKG) defined as follows: $(v_i, v_j) \in E(G) \iff K_i \cap K_j \neq \emptyset$ for all $i \neq j$, i.e., the vertices corresponding to two texts are connected if and only if they have at least one common keyphrase extracted.

As we mentioned in Introduction, the practical importance of TKG is that it describes an alternative way of navigation over texts allowing the user to move on through the semantically related texts of the semistructured text corpus. For user convenience the number of navigation steps required for reaching of a specific text, should be minimized, i.e., the user starting site exploring from one object should be able to find any other object using links between their keyphrases relatively fast.

From these considerations two main requirements to the method of keyphrase extraction arise:

1. (relevance) the extracted keyphrases should be relevant to the main topics of the corresponding text;
2. (connectivity) TKG should be well-connected to make navigation over texts via the graph comfortable for the user.

Quantity and quality of keyphrases affect the connectivity of TKG. The small number of extracted keyphrases does not allow to construct TKG with the edge number that is convenient for navigation. The large number of keyphrases corresponding to a concrete text, or high proportion of irrelevant keyphrases complicates transition from one text to others. For estimation of result quality we chose several numerical characteristics (measures).

To assess how good our method meets the first requirement (relevance) we apply the measures traditionally used for relevance estimation: precision, recall, and F-score [3]. Precision is the number of relevant extracted phrases divided by the total number of extracted phrases, recall is the number of relevant extracted phrases divided by the total number of relevant phrases and F-score is the harmonic mean of them.

The connectivity of TKG is measured using several numerical characteristics in our research. The most simplest ones include the numbers of connected components and isolated vertices of TKG. These measures are useful but they are rather coarse: to consider the result of the keyphrase extraction acceptable it is necessary (but not sufficient) that TKG would have only one connected component, which means that any text can be reached from any other text via navigation over keywords.

More fine-grained measures of connectivity include maximum, average and minimum degrees of TKG vertices and the median of a keyphrase-via-text mapping distribution. The latter measure can be defined as follows. Let $K = K_1 \cup \cdots \cup K_n$ be a set of all extracted keyphrases. Then for a particular keyphrase k define the number of objects that corresponds to the keyphrase as follows:

$$\nu(k) = \#\{v_i \in V | k \in K_i, 1 \leq i \leq n\}.$$

Let the median of the keyphrase-via-text mapping distribution M be a median of the set $\{\nu(k)|k \in K\}$.

The degree of TKG vertices characterizes the user's navigation options. If this measure is large the user can be overloaded with too many possibilities. On the contrary, when it is small (less then 3) the user can spend much time surfing from one text to another.

The median of the keyphrase-via-text mapping distribution estimates usefulness of keyphrases for navigation. If this measure equals 1, most of the extracted phrases are unique in the corpus therefore they do not affect TKG connectivity and the user cannot use them for navigation. If it equals 2 there are many keyphrases either unique or link only two texts therefore they provide too few options for the user choice. So for convenient navigation the median should be at least 3.

3 Related Work

The task of automatic keyphrase extraction appeared quite a long time ago and since that time a number of different approaches to solve it were designed [12]. Existing keyphrase extraction methods usually operate in two steps: extraction of words/phrases, which are considered as candidate keyphrases, using some heuristics; and determination of which of these candidate keyphrases are relevant keyphrases using supervised or unsupervised approaches.

Traditionally supervised approaches consider the task of keyphrase extraction as a binary classification problem. They train a classifier on documents manually annotated with keyphrases. Keyphrases and non-keyphrases are used as positive and negative examples. Different learning algorithms are used to train classifiers, including Naive Bayes, decision trees, bagging, boosting, maximum entropy, multi-layer perceptron, and support vector machines. By-turn, these classifiers use various features including TF*IDF, distance of a phrase, spread, Wikipedia-based keyphraseness, etc.

A well-known supervised keyphrase extraction algorithm is KEA [15]. Firstly, it finds candidate keyphrases using lexical methods and calculates statistical features (e.g., TF*IDF and distance of a phrase) for each candidate. Secondly, KEA builds a prediction model and train it on a set of documents with manually annotated keyphrases. Finally, using the prediction model it applies the Naive Bayes algorithm to determine keyphrases for each document from the corpora.

Another example of supervised keyphrase extraction algorithm is Maui [8]. It is based on KEA and uses the same schema for automatic keyphrase extraction. The difference between the algorithms consists in the fact that Maui calculates more features for candidate keyphrases and uses bagged decision trees instead of Naive Bayes.

In unsupervised keyphrase extraction approaches graph-based ranking methods are state-of-the-art [1,3]. The main idea behind these approaches is to build a graph from the input document and rank its nodes according to their importance using some graph-based ranking method. Each node of the graph corresponds to a candidate keyphrase and edges connect related candidates. The top-ranked candidates from the graph are then selected as keyphrases.

One of the most popular graph-based approach is TextRank [9]. To connect nodes in the graph it uses the co-occurrence relation between words. The idea of the ranking method is that a node is important if there are other important nodes pointing to it. This can be regarded as voting or recommendation among nodes.

Another type of unsupervised keyphrase extraction approaches is a modification of graph-based one. Such algorithms additionally use a stage of grouping of the candidate keyphrases by topics of a document. The example of such an algorithm is Topical PageRank [6]. Just like TextRank this algorithm is based on building a graph of candidate keyphrases and ranking its nodes according to their importance but with some improvements of ranking method that ensure that the extracted keyphrases cover the main topics of the document.

Although the idea of using keyphrases for navigation looks rather straightforward it does not occur often in scientific publications. For example, in [5] the authors describe two systems, Kniles and Phrasies that automatically connect documents within collection based on their keyphrases. These systems use the extracted keyphrases as link anchors inside documents pointing to another related documents and employ some techniques using keyphrases to evaluate document similarity, which is calculated using some standard information retrieval similarity measure. By navigation over keyphrases the user can easily find similar documents within the whole collection. Unlike our research, the authors of this work do not try to increase connectivity of documents by keyphrases.

In [2] the document connectivity is set up using an approach that involves the similarity graph having some common features with TKG. The authors created a framework that visualizes a set of documents to alleviate navigation between them. Visualization is based on the so-called similarity graph—a weighted, undirected graph having individual documents as its nodes that reflects relations between the documents. The relation between two documents is set if documents have some common keyphrases. The difference between this approach and the approach we use in this paper is that improvement of navigation is reached by visualization of the document connectivity graph itself, not by increasing its connectivity.

To retrieve information appropriate for navigation over tourist object descriptions it is possible to use fact extraction techniques [4,11]. However, such techniques can be efficient mostly for extraction of proper names only. For the other types of keyphrases it is difficult to compose appropriate syntactic patterns due to lack of formal structure in majority of tourist texts. That is why the authors decided not to use such techniques in this research.

4   Performance of Existing Methods

For evaluation we used a text corpus extracted from the database of Open Karelia [10]. Open Karelia is an information system about Russian and Finnish museums of the Karelian region. It allows to get information about the museums' history, exhibitions, excursions, work schedule, ticket pricing, contacts, and descriptions of objects stored in the museums.

The corpus we used for experiments contains 986 texts that describe Karelian cultural objects, museum exhibits, and tourist attractions. Each text is a description of a concrete tourist object.

In our experiments we tested four well-known keyphrase extraction methods described in Sect. 3 including two unsupervised and two supervised ones: TextRank, Topical PageRank, KEA, and Maui. For training the supervised methods' model we used a set of 100 texts from the whole corpus accompanied by keyphrases determined by an expert.

For experiments with KEA and Maui algorithms we used their open source implementations in Java from the official websites (https://code.google. com/archive/p/maui-indexer/, http://www.nzdl.org/Kea/index.html). These implementations do not support Russian language out of the box, so we involved the standard Porter stemmer (http://www.algorithmist.ru/2010/12/ porter-stemmer-russian.html). TextRank implementation in Python is available at https://github.com/summanlp/textrank. The Topical PageRank algorithm was implemented based on the TextRank implementation by following of the description of this algorithm from [6] and using Latent Dirichlet Allocation (LDA) implementation from the gensim toolkit (http://radimrehurek.com/ gensim/).

Initially we evaluated the quality of extracted keyphrases for each method. For these experiments we selected a set of 100 texts, which is different from the set of texts used for training. Then we applied the keyphrase extraction method to this set of texts and compared its outcome with the expert's keyphrases. To estimate the quality of the extracted keyphrases we calculated Precision (P), Recall (R), and F-score (F).

The results of these experiments presented in Table 1 show what the Maui algorithm extracted more relevant keyphrases than the other methods and its results look more suitable to meet our first requirement. However, the absolute quality of the results is comparable among the algorithms and is rather poor especially when considering recall: no methods demonstrate recall greater than 25 %.

Table 1. Quality of extracted keyphrases for the set of 100 texts

Algorithm	P	R	F
TextRank	49.6	20.1	28.6
Topical PageRank	51.7	22.0	31.2
KEA	49.1	17.0	25.2
Maui	57.5	24.1	34.0

To evaluate connectivity of TKG resulted from the mentioned algorithms we measured the number of connected components (C) and isolated vertices (I) of TKG, maximum (d_{max}), average $(\bar{d})$, and minimum (d_{min}) degree of vertices in the greatest component of TKG, and also we calculated a median score (M) of distribution of texts that correspond to a concrete keyphrases. The results are shown in Table 2.

Table 2. Connectivity of TKG and median score of distribution of texts corresponding to keyphrases for the set of 100 texts

Algorithm	C	I	$d_{\min}$	$\bar{d}$	$d_{\max}$	M
TextRank	26	25	1	3.71	13	1
Topical PageRank	24	23	1	4.00	15	1
KEA	59	58	1	1.86	7	1
Maui	11	10	1	13.73	41	1

From these results we concluded that the Maui algorithm is also the most suitable for our second requirement. Corresponding TKG is more connected compared to the others—it has minimum number of connected components and greater number of vertex degree. But nonetheless these results are still far from being good enough because the median score of distribution of texts that correspond to keyphrases equals to 1. It means that most of extracted keyphrases are isolated, i.e., each of them corresponds to only one text.

Table 3 shows the result of applying the considered algorithms to the whole corpus containing all 986 texts. The Maui results also remain the best in these experiments providing the most connected TKG and considerably outperforming the other methods in this regard. TKG for Maui in this case links all texts in one connected component, but the average and minimum degrees of vertices are still very low. Therefore, the situation generally does not change for a larger corpus.

Table 3. Connectivity of TKG for the corpus of all 986 texts

Algorithm	C	I	$d_{\min}$	$\bar{d}$	$d_{\max}$	M
TextRank	41	40	1	98.25	438	1
Topical PageRank	31	30	1	108.56	464	1
KEA	158	157	1	10.41	56	1
Maui	1	0	1	241.28	614	1

5   Thesaurus-Based Post-processing Procedure

In order to overcome the insufficiency of quality and TKG connectivity inherent to the standard algorithms of keyphrase extraction we involved external resources, namely a vocabulary of the geographic names and a general purpose thesaurus.

The usage of the geographic names vocabulary was due to the fact that the standard methods rarely extract proper names (including geographic names) as keyphrases whereas in the e-Tourism domain just that names are highly relevant and useful for navigation over the texts of site or information system. In our experiments the vocabulary was used as follows: we identified proper names from all the corpus using the vocabulary and added them to the list of keyphrases during the post-processing procedure. As a result, we found that it improved quality of the extracted keyphrases, i.e., increased their relevance. The quality estimates of this step are in the Table 4. In comparison with the Table 1 we can see that precision and recall increased for all algorithms.

Table 4. Quality of extracted keyphrases with geographic names for the set of 100 texts

Algorithm	P	R	F
TextRank	54.8	27.7	36.8
Topical PageRank	54.9	27.8	36.9
KEA	54.0	23.3	32.6
Maui	59.4	29.6	39.5

The main goal of thesaurus using was to increase the TKG connectivity by adding phrases related to already extracted keyphrases. In this case "related" means "being in some relation defined in the thesaurus with already extracted keyphrases". In our research we used the RuThes general purpose thesaurus (http://www.labinform.ru/pub/ruthes/), which is one of the most popular thesauri for Russian language. It contains 115 000 phrases with relations between them.

It should be noticed that KEA and Maui algorithms per se can be used in conjunction with a thesaurus. They have a special mode in which they extract only phrases contained in the thesaurus. It is different from the use of thesaurus described in the previous paragraph but can also be useful for improvement of relevance, therefore we tested these algorithms in such a mode as well in our experiments.

First of all, we evaluated how much connectivity of TKG can be increased when using the thesaurus-based post-processing procedure and assessed which thesaurus's relations are most effective to improve performance. To figure it out we set experiments where results from the common keyphrase extraction algorithms were supplemented with keyphrases from the thesaurus being in the following relations with them: association ($Assoc$), hypernymy ($Hypm_1$), hypernymy with depth ≤ 2 ($Hypm_2$), hypernymy with depth ≤ 3 ($Hypm_3$), and association with hypernymy ($Assoc + Hypm_1$). The most significant results are shown in Table 5.

From these experiments we concluded what thesaurus-based post-processing procedure significantly improves connectivity of TKG for all algorithms. It increases the degree of vertices in the greatest component and decreases the number of connected components. For example, in comparison with the Table 2 the degree of vertices increased in several times. Specifically, in case of hypernymy with depth ≤ 2 the maximum degree of vertices in the greatest component for Maui increased in 2 times, average—in 6 times and minimum—in 29 times. Also, from these experiments we noticed that the best results again belong to

Table 5. Connectivity of TKG and median score of distribution of texts corresponding to keyphrases for the set of 100 texts with thesaurus-based post-processing procedure

Algorithm	Relationships	C	I	$d_{\min}$	$\bar{d}$	$d_{\max}$	M
Maui	$Assoc$	1	0	1	33.24	70	1
Maui	$Hypm_1$	1	0	9	46.58	78	1
Maui	$Hypm_2$	1	0	29	86.58	99	1
Maui	$Assoc + Hypm_1$	1	0	9	53.42	84	1
Maui with thesaurus	$Assoc$	1	0	2	33.60	66	1
Maui with thesaurus	$Hypm_1$	1	0	5	39.98	75	1
Maui with thesaurus	$Hypm_2$	1	0	5	65.16	87	1
Maui with thesaurus	$Assoc + Hypm_1$	1	0	5	42.18	77	1
Topical PageRank	$Assoc$	6	5	1	23.58	71	1
Topical PageRank	$Hypm_1$	7	6	2	37.87	80	1
Topical PageRank	$Hypm_2$	5	4	2	66.77	92	1
Topical PageRank	$Hypm_3$	5	4	2	84.77	94	2
Topical PageRank	$Assoc + Hypm_1$	5	4	2	42.90	87	1

the Maui algorithm. It has the highest values of degree of vertices and has one connected component for the set of 100 texts.

Unfortunately, this post-processing procedure still does not solve the problem of the small median score. Most of added phrases are unique, they correspond to only one of texts and therefore their presence makes the median score low. Moreover, in this form the post-processing procedure causes serious issues with quality of results (see Table 6): the precision for the most of methods became less than 10 %. The main reason of such significant degradation is due to the

Table 6. Quality of keyphrases extracted by the thesaurus-based post-processing procedure for the set of 100 texts

Algorithm	Relationships	P	R	F
Maui	$Assoc$	5.9	53.9	10.6
Maui	$Hypm_1$	9.8	38.4	11.7
Maui	$Hypm_2$	6.2	85.9	11.6
Maui	$Assoc + Hypm_1$	4.8	78.6	9.1
Topical PageRank	$Assoc$	8.8	44.8	14.6
Topical PageRank	$Hypm_1$	12.6	56.3	20.6
Topical PageRank	$Hypm_2$	8.0	64.5	14.2
Topical PageRank	$Hypm_3$	5.9	66.4	10.8
Topical PageRank	$Assoc + Hypm_1$	6.5	59.2	11.8

large number of resulted keyphrases: its number varies from 90 to 316 per text. Obviously, most of these keyphrases cannot be relevant, so there was necessity of further improvement of the post-processing procedure to retrieve feasible result.

6   Improved Thesaurus-Based Post-processing Procedure with Infrequent Phrase Filtering

To identify possible ways of the post-processing procedure improvement we manually assessed the automatically extracted keyphrases for each set of parameters.

We established the fact that most of the keyphrases added from the thesaurus by the associative relations were contextual synonyms belonging to the foreign domains, so they impaired the quality of keyphrase extraction. Another issue was that in the experiment with addition of hypernyms of depth 3 the maximum and average values of TKG degree were too high, which is undesirable because it leads to the situation when the user would have too much choices during navigation. These findings suggested us not to add associations of the found keyphrases and add hypernyms with depth not greater than 2.

In order to cope with proliferation of unique and rare keyphrases and therefore to increase the median score we added a filter that removes keyphrases that occurred only once or twice in the whole corpus. After repetition of the experiment with these improvements we got the results shown in Tables 7 and 8. The label $Filter_1$ in the second column of Table 8 means using the filter that removes keyphrases occurred once, $Filter_2$—keyphrases that occurred once or twice.

Table 7. Quality of keyphrases extracted by the improved method for the set of 100 texts

Algorithm	Post-processing	P	R	F
Maui	$Hypm_1 + Filter_1$	78.3	46.4	58.3
Maui	$Hypm_1 + Filter_2$	79.1	38.4	51.7
Maui	$Hypm_2 + Filter_1$	70.0	76.8	72.7
Maui	$Hypm_2 + Filter_2$	72.2	70.0	71.1
Maui with thesaurus	$Hypm_1 + Filter_1$	71.8	41.0	52.1
Maui with thesaurus	$Hypm_1 + Filter_2$	74.0	33.0	45.6
Maui with thesaurus	$Hypm_2 + Filter_1$	64.6	60.9	59.2
Maui with thesaurus	$Hypm_2 + Filter_2$	67.6	49.6	57.2
Topical PageRank	$Hypm_1 + Filter_1$	62.6	41.1	49.6
Topical PageRank	$Hypm_1 + Filter_2$	65.8	29.4	40.7
Topical PageRank	$Hypm_2 + Filter_1$	51.8	59.5	55.4
Topical PageRank	$Hypm_2 + Filter_2$	56.6	50.7	53.5

Table 8. Improved method results connectivity of TKG and median score of distribution of texts corresponding to keyphrases for the set of 100 texts

Algorithm	Post-processing	C	I	d_{min}	$\bar{d}$	d_{max}	M
Maui	$Hypm_1 + Filter_1$	1	0	2	35.08	71	3
Maui	$Hypm_1 + Filter_2$	3	2	3	35.18	70	4
Maui	$Hypm_2 + Filter_1$	1	0	4	53.42	89	3
Maui	$Hypm_2 + Filter_2$	1	0	2	53.00	89	5
Maui with thesaurus	$Hypm_1 + Filter_1$	2	1	1	34.59	67	3
Maui with thesaurus	$Hypm_1 + Filter_2$	3	2	3	34.49	67	4
Maui with thesaurus	$Hypm_2 + Filter_1$	1	0	1	42.80	73	3
Maui with thesaurus	$Hypm_2 + Filter_2$	2	1	5	42.91	73	5
Topical PageRank	$Hypm_1 + Filter_1$	8	7	1	27.44	74	2
Topical PageRank	$Hypm_1 + Filter_2$	12	11	2	27.62	71	4
Topical PageRank	$Hypm_2 + Filter_1$	7	6	2	51.26	89	2
Topical PageRank	$Hypm_2 + Filter_2$	8	7	10	51.42	89	4

In comparison with the method that does not involve post-processing, the quality of the extracted keyphrase became significantly higher: the best results show both precision and recall near 70–80 %.

TKG connectivity became lower than in the previous experiments with the post-processing but for the Maui algorithm it is still sufficiently high. In the half of variants for Maui TKG has a single connected component. The minimum, average and maximum degree of vertices reached appropriate values: they are more than 2 and significantly less than 100, so the resulting graph provides the convenient way of navigation. The median is between 3 and 5, which provides the user with the adequate number of choices during navigation.

In summary, all measures for Maui with post-processing satisfy the requirements stated in Sect. 2. The manual overview of the extracted keyphrases made by an expert also witnessed that the proposed method has appropriate performance for practical usage.

7 Conclusion

In this paper we proposed a method of automatic keyphrase extraction targeted at e-Tourism websites and information systems so that the texts linked by common keyphrases form a well-connected graph. The method comprises a combination of a well-known keyphrase extraction algorithm with thesaurus-based post-processing procedure that adds proper names and hypernyms from the thesaurus to the keyphrase set and removes infrequent phrases.

After applying the method descriptions of all tourist objects in the corpus are connected to each other, most extracted keyphrases are relevant and all keyphrases correspond to the significant number of objects. Therefore, the number of navigation steps required for the user to reach information of interest is relatively small. The proposed method is about to be implemented in the Open Karelia tourist information system.

As the result of the proposed method is represented in the form of well-connected TKG, it could be used as a basis for construction of a domain-specific thesaurus. Such a suggestion established on the fact that connectivity is an important quality metrics of vocabularies of different sorts including thesauri [13] can be a subject of future research.

Acknowledgements. The research was supported by the grant of the President of Russian Federation for state support of young Russian scientists (project MK-5456.2016.9).

References

1. Beliga, S., Meštrović, A., Martinčić-Ipšić, S.: An overview of graph-based keyword extraction methods and approaches. J. Inf. Organ. Sci. **39**(1), 1–20 (2015)
2. Berend, G., Farkas, R.: Keyphrase-driven document visualization tool. In: IJCNLP, pp. 17–20. Asian Federation of Natural Language Processing (2013)
3. Hasan, K.S., Ng, V.: Automatic keyphrase extraction: a survey of the state of the art. In: Proceedings of the 52nd Annual Meeting of the Association for Computational Linguistics, pp. 1262–1273. Association for Computational Linguistics (2014)
4. Huang, X., Ng, P.C.: Enabling public access to non-open access biomedical literature via idea-expression dichotomy and fact extraction. In: Workshops at the Thirtieth AAAI Conference on Artificial Intelligence (2016)
5. Jones, S., Paynter, G.: Topic-based browsing within a digital library using keyphrases. In: Proceedings of the Fourth ACM conference on Digital libraries, pp. 114–121. ACM (1999)
6. Liu, Z., Huang, W., Zheng, Y., Sun, M.: Automatic keyphrase extraction via topic decomposition. In: Proceedings of the 2010 Conference on Empirical Methods in Natural Language Processing, pp. 366–376. Association for Computational Linguistics (2010)
7. Liu, Z., Li, P., Zheng, Y., Sun, M.: Clustering to find exemplar terms for keyphrase extraction. In: Proceedings of the 2009 Conference on Empirical Methods in Natural Language Processing EMNLP 2009, vol. 1, pp. 257–266. Association for Computational Linguistics (2009)
8. Medelyan, O., Frank, E., Witten, I.H.: Human-competitive tagging using automatic keyphrase extraction. In: Proceedings of the 2009 Conference on Empirical Methods in Natural Language Processing, vol. 3, pp. 1318–1327. Association for Computational Linguistics (2009)
9. Mihalcea, R., Tarau, P.: TextRank: Bringing order into texts. In: Proceedings of EMNLP, pp. 404–411. Association for Computational Linguistics (2004)
10. Paramonov, I., Mamedov, E., Averkiev, S., Shchitov, I., Krinkin, K., Zaslavskiy, M.: Open Karelia — an informational portal for museums. In: Proceedings of the 17th Conference of Open Innovations Association FRUCT, Yaroslavl, Russia, 20–24 April 2015, p. 331. IEEE (2015)

11. Piskorski, J., Yangarber, R.: Information extraction: past, present and future. In: Poibeau, T., Saggion, H., Piskorski, J., Yangarber, R. (eds.) Multi-source, Multilingual Information Extraction and Summarization. Theory and Applications of Natural Language Processing, pp. 23–49. Springer, Heidelberg (2013)
12. Siddiqi, S., Sharan, A.: Keyword and keyphrase extraction techniques: a literature review. Int. J. Comput. Appl. **109**(2), 18–23 (2015)
13. Suominen, O., Mader, C.: Assessing and improving the quality of SKOS vocabularies. J. Data Semant. **3**(1), 47–73 (2014)
14. Wan, X., Xiao, J.: Single document keyphrase extraction using neighborhood knowledge. In: Proceedings of the 23rd National Conference on Artificial Intelligence AAAI 2008, vol. 2, pp. 855–860. AAAI Press (2008)
15. Witten, I.H., Paynter, G.W., Frank, E., Gutwin, C., Nevill-Manning, C.G.: KEA: practical automatic keyphrase extraction. In: Proceedings of the Fourth ACM Conference on Digital Libraries, pp. 254–255. ACM (1999)

Identifying Product Failures from Reviews in Noisy Data by Distant Supervision

Elena Tutubalina[✉]

Kazan (Volga Region) Federal University, Kazan, Russia
tutubalinaev@gmail.com

Abstract. Product reviews contain valuable information regarding customer satisfaction with products. Analysis of a large number of user requirements attracts interest of researchers. We present a comparative study of distantly supervised methods for extraction of user complaints from product reviews. We investigate the use of noisy labeled data for training classifiers and extracting scores for an automatically created lexicon to extract features. Several methods for label assignment were evaluated including keywords, syntactic patterns, and weakly supervised topic models. Experimental results using two real-world review datasets about automobiles and mobile applications show that distantly supervised classifiers outperform strong baselines.

Keywords: Product defects · Opinion mining · Distant supervision · Problem phrase extraction

1 Introduction

In opinion mining, researchers from both the industry and academia are faced with the challenge of building machine learning approaches. These approaches typically require training examples, which are time-consuming to annotate and cost expensive. At the same time, supervised methods with a set of features ideally should be adapted to a certain opinion mining task. Therefore, the lack of extensively annotated corpora can be seen as a major bottleneck for applying supervised techniques to certain opinion mining problems or certain domains of reviews.

One common solution to avoid lack of manually annotated training data is *distant supervision* which assumes to use noisy signals in a text as positive labels to train a classifier. The intuition behind it is that any text with a noisy class label is more likely to belong to that class. However, this strategy addresses several research questions that are not satisfactorily detailed in previous works and need to be examined to apply for opinion mining tasks.

First, distantly supervised methods rely on automatically labeled data. Recent methods have been explored using information. Recent studies have utilized a manually created set of emoticons (e.g., :-) or :-/) or hashtags (e.g., #happy and #sad) to serve as semantic indicators for sentiment [2,13,20].

© Springer International Publishing Switzerland 2016
A.-C. Ngonga Ngomo and P. Křemen (Eds.): KESW 2016, CCIS 649, pp. 142–156, 2016.
DOI: 10.1007/978-3-319-45880-9_12

Some works have used keywords and trivial patterns (e.g., "I do not like") to represent a class or a topic [17,19]. In the broad field of opinion mining, most methods focus on positive and negative sentiments. The above-mentioned studies have not analyzed how to select proper keywords for detecting labels for tasks beyond sentiment analysis such as identification of product complaints or detection of drug reactions based on user reviews. Therefore, the field lacks a comparative study on the different ways of label assignment.

Second, recent papers on opinion mining have typically applied supervised classifiers in a similar way to those methods that recently applied to sentiment analysis. Several papers have trained a classifier based on noisy class labels to classify testing examples [2,17,19]. The second group of works has applied distant supervision to compute scores for an automatic sentiment lexicon and then train a new classifier based on a small training data [12,24]. Therefore, there are no generally accepted principles on how to use data with noisy labels.

In this paper, we present a comparative study of distant supervision technique for an opinion mining task that focuses on identifying phrases about complaints and product defects from texts. To address the first question about extraction of labels, we employ several approaches including knowledge-based methods and sentiment-aware topic models. To explore the second question about labels' integration into supervised classifiers, we apply several approaches by computing Pointwise Mutual Information (PMI) and learning features' weights from noisy data.

The paper organizes as follows. In Sect. 2, we first give some background on distant supervision and briefly survey recent works, concentrating on sentiment-related ones. Section 3 introduces the problem statement and present our research questions. In Sect. 4, we explain how to incorporate distant supervision assumption and how to assign noisy labels. We then describe several methods for extraction of noisy labels. In Sect. 5, we show the results of an extensive evaluation on two real-life data sets in Russian for distantly supervised methods. Section 6 concludes the paper.

2 Related Work

Our work is related to studies on distant supervision, sentiment analysis, and mining product failures from user reviews.

2.1 Distant Supervision

Recent studies have applied distant supervision for various fields of research including relation extraction [18,21,23], sentiment analysis [2,10,17,24], and topic identification [17]. Formally, distant supervision[1] is a paradigm for using noisy labeled data that was proposed for relation extraction in [18]. In relation extraction, a study focuses on assigning labels to entity pairs. Mintz et al.

---

[1] Also called distant learning in [19].

supposed that any sentence that contains a pair of entities is likely to express that relation in some way if this pair participates in a known relation in a large semantic database Freebase. The method has extracted a large set of possibly noisy features that were combined in a logistic regression classifier. In [21], Poon et al. used an assumption from [23] that some (at least one) sentence in the text expresses a relation known from the database. The authors showed that distant supervision can effectively compensate the lack of annotation of biological pathways based on this assumption.

In [12], authors computed sentiment scores for words from noisy labeled tweets (called pseudo-labeled tweets) marked with hashtagged emotion words such as #joy, #sad, #angry, and #surprised. They computed a sentiment score as a difference between PMI measures in positive and negative corpora. The authors reported that this measure was chosen since it is robust and has been successfully applied in a number of NLP tasks [28]. In [2,13,20], emoticons were used as noisy labels in tweets for distant supervision. In [17], a set of keywords was used to find a target topic. The authors then trained a classifier on a dataset, where each tweet was classified as either positive (i.e., relevant to the target topic) or negative (i.e., irrelevant). In [24], Severyn and Moschitti presented an approach to automatically construct sentiment lexicons using distant supervision. They classified tweets by positive or negative emoticons and used machine learning to find the scores for the lexicon items. Combining created machine-learned lexicon with the state-of-the-art classifiers gave an improvement in F1-measure (from 70.06 % to 71.32 %). In [10], a five-star rating of online reviews was regarded as the overall customer satisfaction. An ordinal classification pairwise approach was used to learn the importance of product characteristics which could benefit product design. However, the mentioned authors did not analyze the best way to assign noisy labels for texts.

We also note a recent work on distant supervision for relation extraction based on piecewise convolutional neural networks to address a wrong label problem [31]. We do not consider neural networks further in this paper but note them as a possible direction for further work for text classification.

2.2   Aspect-Based Sentiment Analysis

The main task of aspect-based opinion mining is identifying aspects and classifying sentiments toward an aspect or a topic group of aspects from a collection of reviews of products. There have been many studies in the area of opinion mining. Liu gave a good overview of the field in [15].

Early works were focused on identifying positive and negative reviews, mostly using linguistic rules, syntactic dependencies, frequency-based methods and supervised classifiers [5,6,12,22]. Mostly, opinion mining approaches use a sentiment lexicon, which contains a list of sentiment words and phrases. Supervised approaches achieve a better performance in the classification task. However, these approaches are limited due to the labeled training dataset's size and lower results after shifting to another domain. Recent works have implemented probabilistic topic models, such as latent Dirichlet allocation (LDA) [1], for multi-aspect analysis tasks [14, 16, 30]. Modifications of LDA from [14, 30] contain additional latent variables to model the latent sentiments of words and sentences, respectively. These models learn topics and sentiment from unlabeled data to predict a review's rating or sentiment polarity.

2.3 Mining Product Defects and User Complaints

Detection of defects and extraction of problem phrases from texts are less studied despite the fact that tasks on sentiment analysis were studied in depth using various automatic techniques. Mostly, works have been focused on creating linguistic rules, based on keywords for expressing major bugs [7], classification rules and patterns to extract noisy examples [19, 25], classifiers [29], and traditional probabilistic models to summarize reviews [8, 19]. The proposed techniques have been applied in four main domain of texts: (i) reports from customer complaint management systems, (ii) short messages on electronic products and services from Twitter, (iii) user reviews from e-commerce websites, and (iv) reviews about mobile applications from app stores' services such as Google Play and App Store.

Mostly authors first manually analyzed a set of user comments to identify patterns and keywords which users use to report a defect or a complaint. In [29], a supervised classifier is used on a set of sentiment and syntactic features without complex feature selection choices. In [25], the authors created classification rules, based on dictionaries. Formally, the authors proposed a simple pattern-matching algorithm to detect whether or not a sentence contains a problem. They explored an approach to the automatic construction of dictionaries using the Google Books Ngram Viewer. They reported the best performance F-measure for the method based on manually created dictionaries. In [9], authors proposed a method based on dictionaries and semantic analysis of sentences with conjunctions. In [4], Galitsky et al. classified complaint scenarios associated with customer-company dialogues, formalizing dialogues as labeled graphs. In [3], conflict scenarios were analyzed using a machine learning approach to capture the similarity between communicative actions.

Moghaddam trained a support vector machine (SVM) classifier based on a bag of words and noisy labels assigned by one of five manually created patterns [19]. The classification results for defect extraction were compared with fully supervised SVM that used manually labeled data. The results showed that both SVMs performed quite similar, comparing the F-measure. However, the author did not analyze more complex classifiers based on a set of statistical, syntactic, and semantic features. To our knowledge, state-of-the-art sentiment classifiers in sentiment analysis have not been fully trained based on distant supervision assumption.

Several studies have presented a manual analysis of bug reports to analyze critical information for developers. Thung et al. [26] categorized bugs that occur in machine learning systems. Tian et al. [27] presented results of manual analysis on the content of software engineering microblogs to better understand microblogging behaviors in Twitter. In [11], Kim et al. created categories of usability problems to analyze user experience with electronic products and correlation between product categories and difficulties.

3   Problem Statement and Contribution

Let $P = \{p_1, ..., p_m\}$ be a set of products or services provided by a company and collected from websites, social comments, technical support, etc. We assume as input a collection of reviews $D = \{d_1, ..., d_n\}$ of each product $p_i \in P$. Each product $p_i \in P$ consists of opinion targets $T_i = \{t_1, ..., t_k\}$. In some of the reviews, customers report a defect and complain about the mechanism of the products, their performance of a product, and some difficulty in understanding the functions. Further, we define research terms more formally:

Problem phrase: A problem phrase is a text fragment of the user's response (called *opinion unit*) containing an indication of difficulty in the use of the product or inability to use the product due to a failure (bug, defect, or crash).

Problem indicator: A problem indicator is a single-word or multiword expression that points out on a product's problem explicitly or indirectly, for example, "reload", "garbage", and "trouble".

The task determines whether a given text contains a problem or not. To support the extraction of problem phrases in a general and robust manner, we treat each sentence of a review or a short comment as an opinion unit. This assumption is reasonable due well-defined meaning of a sentence. For example,

– "I installed it from the HP and received an error code '*the following printhead has a problem*'."
– "*The store would not take* it back because I bought it too long ago."
– "*Still missing fundamental basic options* for a mail app."
– "*Hall of shame product by mnc microsoft* few spits to coder and so called mnc running the circus by employing *useless banana eating monkeys*."

The first and third sentences indicate a difficulty with the product and the app, while the second sentence about a local shop is not domain specific and not related to an HP product. The fourth sentence about the Skype app is useless for Android developers even if it contains negative opinion.

Mostly, opinion mining approaches are based on *the main distant supervision assumption* that all sentences that contain keywords or patterns are likely to belong to a certain class.

The main goal of our research is to investigate the distant supervision assumption to identify problem phrases in unlabeled texts. Thus, the research questions that guided our work:

RQ1: What is the best way to extract noisy labels from unlabeled data based on the distant supervision assumption?
RQ2: How to utilize noisy labeled data in supervised methods to improve text classification?

4   Distant Supervision Approach

In this section, we describe the research approach we followed to answer our research questions and methods that further used for distant supervision. We examine different classifiers: knowledge-based, weakly supervised topic models, and support vector machines (SVM) with a set of commonly used features including lexicon features.

4.1   Approach Overview

While much work on the extraction of problem phrases has studied fully supervised approaches or knowledge-based methods, we focus on unlabeled data. We build a simple distant supervision framework that treats classification methods and feature extractors as two distinct components. In this context, this work utilizes the following methods to decide whether an opinion unit is related to a required class (contains a problem phrase) and assign a corresponding label:

- a dictionary-based approach that uses a simple pattern-matching algorithm based on dictionaries [25];
- a clause-based approach that analyzes a sentence according to conjunctions [9];
- sentiment-based topic models: JST and Reverse-JST;
- star rating from reviews that serves as an indicator of a product's quality. Each review in corpora is associated with an overall rating, which is a numeric score between 0 (lowest) and 5 (highest). If the score for a given text is equal or less than 3, the text is considered as a problem phrase; otherwise, the text is considered as "negative" (without user complaints).

Since methods based on distant supervision assumption suffers from false negatives (some negative examples may be positive due to the limited size of dictionaries), opinion mining methods can use (i) a small set of manually labeled training data, and (ii) a large corpus of texts with noisy labels to create opinion lexicons and extract indicator's scores. Therefore, we evaluate the classification performance of the various state-of-the-art models using the following methods:

1. We evaluate classifiers trained on noisy labels.
2. We evaluate classifiers with an extended set of lexicon features. The lexicon features were calculated for an automatically created lexicon where word scores are estimated on data with noisy labels using pointwise mutual information (PMI) or classifier's feature weights.

In Sect. 5, we present a performance evaluation of classification methods in a distant supervision framework based on two different real-world datasets[2].

---

[2] Datasets are available at https://github.com/tutubalinaev/noisy-data.

4.2 Knowledge-Based Methods

To automatically extract noisy labels, we apply two knowledge-based approaches: (i) the dictionary-based approach further called **DbA** [25] and (ii) the more complex clause-based approach called **CbA** as an extension of **DbA** [9]. Note that we employ these methods due to a set of various dictionaries to minimize the number of false negatives. Statistics and examples of entries from the dictionaries are presented in Table 1. Please refer to previous papers for a detailed description.

Table 1. Summary of statistics of the considered dictionaries.

Dictionary	Size	Examples of Dictionary Entries
ProblemWord	940	неисправность [malfunction], ухудшиться [to worsen], ошибка [error], неправильно [incorrectly], повреждение [damage]
NotProblemWord	70	наладить [to adjust], удобно [easy-to-use], комфортно [comfortably], выгодно [favourably], доступный [available]
Action	7800	хлопать [to clap], работать [to work], печатать [print], рекомендовать [to recommend], испечься [to bake]
AddWord	30	чересчур [too], излишне [unwarranted], пришлось [have to], слишком [too], чрезмерно [excessively], больно [a bit]
Negation	14	ни один [none], не [not], ни [non], нельзя [be impossible], не мочь [can not], никогда [never], нечего [nothing]
PositiveWord	1078	прекрасный [fine], уютный [cozy], привлекательный [attractive], замечательный [remarkable], респект [respect]
NegativeWord	1476	грубый [rough], ломкий [fragile], некрасивый [ugly], небезопасный [unsafe], гнилой [rotten], жестокий [cruel]
ImperativePhrases	26	сделайте [make], почините [repair], откорректируйте [modify], реализуйте [realize], проверьте [check], исправьте [correct]

DbA considers related negations of words and uses matching algorithm based on several dictionaries: *Action*, *Negation*, *ProblemWord*, *NotProblemWord*, and *ImperativePhrases*. The improved **CbA** uses eight dictionaries from Table 1 and determines the class of the sentence using 28 manually created rules for texts in Russian based on decomposed clauses and seven different conjunctions. For example:

(a) $< neg\_action >$, но [but] $< neg\_problem > \rightarrow \neg < problem >$.
(b) $< positive >$, но [but] $< addWord > \rightarrow < problem >$.
(c) $< problem >$, хотя [though] $< neg\_problem > \rightarrow < problem >$.

Formally, the clause-based approach consists of decomposing the extraction of problem phrases into a combination of simple decisions for sentences with conjunctions.

4.3   Supervised Machine Learning for Classification

We utilize a classifier, called **SentSVM**, with a set of features that is often used for approaches in SemEval sentiment analysis tasks and described in detailed in [6,12,22]. The set of features includes term frequency features, syntactic features, and lexicon-based features. We train an SVM with linear kernel[3] using the following set:

- **N-grams.** The occurrences of contiguous sequences of 1-grams to 2-grams and non-contiguous n-grams (n-grams with one token replaced by *). Similarly, character n-grams: the occurrences of character 3-grams up to word 5-grams are used.
- **All-caps and elongated words.** The feature counts the number of words which contain all capitalized characters and the number of words with one letter repeated more than twice.
- **Parts of speech.** We use the occurrences of each part-of-speech tag.
- **Punctuation marks.** The features count the number of marks in sequences of exclamation marks, question marks, or a combination of these marks and the number of marks in contiguous sequences of dots. Sequences that consist of more than one mark (e.g., !! or ???) are used as features.
- **Negation.** The number of negated words in negated contexts (defined as text spans between a negation word and a punctuation mark) is used. We add a suffix NEG to every token in a negated context (e.g., perfect_NEG). As negation words, we use *никакой* [any], *ничто* [nothing], and *нисколечко*.
- **Emoticons.** The following features are extracted: the emoticons contained in a text and whether the last token of a text is an emoticon.
- **Lexicon features.** For each of the created dictionaries, we compute the following eight features: the number of entries with a positive score and the number of entries with a negative score, the sum of the positive scores and the sum of the negative scores of the tweets dictionary entries, the maximum positive score and the minimum negative score of the tweets dictionary entries, and the last positive and negative scores.

In opinion mining, lexicon features are usually based on word polarity dictionaries *PositiveWord* and *NegativeWord*, where each dictionary entry is marked either positive (with a score equals 1.0) or negative (with a sentiment score equals -1.0). In order to identify problem phrases, we apply the task-specific dictionaries with problem indicators described above and used for **DbA**.

For an automatically generated lexicon, we induce the scores as a difference between PMI measures or from a linear model trained on large corpora, following state-of-the-art approaches in sentiment analysis (e.g., [12]). First, the sentiment score is computed by measuring the PMI between the term w and labeled texts:

$$score(w) = PMI(w, pt) - PMI(w, nt), PMI(w, pt) = \log \frac{p(w, pt)}{p(w) * p(pt)}, \quad (1)$$

where pt and nt denote texts with positive and negative labels, respectively.

---

[3] We use an implementation of the NRC-Canada classifier from [6]: https://github.com/webis-de/ECIR-2015-and-SEMEVAL-2015.

4.4   Topic Extensions of LDA for Sentiment Analysis

To extract noisy labels considering the major topics of the text, we apply basic topic models for sentiment analysis which extract certain topics (aspects) from a corpus of documents and distinguish words' sentiment-related labels. In [14], the authors proposed joint sentiment-topic (JST) and reverse joint sentiment-topic (Reverse-JST) models. The graphical models are shown in Fig. 1.

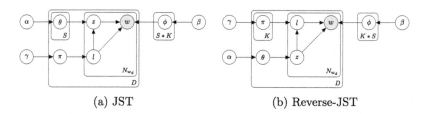

(a) JST   (b) Reverse-JST

Fig. 1. Graphical representation for (a) JST and (b) Reverse-JST.

Formally speaking, for S different possible sentiment labels (usually S is small), it extends the ϕ word-topic distributions, generating separate distributions ϕ_{lt} for each sentiment label $l \in \{1, \ldots, S\}$ (with a corresponding Dirichlet prior λ) and a multinomial distribution on sentiment labels π_d (with a corresponding Dirichlet prior γ) for each document $d \in D$. Each document now has S separate topic distributions for each sentiment label, and the generative process operates as follows: for each word position j, (1) sample a sentiment label $l_j \sim \text{Mult}(\pi_d)$; (2) sample a topic $z_j \sim \text{Mult}(\theta_{d,l_j})$; (3) sample a word $w \sim \text{Mult}(\phi_{l_j,z_j})$. The work [14] derives Gibbs sampling distributions for JST by marginalizing out π_d. Denoting by $n_{w,k,t,d}$ the number of words w generated with topic t and sentiment label k in document d and extending the notation accordingly, a Gibbs sampling step can be written as

$$p(z_j = t, l_j = k \mid \nu) \propto \frac{n_{*,k,t,d}^{\neg j} + \alpha_{tk}}{n_{*,k,*,d}^{\neg j} + \sum_t \alpha_{tk}} \cdot \frac{n_{w,k,t,*}^{\neg j} + \beta_{kw}}{n_{*,k,t,*}^{\neg j} + \sum_w \beta_{kw}} \cdot \frac{n_{*,k,*,d}^{\neg j} + \gamma}{n_{*,*,*,d}^{\neg j} + S\gamma},$$

where α_{tk} is the Dirichlet prior for topic t with sentiment label k, and $\nu = (z_{-j}, w, \alpha, \beta, \gamma, \lambda)$ is the set of all other variables and model hyperparameters.

In JST, topics were generated conditional on sentiment labels drawn from π_d. In Reverse-JST, it works in the opposite direction: for each word w, first we draw a topic label $z_j \sim \text{Mult}(\theta_d)$ and then draw a sentiment label k conditional on the topic and a word conditional on both topic and sentiment labels. Again, inference is a modification of Gibbs sampling:

$$p(z_j = t, l_j = k \mid \nu) \propto \frac{n_{*,*,t,d}^{\neg j} + \alpha_t}{n_{*,*,*,d}^{\neg j} + \sum_t \alpha_t} \cdot \frac{n_{w,k,t,*}^{\neg j} + \beta}{n_{*,k,t,*}^{\neg j} + W\beta} \cdot \frac{n_{*,k,t,d}^{\neg j} + \gamma}{n_{*,*,t,d}^{\neg j} + S\gamma},$$

and in the sentiment estimation step, Reverse-JST evaluates document sentiment as $p(k \mid d) = \sum_z p(k \mid z, d)p(z \mid d)$.

We set S = 2, hence a sentiment label can either be "positive" (i.e., relevant to the product defects and user complaints) or "negative" (i.e., irrelevant). We incorporate knowledge from the *ProblemWord* and *NotProblemWord* dictionaries into asymmetric Dirichlet priors β_{lw} using words and words' negations.

5   Experiments and Evaluation

Experimental evaluation of all models was carried out on two real-world datasets. We crawled play.google.com and otzovik.com to collect 123,014 comments and 8,271 reviews in Russian about financial applications and automobiles, respectively. In order to identify problem phrases, we split reviews into 134,359 sentences using the TextoKit library[4] and assign reviews' rating to these sentences. During a preprocessing step, we converted word tokens to lowercase and lemmatized all words using the Mystem library[5].

To train topic models, we removed all the stopwords and punctuation marks, and reduced datasets' words that occurred less than five times. In order to evaluate the methods, we created a testing corpus of 5,888 opinion units. Three annotators independently labeled sentences using the annotation scheme described in [25]. Fleiss' kappa was 0.58, which may be treated as fair interexpert agreement. Statistics are presented in Table 2.

Table 2. Summary of statistics of review datasets.

Product domain	#opinion units with rating r						#labeled examples (test)	
	$r = 1$	$r = 2$	$r = 3$	$r = 4$	$r = 5$	—V—	Problem	No-problem
Automobiles	450	2552	10773	48320	72259	58810	534	2285
Applications	26019	9337	10432	16288	60938	42393	1653	1216

We compare the following methods in the distant supervision framework:

- **SentSVM** with a set of standard features for sentiment analysis including features based on PositiveWord and NegativeWord;
- **SentSVM+PWs** with an extended set of lexicon features described below.
- Two standard baselines, Maximum Entropy classifier (MaxEnt) and Support Vector Machines (SVM) trained with both unigrams and bigrams;
- Dictionary-based and clause-based methods, **DbA** and **CbA**;
- **Naive Bayes** classifier that described in [29] and used for categorizing reviews into various types (e.g., bug reports and feature requests);
- **JST** and **Reverse-JST** (further called RJST).

---

[4] http://textocat.ru/textokit.html.
[5] https://tech.yandex.ru/mystem/.

We used manually created dictionaries to compute the feature set based on problem indicators (32 features in total) for SentSVM+PWs: (i) *ProblemWord* (8 features considering words without a negation as "positive", otherwise, "negative"), (ii) *Action* and *NotProblemWord* (16 features considering words with a negation as "positive", otherwise, "negative"), and (iii) *ImperativePhrases* and *AddWord* (8 "positive" features).

The Naive Bayes classifier depends on a set of features based on a bag of words, lemmatized unigrams, star rating, the length of the review, and a number of verbs in past, present, and future tenses. The Weka library[6] is used for the implementation of classifiers (with default parameters).

For topic models, we first set the priors for all words to $\beta_{*w} = 0.01$. Then for a problem indicator presented in the dictionary, we set $\beta_{rw} = (1, 0.001)$ (1 for "positive", 0.001 for "negative") and for an indicator with the negation, $\beta_{rw} = (0.001, 1)$. Posterior inference for the models was drawn using 1,000 Gibbs iterations and set $K = 20$, $\alpha = \frac{50}{K}$, and priors $\gamma = \frac{0.01*AvgL}{S}$, where $AvgL$ is the average review length. The topic models were trained on unlabeled corpora.

We validate the utility of the distant supervision technique by conducting an extensive set of experiments. First, we trained SentiSVM on noisy labeled corpora and evaluate the classifier on the testing data. We marked these methods with "DS$_{apr}$", where apr is an approach for assigning noisy labels. Second, following [12,24], we trained a linear SVM model (with IF-IDF) or computed PMI on noisy labeled corpora, extracting words' and bigrams' scores for an automatic lexicon. We evaluate SentiSVM with the extended set of lexicon features on the testing data using tenfold cross-validation. Classifiers were marked with "PMI$_{apr}$" and "F$_{apr}$" if noisy labels were used to compute PMI or features' weights, respectively.

Table 3 presents the classification results on the test dataset. We used macro-averaged precision, recall, and F1-measure across the two classes of the dataset to evaluate classifiers.

Several observations can be made based on Table 3. First, comparing the results of CbA and SentSVM+PWs, we can see in both tasks that SentSVM+PWs outperformed CbA using an additional knowledge from task-specific dictionaries. Second, comparing the results of DS$_{CbA}$+SentSVM+PWs and the baselines, we can see for both domains that the method based on distant supervision achieved higher F1-measure than JST, Reverse-JST, MaxEnt, SVM, and NaiveBayes. Note that no manually labeled data is used for training the distantly supervised method. DS$_{CbA}$+SentSVM+PWs and DS$_{DbA}$+SentSVM+PWs performed better than DS$_{rating}$+SentSVM+PWs. Hence, the distant supervision framework requires the knowledge-based assignment of labels rather than using trivial methods (e.g., rating, emoticons) to classify texts. Third, SentSVM+PWs+PMI$_{rating}$ performed slightly better than SentSVM+PWs+PMI$_{CbA}$. A possible explanation of these results is that scores from a lexicon PMI$_{rating}$ were computed by measuring the PMI between a token and two categories of user texts (with the ratings ≤ 3 or ≥ 4). In this case, we did

---

[6] http://www.cs.waikato.ac.nz/ml/weka/.

Table 3. Classification results.

Method	Automobiles				Mobile applications			
	Acc	P	R	F	Acc	P	R	F
JST	.808	.620	.510	.559	.423	.453	.498	.474
Reverse-JST (RJST)	.655	.566	.597	.581	.738	.732	.728	.730
MaxEnt 2-grams	.704	.581	.607	.586	.689	.684	.687	.685
SVM 2-grams	.817	.695	.670	.680	.805	.800	.801	.800
NaiveBayes	.754	.624	.645	.634	.791	.786	.783	.784
DbA	.814	.708	.746	.726	.806	.802	.803	.802
CbA	.814	.709	**.751**	.730	.820	.816	.815	.816
SentSVM	.840	.742	.701	.720	.821	.818	.812	.815
SentSVM+PWs	.847	.754	.712	.732	.831	.829	.824	.826
DS_{DbA}+SentSVM+PWs	.782	.679	.734	.706	.795	.801	.778	.789
DS_{CbA}+SentSVM+PWs	.781	.681	.742	.710	.809	.828	.787	.807
DS_{rating}+SentSVM+PWs	.813	.725	.511	.600	.774	.770	.774	.772
DS_{JST}+SentSVM+PWs	.812	.724	.506	.596	.465	.547	.522	.534
DS_{RJST}+SentSVM+PWs	.742	.593	.601	.597	.711	.704	.702	.703
SentSVM+PWs+PMI_{DbA}	.852	.762	.723	.742	.833	.831	.825	.828
SentSVM+PWs+PMI_{CbA}	**.853**	**.766**	.721	.743	.834	.832	.826	.829
SentSVM+PWs+PMI_{rating}	**.853**	.765	.726	**.745**	.839	.836	.832	.834
SentSVM+PWs+PMI_{JST}	.847	.755	.710	.732	.836	.834	.829	.831
SentSVM+PWs+PMI_{RJST}	.846	.752	.714	.732	.834	.832	.826	.829
SentSVM+PWs+F_{DbA}	.850	.759	.712	.735	.839	.837	.832	.834
SentSVM+PWs+F_{CbA}	.850	.760	.716	.737	.842	.840	.835	.838
SentSVM+PWs+F_{rating}	.847	.755	.711	.732	.837	.833	.831	.832
SentSVM+PWs+F_{JST}	.846	.754	.704	.728	.835	.832	.829	.831
SentSVM+PWs+F_{RJST}	.845	.750	.707	.728	.832	.830	.825	.828
SentSVM+PWs+$PMI_{rat.}$+F_{CbA}	.852	.763	.722	.742	**.843**	**.841**	**.837**	**.839**

not employ a machine learning method to address our classification problem. Therefore, it is more robust, which is important in the case of noisy labeled data. SentSVM+PWs+F_{CbA} and SentSVM+PWs+F_{Dba} obtained higher F1-measure than SentSVM+PWs+F_{rating} due to more accurate prediction of noisy labels (also see results of DS_{CbA}+SentSVM+PWs and DS_{rating}+SentSVM+PWs). Depending on corpora, the best results were achieved by SentSVM+PWs+F_{CbA} and SentSVM+PWs+PMI_{rating} or by the model with features based on both automatically constructed sentiment lexicons.

6 Conclusion

In this paper, we have examined a problem of assigning labels to user texts for use in machine learning. We have considered recent studies on distant supervision by applying several techniques to two corpora in Russian. Two key issues have been discussed: (i) the problem of extracting noisy labels from data and (ii) the problem of selecting a reasonable way to utilize noisy labeled data in existing machine learning approaches. Experimental results on two real-world datasets from reviews about automobiles and mobile applications have demonstrated the importance of knowledge-based label assignment in accurately classifying reviews' sentences and short comments. We have shown that assigning noisy labels obtained by a knowledge-based method on a set of dictionaries and hand-crafted rules is a more effective way to perform distantly supervised learning than trivial ways such as rating. Our results have demonstrated that the distantly supervised technique can achieve comparable results to the fully supervised SVM model with no manual annotation cost. Possible future directions for this work include applying deep learning to reduce mislabeled samples in corpora.

Acknowledgements. We thank the anonymous reviewers for their valuable comments. This work was supported by the Russian Science Foundation grant no. 15-11-10019.

References

1. Blei, D.M., Ng, A.Y., Jordan, M.I.: Latent dirichlet allocation. J. Mach. Learn. Res. **3**(Jan), 993–1022 (2003)
2. Davidov, D., Tsur, O., Rappoport, A.: Enhanced sentiment learning using twitter hashtags and smileys. In: Proceedings of the 23rd International Conference on Computational Linguistics: Posters, pp. 241–249. Association for Computational Linguistics (2010)
3. Galitsky, B., de la Rosa, J.L.: Learning adversarial reasoning patterns in customer complaints. In: Applied Adversarial Reasoning and Risk Modeling (2011)
4. Galitsky, B.A., González, M.P., Chesñevar, C.I.: A novel approach for classifying customer complaints through graphs similarities in argumentative dialogues. Decis. Support Syst. **46**(3), 717–729 (2009)
5. Günther, T., Furrer, L.: Gu-mlt-lt: Sentiment analysis of short messages using linguistic features and stochastic gradient descent (2013)
6. Hagen, M., Potthast, M., Büchner, M., Stein, B.: Twitter sentiment detection via ensemble classification using averaged confidence scores. In: Hanbury, A., Kazai, G., Rauber, A., Fuhr, N. (eds.) ECIR 2015. LNCS, vol. 9022, pp. 741–754. Springer, Heidelberg (2015)
7. Iacob, C., Harrison, R., Faily, S.: Online reviews as first class artifacts in mobile app development. In: Memmi, G., Blanke, U. (eds.) MobiCASE 2013. LNICST, vol. 130, pp. 47–53. Springer, Heidelberg (2014)
8. Iacob, C., Harrison, R.: Retrieving and analyzing mobile apps feature requests from online reviews. In: 2013 10th IEEE Working Conference on Mining Software Repositories (MSR), pp. 41–44. IEEE (2013)

9. Ivanov, V., Tutubalina, E.: Clause-based approach to extracting problem phrases from user reviews of products. In: Ignatov, D.I., Khachay, M.Y., Panchenko, A., Konstantinova, N., Yavorsky, R.E. (eds.) AIST 2014. CCIS, vol. 436, pp. 229–236. Springer, Heidelberg (2014)

10. Jin, J., Ji, P., Liu, Y.: Product characteristic weighting for designer from online reviews: an ordinal classification approach. In: Proceedings of the 2012 Joint EDBT/ICDT Workshops, pp. 33–40. ACM (2012)

11. Kim, C., Christiaans, H., Van Eijk, D.: Soft problems in using consumer electronic products. In: IASDR Conference (2007)

12. Kiritchenko, S., Zhu, X., Cherry, C., Mohammad, S.M.: Nrc-canada-2014: detecting aspects and sentiment in customer reviews. In: Proceedings of the 8th International Workshop on Semantic Evaluation (SemEval 2014), pp. 437–442 (2014)

13. Kouloumpis, E., Wilson, T., Moore, J.D.: Twitter sentiment analysis: the good the bad and the omg! Icwsm **11**, 538–541 (2011)

14. Lin, C., He, Y., Everson, R., Ruger, S.: Weakly supervised joint sentiment-topic detection from text. IEEE Trans. Knowl. Data Eng. **24**(6), 1134–1145 (2012)

15. Liu, B.: Sentiment Analysis: Mining Opinions, Sentiments, and Emotions. Cambridge University Press, Cambridge (2015)

16. Lu, B., Ott, M., Cardie, C., Tsou, B.K.: Multi-aspect sentiment analysis with topic models. In: 2011 IEEE 11th International Conference on Data Mining Workshops, pp. 81–88. IEEE (2011)

17. Marchetti-Bowick, M., Chambers, N.: Learning for microblogs with distant supervision: political forecasting with twitter. In: Proceedings of the 13th Conference of the European Chapter of the Association for Computational Linguistics, pp. 603–612. Association for Computational Linguistics (2012)

18. Mintz, M., Bills, S., Snow, R., Jurafsky, D.: Distant supervision for relation extraction without labeled data. In: Proceedings of the Joint Conference of the 47th Annual Meeting of the ACL and the 4th International Joint Conference on Natural Language Processing of the AFNLP, vol. 2, pp. 1003–1011. Association for Computational Linguistics (2009)

19. Moghaddam, S.: Beyond sentiment analysis: mining defects and improvements from customer feedback. In: Kazai, G., Rauber, A., Fuhr, N., Hanbury, A. (eds.) ECIR 2015. LNCS, vol. 9022, pp. 400–410. Springer, Heidelberg (2015)

20. Pak, A., Paroubek, P.: Twitter as a corpus for sentiment analysis and opinion mining. LREc. **10**, 1320–1326 (2010)

21. Poon, H., Toutanova, K., Quirk, C.: Distant supervision for cancer pathway extraction from text. In: Pacific Symposium on Biocomputing, pp. 120–131 (2015)

22. Proisl, T., Greiner, P., Evert, S., Kabashi, B.: Klue: simple and robust methods for polarity classification. In: Second Joint Conference on Lexical and Computational Semantics (*SEM), vol. 2, pp. 395–401 (2013)

23. Riedel, S., Yao, L., McCallum, A.: Modeling relations and their mentions without labeled text. In: Balcázar, J.L., Bonchi, F., Gionis, A., Sebag, M. (eds.) ECML PKDD 2010, Part III. LNCS, vol. 6323, pp. 148–163. Springer, Heidelberg (2010)

24. Severyn, A., Moschitti, A.: On the automatic learning of sentiment lexicons. In: Proceedings of the Conference of the North American Chapter of the Association for Computational Linguistics (NAACL HLT 2015) (2015)

25. Solovyev, V., Ivanov, V.: Dictionary-based problem phrase extraction from user reviews. In: Sojka, P., Horák, A., Kopeček, I., Pala, K. (eds.) TSD 2014. LNCS, vol. 8655, pp. 225–232. Springer, Heidelberg (2014)

26. Thung, F., Wang, S., Lo, D., Jiang, L.: An empirical study of bugs in machine learning systems. In: 2012 IEEE 23rd International Symposium on Software Reliability Engineering (ISSRE), pp. 271–280. IEEE (2012)

27. Tian, Y., Achananuparp, P., Lubis, I.N., Lo, D., Lim, E.P.: What does software engineering community microblog about? In: 2012 9th IEEE Working Conference on Mining Software Repositories (MSR), pp. 247–250. IEEE (2012)

28. Turney, P.D., Littman, M.L.: Measuring praise and criticism: inference of semantic orientation from association. ACM Trans. Inf. Syst. (TOIS) **21**(4), 315–346 (2003)

29. Walid, M., Hadeer, N.: Bug report, feature request, or simply praise? on automatically classifying app reviews. RE 2015 (2015)

30. Yohan, J., H., O.A.: Aspect and sentiment unification model for online review analysis. In: Proceedings of the 4th ACM International Conference on Web Search and Data Mining, WSDM 2011, pp. 815–824 (2011)

31. Zeng, D., Liu, K., Chen, Y., Zhao, J.: Distant supervision for relation extraction via piecewise convolutional neural networks. In: Proceedings of the 2015 Conference on Empirical Methods in Natural Language Processing (EMNLP), Lisbon, Portugal, pp. 17–21 (2015)

Elicitation Taxonomy for Acquiring Biodiversity Knowledge

Andréa Corrêa Flôres Albuquerque[1(✉)], José Laurindo Campos dos Santos[2], and Alberto Nogueira de Castro Júnior[1]

[1] Instituto de Ciência da Computação (IComp), Universidade Federal do Amazonas (UFAM), Manaus, AM, Brazil
andreaalb.1993@gmail.com, alberto@icomp.ufam.edu.br
[2] Laboratório de Interoperabilidade Semântica (LIS), Instituto Nacional de Pesquisas da Amazônia (INPA), Manaus, AM, Brazil
laurindo.campos@inpa.gov.br

Abstract. Traditionally, knowledge is kept by individuals and not by institutions. This weakens an institution's ability to survive and be competitive. Questions such as, *how to maintain and manage knowledge of institutional interest regarding employee turn over and the information entropy problem*, must be addressed. Evidence in the literature suggests gaps in the process of knowledge elicitation and acquisition. For a complex domain such as biodiversity, mechanisms are needed to acquire, record and manage knowledge with high level of expressiveness, which includes tacit knowledge. This paper presents techniques for eliciting tacit knowledge in the biodiversity domain to be used in a conceptual framework that integrates scientific knowledge. The main issues related to the knowledge elicitation process are presented, as well as the knowledge elicitation techniques used. As a case study, knowledge was elicited from INPA's ichthyology group.

1 Introduction

This research is part of a conceptual framework that integrates Experts' Mental Models (EMMs) adopted to map semantic components, of attachable structures to formal ontologies, and semantic annotation for dissemination and reuse in the Semantic Web environment [1]. The purpose of the conceptual framework is to provide a technological environment for knowledge integration, with features to add semantic expressiveness to formal ontologies and others structuring instruments of knowledge through EMMs. Also, the framework provides mechanisms to deal with issues not covered by the ontology used (OntoBio, a formal biodiversity ontology [2]), since knowledge is managed in different levels of granularity.

An important fact regarding biodiversity is the lack of unstructured data, such as the legacy knowledge of the researcher, the bushman, fisherman, local guide, etc., potential to add value to data available. These data are found at the EMM and tend to be lost in the process of information entropy.

The framework incorporates to the ontology and databases, as well as syntax and semantic structure, a domain experts' model of conceptual and cognitive structure.

© Springer International Publishing Switzerland 2016
A.-C. Ngonga Ngomo and P. Křemen (Eds.): KESW 2016, CCIS 649, pp. 157–172, 2016.
DOI: 10.1007/978-3-319-45880-9_13

The result can be mapped to the desired ontology to guide the process of evolution and/or provide a Progressive Formalization Scheme (PFS) for increasing semantic expressiveness of structural elements of knowledge. The framework designed is comprised of five steps: (E1) Elicitation of the EMMs; (E2) Formalization of different EMMs, through a PFS; (E3) Composition of knowledge to determine the correspondence and new concepts in the ontologies; (E4) Ontology evolving process; (E5) Validation of the conceptual framework by implementing the changes suggested at *E4*. This paper describes the whole process of E1 and main issues.

2 Theoretical Background

With Web evolution, the emphasis of many knowledge engineering efforts have increased. The development of computational formal ontologies concentrated on the elicitation, representation and exploitation of human knowledge [3, 4]. Knowledge management aims to provide functionalities for users to explore knowledge in a full spectrum.

Knowldege elicitation is the process of collecting human knowledge that is considered relevant. It is part of the larger process of knowledge acquisition, which is a component of the knowledge engineering discipline [5, 6]. Knowledge acquisition is more comprehensive because it also involves the explanation and formalization of the acquired knowledge. In knowledge engineering, the data obtained are processed in a computational model.

There is academic consensus, that tacit knoweldge aggregates adictional semantics to structural instruments of knowledge (knowledge representation techniques). This justifies the interest in the use of Knowledge Elicitation Techniques (KETs) to support the transformation of tacit knowledge into explicit knowledge as part of the organizational knowledge creation cycle [3, 8–10]. In this research, the tacit knowledge [7] considered is scientific. This knowledge is not necessarily formalizable, but must be capable of systematization, associated with a logical process, subjected to the laws of knowledge. Additionally, it is not intuitive and not heuristic in essence, but it must be in agreement with the fundamental conceptual structure of the domain.

Knowledge engineering identifies two roles in the Knowledge Elicitation (KE) process: expert and analyst [11, 12]. The expert is an individual who has valuable knowledge that is of interest to the organization. The analyst is performed by the knowledge engineer and is responsible for eliciting the knowledge of the expert; which is a time-consuming and expensive activity.

Although the first conceptualizations of KE define the process as extraction or expert knowledge mining, researchers recognize the following: complexity of knowledge, difficulties in its elicitation, and uncertainties associated with the final product of KE.

The problem of knowledge communication and transference amongst individuals within an organization must be dealt. The open question is *how to establish the ideal conditions that allow experts to communicate their knowledge?*. Much of the power of human expertise is the result of experience, gained through years, and represented as heuristics. Often the expertise becomes so routine that the experts have difficulty to describe specific tasks. In other cases, the knowledge is distributed throughout the organization and most of the time resides in the minds of a number of experts. Furthermore, also involves the abstraction of expert's knowledge models of data collected during elicitation. These data are limited in expressivity because much of the knowledge is tacit and it is not subjected to conscious introspection and subsequent verbalization. In many cases, experts act automatically and knowledge is not easily verbalized or even considered. An additional challenge is that, knowledge can be articulated, but the expert may be unavailable. When available, some situations need to be considered: experts can simplify or distort knowledge when transmitting it to a non-expert with limited knowledge of the domain; in other cases, since it is tacit, the elicited knowledge is a note or information generated on the spot. The expert may provide inaccurate information unintentionally, but in an attempt to satisfy the knowledge engineer, transfers any knowledge possible available.

Due to its tacit nature, the validity of verbal reports as a way to elicit knowledge has been controversial since they are considered to be incomplete and inaccurate [13]. Ericsson and Simon consider verbal reports as a valid elicitation method. Verbal reports provide interesting information, but are informal, requiring verification of data, affecting the ways in which verbalizations are collected and analyzed [14].

During KE at INPA, the following are considered: the techniques' description; the characteristics of domain experts and their associated knowledge that can directly affect the process; and the issues surrounding the proper selection of KETs. In this work are considered dialogue-based communication, and also other types, such as non-verbal, and visual.

2.1 Expert and Knowledge Relationship: Main Issues

The lack of attention to the differences between experts and the level of knowledge they possess, can affect the efficiency of the process of KE, and the quality of the knowledge.

One of the first challenges that must be addressed in any knowledge engineering project is to identify individuals with relevant experience. In some cases it may be obvious who the experts are in a given field; in others, however, it may not be clear how the experts must be identified. For this issue, factors such as the professional qualifications, experience, occupational position, results of tests and screening processes, can be used for identify experts. For example, the official position held by an individual in an organization is usually one of the criterion for the selection of experts; although, the reasons for an individual to be awarded a position within a given occupational context may not be related to actual experiences [15]. Since this research aims to capture the individual tacit knowledge of ichthyology professionals, their daily practice experience is very important. Those considered experts are: researchers (trained or in training), technicians with practical work at the Institute, fisherman, jungle guides and natives associated with INPA, and those whose activities and expertise are recognized by researchers.

Once the experts have been identified, it is important to consider the differences between the experts, and the nature of the experience of each one. A classification considered useful distinguishes three types of experts: the academic (organized structure of the domain logically), the practitioner (daily troubleshooting of the domain), and the samurai (optimum performance) [16]. Each of them differs in their deliberations, the problem-solving environment in which they work, in the state of knowledge they possess (both its internal structure and its external manifestation), position and responsibilities, source of information, and the nature of training. These distinctions amongst experts can be found in any domain. In practice, experts do not fit into either category; rather, they incorporate elements of the three types mentioned.

A number of expertise development models were proposed by cognitive science and human communities factors, and can be used as the basis for a second dimension in which the experts can be classified according to the level of expertise. The model proposed by Dreyfus and Dreyfus [17], suggests that experience is developed through the progression of five sequential stages: novice, advanced beginner, competent, proficient and expert. The transition between these phases depends on the accumulation of practical experience.

Identifying the expertise of an individual classified as expert is important for the purpose of eliciting knowledge. Clearly, individuals with well-developed expertise levels are ideal targets for KE, since they are likely to possess the knowledge relevant to the domain. The development of expertise tends to be associated with tacit knowledge, and therefore, individuals in different parts of the development path, from learner to master can be differentially sensitive to certain types of KETs. In some domains, for example, a proficient individual may have direct access to relevant knowledge of a domain than an expert. This suggests that, techniques such as interviews may produce more information from those at intermediate levels of expertise, compared to those at the highest level.

In E1, the description of the main techniques of knowledge elicitation, the characteristics of domain experts and their associated knowledge that can directly affect the process of eliciting knowledge, the issues surrounding the proper selection of KETs are considered when eliciting knowledge at INPA. In the scope of this work, it was not considered only dialogue-based communication, but also other types of communication as non-verbal, visual, etc.

3 Methods for Eliciting Scientific Tacit Knowledge

From 2000, Knowledge Management (KM) practices and intelligent system development joined strategies for knowledge engineering research towards elicitation [16]. There are several methods, techniques and tools for eliciting tacit knowledge [3, 10, 16, 18, 19]. The results of research done by Shadbolt and Smart [16], Cooke [3, 18] and Gavrilova and Andreeva [10] provided insights to guide the study presented in this paper.

Likewise, a number of KE taxonomies have been proposed [3, 10, 16]. Cooke [3], presents three groups of KE methods that can be differentiated based on the degree of specification of the methods and analysis: observation and interviews are relatively informal, tracking process methods are better specified than the previous ones and formal conceptual and well specified techniques. Shadbolt and Smart [16] define a taxonomy based on the nature of the techniques: natural and artificial, depending on the level of formality of the method. A perspective of a KM taxonomy highlights distinct roles of analyst (knowledge engineer) and expert and potential variations in these roles are proposed by Gavrilova and Andreeva [10], illustrated in Fig. 1. This approach introduces a new element based on the leadership role played by the analyst, that performs the task of facilitator and organizer of the process of KE. The analyst works as an interface between the holder of the knowledge and the knowledge receiver. The authors suggest that this taxonomy method can enrich the KM practice, by introducing the role of a knowledge analyst and structuring of elicitation methods in accordance with the different functions that the analyst can perform.

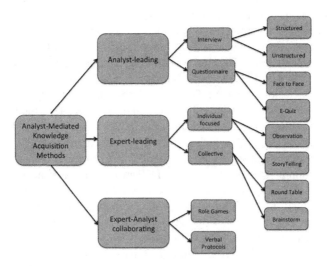

Fig. 1. Taxonomy of knowledge elicitation techniques. Adapted from Gavrilova and Andreeva [10].

Since this work requires the presence of the analyst to conduct the process of acquiring knowledge, indirect methods (such as data mining and other techniques assisted by computers) are excluded from the taxonomy for not allowing the interaction between the expert and analyst; similarly, secondary methods (such as repertory grids, classification, concept mapping and card sorting) are also excluded since they can only be used after the application of the methods presented in Fig. 1. Also, the methods range from informal techniques such as verbal reports and observations, interviews and questionnaires, to more formal techniques used in the development of knowledge-based systems [9]. The methods are divided into three categories using criteria such as the level of involvement of the expert and analyst, and type of interaction/collaboration between them.

From the three categories, two may be classified as *passive methods* and *active methods*, respectively (from the degree of engagement of an analyst in comparison with the efforts of an expert), and the third category demands equal involvement of both parties. *Active methods* (guided by the analyst), imply the techniques that require the active position of an analyst, that *extracts* the expert's knowledge with the help of specially defined questions. *Passive methods* (guided by the specialist) are techniques in which the involvement of the analyst in the process is very limited. *Observation* is a good example of a *passive method*, where the role of the analyst is to watch, listen and analyse the activity.

4   Selection Criteria for Elicitation Methods of Knowledge

The choice of which KE technique to use in any particular situation is guided by a list of criteria, including domain characteristics, the nature of the domain expert and requirements associated with the knowledge system proposed. Moreover, it is clear that some techniques will be greater time cost to the expert, or in terms of the effort required to analyze the elicited material. To select a suitable KE technique, it is necessary to understand the method that best fits the particular situation and problem. Literature provides some general conclusions about KETs' relative effectiveness [19–21].

One of the main criteria for choosing between different KETs is the kind of knowledge that needs to be elicited. Hoffman and Lintern [22] suggest that the different KETs establish different conditions in which the tacit KE is more or less probable: KETs should be seen as resource to support the expression or the knowledge communication. The conditions of communication in the process of KE are influenced by the type of technique used, since each technique is associated with several forms of social interaction, access to mnemonic signals, use of different schematic representations, and so forth. It can be argued, therefore, that tacit knowledge should not be viewed as a knowledge form that can never, in principle, be verbalized by experts; instead, it should be viewed as a knowledge form that is more easily represented in certain situations, than in anothers.

To browse through the variety of KE methods, it is necessary to identify the most appropriate method for a particular situation. Milton et al., [23] and Gavrilova and Andreeva, [10] formulated principles of KE, which are: there are different kinds of knowledge, of experts and expertise; different ways of representing knowledge, which can help elicitation, validation and reuse of knowledge; different ways to use knowledge, so that the elicitation process can be guided by the use purpose of the elicited knowledge; and therefore, KE methods should be chosen appropriately to meet the contingencies.

From the perspective of knowledge management, the types of knowledge can guide the choice of KE method. Another important feature of knowledge is who masters this knowledge - the individual or the collectivity [24]. In this research, only tacit KE methods associated with individual KE methods are considered. Based on that, the elicitation techniques selected are: interview (semi-structured and unstructured); observation (real time); storytelling (ability to prepare and understand lectures); and verbal protocols (sets the decision making logic).

From the four methods listed, except for the interview, the others do not require or even allow the intervention or mediation of the KE process by the analyst. Registration ensures fidelity to the mental model of the expert[1]. In the interview, the action of the analyst can guarantee success or failure of the method, and this is the method used in this work.

---

[1] The expert's mental model represents the thinking process of an individual about how something works (the understanding of the world around); it is based on incomplete facts, past experiences and even intuitive perceptions [36].

4.1 Interview

Interview is the KET most commonly used [3, 9, 10, 16]. It is a specific form of communication between the analyst and expert, in which the analyst poses questions prepared in advance to get a better understanding of a specific area of expertise [10]. The interview may have different levels of organization offering the analyst different levels of freedom; in all cases, the main purpose of the interview is to obtain information about how a task is performed or how a decision making process is undertaken. It can be direct or indirect and may contain explicit or implicit questions. Interviews are generally retrospective since they require the specialist to retrieve information based on past experience [25].

Interviews can be unstructured and structured. Unstructured interviews are format-free, in which neither the content nor the sequence of topics are predetermined. The structured interview is a formal version in which the analyst plans and directs the session. Structured interviews differ from unstructured since they follow a predetermined format [9]. They range from highly structured, in which the content and order of events are predefined to semi-structured, in which the content is predetermined, although the sequence may vary. Questions can be either open (what, how, why questions), imposing minimal restriction in response, or closed (who, where, when questions), imposing greater restrictions [26]. Specific types of questions vary and are the major source of difference between the structured interview techniques. In interviews, the action of the analyst can ensure the success or failure of the method.

An example of a semi-structured interview is the *knowledge acquisition grid* [27] - an array of types and forms of knowledge. Examples of knowledge forms are layouts and stories, while some examples of types of questions are *grand tour* and *cross-checking*. A *grand tour* involves aspects such as distinction of domain boundaries and the overall organization goals; *cross-checking* involves the validation of the knowledge acquired by the engineer, for example, by cross-reference.

Another form of semi-structured interview technique is *teachback* [28]. In this technique, the expert explains something to the analyst who tries to explain it back to the expert – the knowledge is actually "taught back" to the expert. The specialist has the opportunity to check and, if necessary, change the information.

Studies indicate that interviews are the most effective technique of KE; unstructured interviews are more recommended than introspective techniques [29, 30].

Interviews, particularly the semi-structured, are one of the most used methods of qualitative research, along with self-reports, observations, protocols analysis and narratives. Those information gathering are open process; since they focus on individuals or situations.

5   Application of KE Methods Guided by the Analyst

The general method of conducting a semi-structured interview includes the following steps:

(a) Minimize the hierarchical status of discussion to a comfortable level for the respondent;

(b) Use the interview guide to make open-ended questions; it is the working tool for the interviewer and contains the set of issues to be explored during the discussion;

(c) Explore the issues in depth, following the interviewee's thought process. The goal is to capture as much information as possible (both, the thinking strategy and the content of the thoughts of the interviewee).

To make sure that the interviewee speaks openly and freely, the following must be followed: have clear formulation; be concised and focused; be more general at the beginning and more specific and detailed at the end; include examples to improve understanding.

Recomendations for performing a quality interview and formulating understandable questions are: avoid using composite questions, it is best to divide and ask; avoid using poor language, jargon, slang; avoid multiple choice questions; specific questions can result in dissonant answers; generic questions may be better accepted; use confirmations of understanding to keep the focus of the respondent; use questions to establish bond and compliance; use persistent questions like *"Is there anything more?"*, *"What else?"*, *"What next?"*.

The interviews were face-to-face, individual and no time limit. The questions prepared and used in the semi-structured interview regarding this case study are listed:

1. *What environmental factors, determine the occurrence of specimen?*
2. *What kind of observation/information is important in the collection process?*
3. *What is the influence of the collection method when capturing specimens from distinct taxonomic classifications?*
4. *Have you used a non-standard method for collecting with better results?*
5. *Cite facts observed during your professional life that led you to acquire new knowledge? Ex.: Occurrence of swamp rice grass, usually indicates outbreak of malaria; the vector finds favorable conditions to be spread in the environment.*
6. *What are the preferred habitat of a particular group?*
7. *What types of biotic and abiotic factors indicate the presence or absence of individuals or groups of individuals in a given place/environment?*
8. *What are the actions incorporated into the individuals/groups of individuals collection process who come from the researcher's experience?*
9. *What are the environmental characteristics that define an event?*
10. *What is the empirical knowledge acquired during your research?*
11. *Have you noticed any aspect/event that differs from standard literature?*

5.1 Elicitation Results: Mental Models

The elicitation schema developed is comprised of four steps that must be managed and recorded: the KE medias, the KE transcriptions, the expert EMMs and their conceptual maps.

Figure 2 illustrates a sample of an elicitation schema. A record of the KE media is kept for further use. The next step of the schema is to transcript the knowledge elicited. This is followed by a description of the EMM and its conceptual map. This description is subjected to the analyst understanding of the EMM and decision of which knowledge and how it is going to be modelled. As illustrated, the use of casting net as a collection method in waterfalls is not considered. It is an explicit knowledge in the collection protocols of INPA, already consolidated. This is the choice of modelling and representation of knowledge by the analyst. The adoption of hand fishing as a collection method is innovative, result of specialist's expertise, and was considered by the analyst when modelling knowledge. It's clear that during the transition between the transcription of interview and the description of an EMM, and semantic losses may occur.

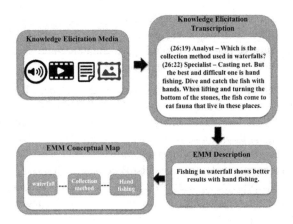

Fig. 2. A sample of an elicitation schema of an ichthyology's EMM.

Examples of the EMMs descriptions identified at the the interview with ichthyology experts are presented:

EMM1 - Cangati (family Auchenipteridae, genus *Centromochlus*) is a catfish easier to be collected at sunset using dip net. The fish come to the surface to catch insects that fall into the water.

EMM2 – Carataí (family Auchenipteridae, genus *Tatia*) is a catfish easier to be collected at sunset using dip net. The fish come to the surface to catch insects that fall into the water.

EMM3 - Fishing in waterfall shows better results with manual fishing.

EMM4 – Four-eyed fish (family Anablepidae, genus *Anableps*) is easier to fish at night, with dip net in the water-land transition zone.

EMM5 - Occurrence of cará (Family Ciclhidae) when there is the presence of floating macrophyte (Neptunia oleracea).

EMM6 - Occurrence of bodó (catfish) (Loricariidae family, Hypoptopomatinae sub-family), in old soaked trunks with recesses.

EMM7 - To fish jatuarana (genus *Brycon*), use jauari seed (genus *Astrocarium*).

EMM8 - The favorite manatee (*Trichechus manatus*) macrophyte is the grass Canarana (*Echinochloa polystachya*).

EMM9 - Well catfish (*Phreatobius cisternarum*) is found in wells and deep banks with submerged leaves.

EMM10 - Occurrence of bodó (catfish) (Loricariidae family, sub-family Hypoptopomatinae) in rapids, seek submerged branches behind and underneath.

This EMMs demonstrate the importance of eliciting the same knowledge from several experts. The purpose is to elicit the mental models of different specialists and register the different world views on the same subject, pointing out the different profiles of experts. Adverbs of frequency must be considered (always, often, frequently, sometimes, occasionally, hardly ever, never, etc.), since they are indicators that guide the search/generation of new knowledge and need to be registered. Conflicting knowledge is not treated, only consensual knowledge; it is not in the scope of this research to judge the specialist's expertise.

The EMMs are the result of the interviews with experts. A validation process consisting of additional interviews and analysis of the primary EMMs to guarantee semantic integrity must be performed, to mitigate semantic interoperability issues.

Observing the EMMs presented above, it can be inferred that:

- MM1 to MM4 and MM9 and EMM10 were elicited with specialist A and constitute more scientific information and practical techniques, featuring an expert with great scientific experience, but "get hands-on experience". This specialist despite his scientific training, is the kind of expert who participates in the entire body of the collection process. It is the practitioner specialist.
- MM5 to MM8 were elicited with specialist B and are composed of scientific information, and a little practice. This expert can be classified as academic.
- After concluding the experiment in totality, all kinds of specialists classified as Shadbolt and Smart [16] were identified.

5.2   Conceptual Map Tools for Mental Models' Representation

Some computational tools are used to help the process of KE such as xLine[2], IThought[3], and SimpleMind[4], among others. These tools produce conceptual maps used to organize and represent knowledge. Special attention is devoted to some tools influenced by the SW and ontology. Recent versions of PCPACK[5] support the export of RDF, while plug-ins for KE in Protégé[6] interoperate with the Protégé-OWL plug-in [31]. There are also Cmap-Tools[7] extending initiatives to provide support for viewing and editing OWL ontologies.

Any of these tools are suitable to represent EMMs as conceptual maps, since no semantic support is demanded at this phase of the research and they are only used as a graphical source to view the EMMs. Figure 3a and b presents the EMMs elicited using SimpleMind.

The analyst when defining the EMMs conceptual map did not consider some aspects of EMMs 1, 2 and 4: the *period* of time indicated to the *collection* and the *justification* of the use of a specific *collection method*. In EMM3 *hand fishing* is a *collection method* for *waterfalls*. At this EMM, *collection method(hand fishing)* and *collection environment(waterfall)* should be related to a *biotic entity* (a living organism, precisely, a fish); in fact, *biotic entity* is associated to a *collection method*, *biotic entity* has a *collection*

---

2 https://apps.adnx.com/en/apps/xLine.
3 http://toketaware.com/ithoughts-new-home/.
4 http://www.simpleapps.eu/simplemind/desktop/osx/features.
5 http://www.tacitconnexions.com/PCPACK%20download%20promo%20page.htm.
6 http://protege.stanford.edu/.
7 http://cmap.ihmc.us/.

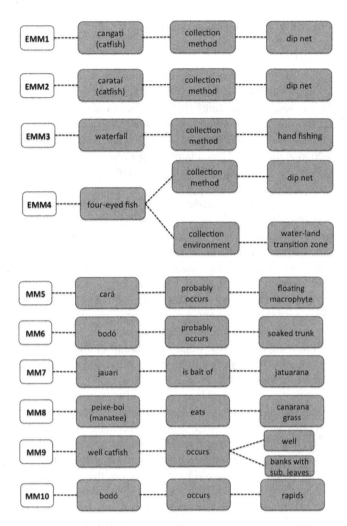

Fig. 3. a. EMM 1 to 4 elicited and represented as conceptual maps. b. EMM 5 to 10 elicited and represented as conceptual maps.

environment and the *collection environment* has a *collection method* associated to a specific *biotic entity*. During the elicitation, the expert did not specify an organism. The analyst cannot affirm that *all* fish found in the *collection environment waterfall* can be catch by *hand fishing*, but it is a possibility. *Collection method* and *collection environment* should be related when modeling the domain. Graphical resources to visualize knowledge are limited and may allow semantic losses of knowledge. When the analyst is familiar with domains' knowledge, the elicitation schema correspondent to the interview is more efficient. In EMM6 the information that the soaked trunks must be "old with recesses" was abstracted. At EMM7 jauari could be understood as food and not as bait (a collection method).

6   Management of Elicited Knowledge

The elicited knowledge requires management for future reference. To record knowledge is important because they are normally handled by the analyst, that in turn, has a personal mental model of what was elicited, with his/her particular view that can express misconceptions. In some cases, not everything that was elicited is considered by the analyst, demanding that this knowledge should be revisited. Sometimes, aspects of what is elicited is only considered in a specific scenario; or, certain knowledge is not yet formalizable and is used as annotation. Thus, maintaining a repository with the record of elicitation is mandatory. This demand, guides the decision about which architecture must be used to manage and store the records.

Individual management of research data tended to use scientific data repositories and transparency is required in scientific data treatment. The use of repositories and digital libraries is strongly associated with the preservation of information and knowledge, cultural, scientific and historical heritage, memory and dissemination.

Repository is a computer system used to store collections of a digital library and disseminate them to users. The type of repository is determined by the application and the goals to which it applies, as well as technological tool that will be adopted.

Digital libraries are defined as an environment where joined collections, services and staff that support the full cycle of creation, dissemination, use and preservation of data, information and knowledge. A digital library consists of multimidia content, interconnections and software [32]. They are comprised of systems for organization, labeling, navigation and search engines [33].

Semantic technologies offer a new level of flexibility, interoperability, and relationships for digital repositories. The use of a semantic digital library is recommended [34] to manage the elicited knowledge, since it is an information retrieval system through semantic metadata. It differs from traditional digital libraries because the user can navigate the metadata more easily, once they are no loose concepts, but related by the underlying ontology. The digital library would keep memory on the topic, but with semantics. An architecture that can be adopted for this purpose should use semantic digital libraries. The semantic digital libraries have informational contents in digital formats that are stored and made available for access, according to standardized processes, in local servers or distributed, and accessed via computer networks in other libraries. The idea is to organize, categorize and structure information resources - text, image, sound, etc. - so they can be stored, published edited and reused with flexibility [35].

Further, there are many projects such as SIMILE[8], Greenstone[9], DELOS[10], BRICKS[11], DuraSpace (FEDORA, DSpace, Mulgara)[12], JeromeDL[13] using semantics to organise and manage data similarly to the way suggested in this section.

---

[8] http://simile.mit.edu/.
[9] http://www.greenstone.org/.
[10] http://delos.info/.
[11] http://www.brickscommunity.org/.
[12] http://www.fedora-commons.org/.
[13] http://www.jeromedl.org/.

6.1   Architecture for Use of Repositories and Digital Libraries in the Conceptual Framework

Figure 4 illustrates an architecture for use of repositories and digital libraries for this purpose and is based on the definition of Toutain [35]. Semantic Digital Libraries (SDL) use informational contents in digital formats, in dedicated servers (Reps) or distributed and accessed (interface/public site) via the Internet in other libraries or library networks of the same nature. The objective is the organization, structuring and categorization of information resources (multimedia data) so they can be stored, published, edited and reused with greater flexibility.

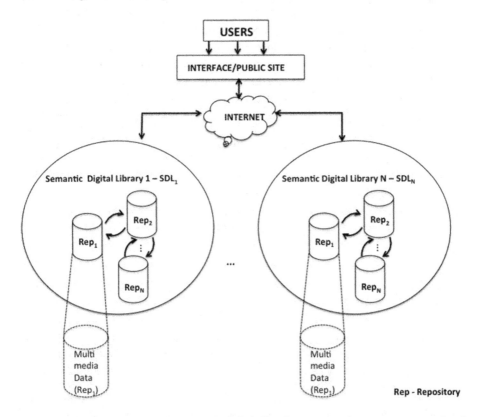

Fig. 4. Architecture of repositories and digital libraries suggested to manage records of knowledge elicitation.

This architecture of semantic digital libraries is part of the proposed framework to answer questions that OntoBio can not, since it keeps the records of the elicited knowledge. Thus, in addition to guiding the process of evolution of the reference ontology (OntoBio), updating and improving it, the proposed framework also provides resources to answer more questions about the domain and better than answered before and also answer questions not considered in the scope of the ontology.

7   Conclusions

The integration of scientific tacit knowledge performed by the conceptual framework can be jeopardized by knowledge elicitation if the process is misconducted. The paper demonstrates that the analyst should elicit knowledge from experts. It allowed the identification of the most adequate methods for tacit KE, whereas the domain experts of biodiversity may not have the necessary skills to carry out the process of acquiring knowledge without the participation of ontology engineer.

This research clarified major issues about scientific tacit KE. For each KE instance, a scenario, an expert with expertise, a knowledge is considered. Elicit tacit knowledge in biodiversity domain lead us to singular issues and requires to understand the KE methods that best fits a particular situation and problem.

This work, also demonstrates the important role that the analyst plays in KE for tacit knowledge. The definition of an EMM is analyst dependent, his/her comprehension of the domain and the understanding of the domains scope to be modelled. Semantic knowledge loss may occurs during the definition of EMM and its representation as a conceptual map. The use of semantic digital libraries are recommended to manage and store the knowledge elicited. It is easier to recover knowledge for further use.

The next step of this research is the formalization of the elicited knowledge. The definition of EMM and its conceptual map is a simplified form towards knowledge formalization.

Acknowledgments. We would like to thank ICOMP-UFAM, LIS-INPA, FAPEAM (Foundation for the State of Amazonas Research), Grant Number 021/2011 062.03101 / 2012-DO and CNPq (National Council for Scientific and Technological Development) Grant Number 486333 / 2011-6 for partially funding this research.

References

1. Albuquerque, A.C.F., Santos, J.L.C., Castro JR, A.N.: Elicitation process and knowledge structuring: a conceptual framework for biodiversity. In: Proceedings of 14th Mexican International Conference on Artificial Intelligence, MICAI 2015, Cuernavaca, Guerrero, Mexico, October 25–31. (IEEECPS) IEEE Computer Society Order Number P5721, pp. 67–72 (2015). ISBN: 978-1-5090-0323-5
2. Albuquerque, A.C.F., Santos, J.L.C.; Castro JR, A.N.: OntoBio: a biodiversity domain ontology for Amazonian. In: Proceedings of 48th Hawaii International Conference on System Sciences, Kauai, Hawaii, 5–8 January 2015. ISBN: 978-1-4799-7367-5
3. Cooke, N.J.: Varieties of knowledge elicitation techniques. Int. J. Hum. Comput. Stud. **41**(6), 801–849 (1994). Gaines, B.R. (ed.). Academic Press, Inc. Duluth, MN, USA. ISSN: 1071-5819
4. Gil, Y.: Interactive knowledge capture in the new millennium: how the semantic web changed everything. Knowl. Eng. Rev. **26**(1), 45–51 (2011)
5. Schreiber, G.: Knowledge acquisition and the web. Int. J. Hum. Comput. Stud. **71**(2), 206–210 (2013)

6. Regoczei, S.B., Hirst, G.: Knowledge and knowledge acquisition in the computational context. In: Hoffman, R.R. (ed.) The Psycology of Expertise: Cognitive Research and Empirical AI, pp. 12–25. Springer Verlag, New York (1992)

7. Botha, A., Kourie, D., Snyman, R.: Coping with Continuous Change in the Business Environment, Knowledge Management and Knowledge Management Technology. Chandice Publishing Ltd, London (2008)

8. Nonaka, I., Takeuchi, H.: The Knowledge-Creating Company: How Japanese Companies Create the Dynamics of Innovation. Oxford University Press, New York (1995)

9. Cordingley, E.S.: Knowledge elicitation techniques for knowledge-based systems. In: Diaper, D. (ed.) Knowledge Elicitation: Principles, Techniques, and Applications, pp. 89–175. Wiley, New York (1989)

10. Gavrilova, T., Andreeva, T.: Knowledge elicitation techniques in a knowledge management context. J. Knowl. Manage. 16(4), 523–537 (2012)

11. Waterman, D.: A Guide to Expert Systems. Pearson Education, London (2004)

12. Kendal, S., Creen, M.: An Introduction to Knowledge Engineering. Springer, London (2006)

13. Nisbett, R.E., Wilson, T.D.: Telling more than we can know: verbal reports on mental processes. Psychol. Rev. 84, 231–259 (1977)

14. Ericsson, K.A., Simon, H.A.: Protocol Analysis: Verbal Reports as Data. Bradford Books/MIT Press, Cambridge (1984)

15. Farrington-Darby, T., Wilson, J.R.: The nature of expertise: a review. Appl. Ergon. 37(1), 17–32 (2006)

16. Shadbolt, N., Smart, P.R.: Knowledge elicitation: methods, tools and techniques. In: Wilson, J.R., Sharples, S. (eds.) Evaluation of Human Work, Boca Raton, Florida, USA, pp. 163–200. CRC Press (2015). IBNS: 9780415267571

17. Dreyfus, H.L., Dreyfus, S.E.: Mind over Machine: The Power of Human Intuition and Expertise in an Era of the Computer. Free Press, New York (1986)

18. Cooke, N.J., Nickerson, R.S.: Knowledge elicitation. In: Durso, F.T., Nickerson, R.S., Schvaneveldt, R.W., Dumais, S.T., Lindsay, D.S., CHI, E.M.T.H. (eds.) Handbook of Applied Cognition, pp. 479–510. Wiley, Chichester (1999)

19. Hoffman, R.R., Shadbolt, N.R., Burton, A.M., Klein, G.: Eliciting knowledge from experts: a methodological analysis. Organ. Behav. Hum. Decis. Process. 62(2), 129–158 (1995)

20. Shadbolt, N.R., O'Hara, K., Crow, L.: The experimental evaluation of knowledge acquisition techniques and methods: history, problems and new directions. Int. J. Hum. Comput. Stud. 51(4), 729–755 (1999)

21. Burton, A.M., Shadbolt, N.R., Rugg, G., Hedgecock, A.P.: The efficacy of knowledge elicitation techniques: a comparison across domains and levels of expertise. Knowl. Acquisition 2(2), 167–178 (1990)

22. Hoffman, R.R., Lintern, G.: Eliciting and representing the knowledge of experts. In: Ericsson, K.A., Charness, N., Feltovich, P., Hoffman, R.R. (eds.) Cambridge Handbook of Expertise and Expert Performance. Cambridge University Press, New York (2006)

23. Milton, N., Clarke, D., Shadbolt, N.: Knowledge engineering and psychology: towards a closer relationship. Int. J. Hum. Comput. Stud. 64(12), 1214–1229 (2006)

24. Spender, J.C.: Making knowledge the basis of a dynamic theory of the firm. Strateg. Manag. J. 17, 45–62 (1996)

25. Gavrilova, T.: Choice of knowledge elicitation technique: the psychological aspect. Int. J. Inf. Theor. Appl. 1(8), 20–26 (1993)

26. Shaw, M.L.G., Woodward, J.B.: Modeling expert knowledge. Knowl. Acquisition 2, 179–206 (1990)

27. Lafrance, M.: The knowledge acquisition grid: a method for training knowledge engineers. Int. J. Man Mach. Stud. **26**(2), 245–255 (1987)
28. Johnson, L., Johnson, N.: Knowledge elicitation involving teachback interviewing. In: Kidd, A. (ed.) Knowledge Elicitation for Expert Systems: A Practical Handbook. Plenum Press, New York (1987)
29. Davis, A., Dieste, O., Hickey, A., Juristo, N., Moreno, A.M: Effectiveness of requirements elicitation techniques: empirical results derived from a systematic review. In: 14th IEEE International Requirements Engineering Conference (RE 2006), Minneapolis/St. Paul, Minnesota, USA, pp. 179–188 (2006)
30. Dieste, O., Hickey, A., Juristo, N.: Systematic review and aggregation of empirical studies on elicitation techniques. IEEE Transactions on Software Engineering, vol. 99 (2010)
31. Wang, Y., Sure, Y., Stevens, R., Rector, A.: Knowledge elicitation plug-in for protégé: card sorting and laddering. In: 1st Asian Semantic Web Conference, Beijing, China (2006)
32. Cunha, M.B.: Das Bibliotecas Convencionais às Digitais: Diferenças e convergências. Perspectivas em Ciência da Informação **13**, 2–17 (2008)
33. Vidotti, S.A.B.G., Sant'Ana, R.G.: Infra-Estrutura Tecnológica de uma Biblioteca Digital: Elementos Básicos. Em C.H. Marcondes et al. (Orgs.). Bibliotecas digitais: saberes e práticas. Salvador: UFBA, pp. 77–91 (2006)
34. Kruk, S.R., McDaniel, B.: Semantic Digital Libraries. Springer, Heidelberg (2009). ISBN 3540854339
35. Toutain, L.M.B.B.: Bibliotecas Digital: Definição de Termos. Em C.H. Marcondes et al. (Orgs.). Bibliotecas digitais: saberes e práticas. Salvador: UFBA, pp. 77–91 (2006)
36. Carey, S.: Reorganization of knowledge in the course of acquisition. In: Strauss, S. (ed.) Ontogeny, Phylogeny and Historical Development. Ablex Publishing Corporation, Norwood (1988)

Semantic Knowledge Base: Quantifiers and Multiplicity in Extended Semantic Networks Module

Marek Krótkiewicz, Krystian Wojtkiewicz$^{(\boxtimes)}$, Marcin Jodłowiec, and Waldemar Pokuta

Institute of Control and Computer Science, Opole University of Technology, Opole, Poland
mkrotki@mkrotki.com, krystian.wojtkiewicz@gmail.com, marcin.jodlowiec@gmail.com, wpokuta@gmail.com

Abstract. This article is a concise description of solutions for supplementing classic semantic networks with new functionality developed by the authors. Ths study is based on the Semantic Knowledge Base modeled in association-oriented database metamodel. It was the Extended Semantic Networks Module that has been introduced with modifiers, quantifiers, multiplicities and certainity factor extending classical approach towards semantic networks and expanding theirs knowledge representation capacity.

Keywords: Semantic networks · Semantic knowledge base · Knowledge representation · Quantifiers · Multiplicity

1 Introduction

When solving specialized problems requiring professional expertise, computer programs called expert systems are used. At the end of the 1970s it was noticed that the effectiveness of an expert program in solving a problem depends on the knowledge coded into it and not on the schemes of reasoning contained therein. It can therefore be said that the more complete the knowledge, the quicker a solution can be achieved. Expert systems are essentially organized in such a way that the knowledge base is separated from the rest of the system containing knowledge processing e.g. mechanism of reasoning. Over the last forty years many projects contributed the next steps of development of expert systems and of artificial intelligence as a whole. Many others, however, were discontinued. Following the publishing of the Lighthill report the progress in this field was temporarily halted. In computer information systems data processing is increasingly abandoned in favor of knowledge processing (knowledge-based systems) [3–5,10]. It is therefore important to properly formalize the structure of the knowledge base (KB) in newly developed solutions. It may in future result with application of knowledge-based systems in many fields of engineering, such as described in [8,11].

© Springer International Publishing Switzerland 2016
A.-C. Ngonga Ngomo and P. Křemen (Eds.): KESW 2016, CCIS 649, pp. 173–187, 2016.
DOI: 10.1007/978-3-319-45880-9_14

A knowledge-based system (KBS) can be defined in the following way: it is a system consisting of facts and rules concerning the world that we wish to represent for some purpose [6]. KBS has a built-in mechanism of reasoning which, on the basis of knowledge contained within the base, draws conclusions, derives new facts and identifies inconsistencies [16]. The knowledge in the knowledge base can be represented in various ways. Based on how the knowledge is organized the following approaches to solving the problems in this area can be discerned:

- Ontologies,
- Agent-based approaches,
- Facts and rules.

Based on how the knowledge is represented the following, among others, can be identified:

- Rules,
- Frames,
- Semantic networks.

There are also hybrid approaches [17]. They involve using two or more methods of knowledge representation within a single system. This is the type of solution adopted by the authors. Within the Semantic Knowledge Base (SKB) [13,14] research project a modular system of knowledge representation has been developed, consisting of:

- Structural Module (SM^{SKB})
  - Ontological Core Module (OCM^{SKB})
  - Relationships Module (RM^{SKB})
  - Cyclic Value Ranges Module ($CVRM^{SKB}$)
- Dimension & Space Module (DSM^{SKB})
- Extended Semantic Network Module ($ESNM^{SKB}$)
- Behavioral Module (BM^{SKB})
- Linguistic Module (LM^{SKB})

Its key characteristic is a synergy effect involving individual modules mutually supporting each other. It means using multiple modules at once to represent knowledge of various types which nevertheless constitute a logical whole.

In this work the authors pay particular attention to the problems of knowledge representation in the structures of semantic networks. In further parts of this article the classic approach based on semantic networks as well as its extensions proposed over the last few decades will be introduced. Next, the authors will present their own solution in which the main focus will be placed on the issue of adopting quantifiers for describing individual network nodes; they will also demonstrate how they made it possible to determine the number of elements participating in a given node for a given role. Those extensions to semantic networks are essential for them to be considered as reliable structures for knowledge representation and processing, as they introduce greater precision and enable decreasing of the network semantics ambiguity.

2 Semantic Networks as a Method of Knowledge Representation

2.1 Basic Design of a Semantic Network (SN)

The memory model developed by Collins and Quillian, known as the Teachable Language Comprehender (TLC) model [15], assumes that there are three types of information encoded in memory: **units**, their **properties** and **links** between units and properties [2]. The first two types take the form of nodes, wherein each property consists of an **attribute** and its defined **value** (e.g. the property "is white" consists of an attribute - the color - and the value "white"). Links in the network correspond to both the attribute and its value. Concepts are the basic memory units. Each concept is defined by a set of relevant properties. A given concept is linked to concepts one level (category) higher and one level (category) lower on the grid; therefore, a list of properties directly characterizes the concept to which it was assigned and indirectly characterizes concepts lower in hierarchy. The model also assumes that the meaning of a concept can be expressed through its relations with other concepts, hence a concept can have multiple meanings depending on current context.

The **spreading-activation model** developed by Collins and Loftus is a more advanced semantic memory model [1] - an improved version of the model developed by Collins and Quillian. Here, contrary to the previous model, it is assumed that associations can be of varying strength: concepts more strongly associated with one another have stronger (or closer) links than concepts with weaker associations. The same holds true for properties - the more characteristic ones are placed closer to the node representing a given category than the those less characteristic. Representations of objects are linked not only by category, but also by common properties. The more properties they have in common, the stronger (closer) the relationship between two nodes. There are also negative links in the network. The strength of links between the nodes influences the spread of activation. Once a specific node is activated, this activation spreads along associative links and gradually weakens. That way the closer nodes are activated more quickly and strongly than farther nodes.

2.2 Hendrix's Partitioned Semantic Networks

In the mid-1980s Gary Hendrix introduced a new way of looking at semantic networks through a proposed idea of partitioning (isolating subnetworks) [7]. It is based on grouping various nodes and arcs into units called spaces so that every node and arc is assigned to exactly one space and all nodes and arcs within a single space can be differentiated from those in other spaces. The main reason for partitioning semantic networks into spaces is to separate the scopes of quantified variables. More than a solution based on simply grouping coherent semantic areas, the fact that Hendrix established an approach wherein a mathematical apparatus is used to describe semantic networks is far more important. That way this method of representation, which has its origin in linguistics, was enriched with a formal foundation based on, among others, graph theory.

3   Main Features of the Extended Semantic Nets Module (ESNM)

3.1   Basic Assumptions and Primitive Notions of ESNM

The basic primitive notions of ESNM include: an *operator*, an *operand* and a *role*. An *operator* links *operands* using *roles*. *Operators* can be of widely varied characters. They can be, for example, actions being performed. In the sentence *"John goes to New York City by car."* the action represented by the verb (to) *"go"* is the operator. It links operands designated by *"John"*, *"New York City"* and *"car"* through appropriate roles termed *"actor"*, *"destination"* and *"tool"*. This is demonstrated in a graph in Fig. 1.

Fig. 1. Graph representing the sentence *"John goes to New York City by car."*

Operators may be of various types depending on their behavioral character. A list of operator types is shown in Table 1:

Table 1. List of operator types.

Operator type	Symbol	Question
Activity	A	What is it doing?
State	S	What state is it in?
Property	P	What property does it have?

An *activity* type denotes an action/operation which is being performed by the *actor* or actors. *State* is the state of the described *object* or objects. *Property* means defining a property of the *object* or objects and is used to e.g. define the values of *attributes*. A list of predefined operand *roles* is shown in Table 2.

A relationship can also be an operand. For example, the term *"marriage"* could be represented in semantic networks as an operator linking two operands via roles termed *"husband"* and *"wife"*. Figure 2 shows a graph corresponding to this relationship.

An SKB has a far more complex way of building semantic networks than what is shown above. It was termed Extended Semantic Network Module because of its various features significantly expanding them in terms of both grammar and semantics.

Table 2. List of predefined operand roles.

Operand role	Question
Actor	*Who/What?*
Co-operator	*With whose participation?*
Object	*Whom/What?*
Owner	*Whose?*
Modality internal	*Is he able to? Does he want to? Does he need to?*
Modality external	*Is he allowed to? Should he? Does he have to?*
Relationship	*In what relation/In what kind of relationship?*
Adverbial of manner	*How?*
Source	*Where from?*
Target	*Where to?*
Tool	*Using what?*

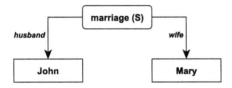

Fig. 2. Graph representing a relationship termed marriage.

3.2 Extensions of the Classic Semantic Networks in ESNM

Among the most important extensions of classic semantic networks are such concepts as: modifiers, quantifiers, cardinality and certainty factor.

Modifiers. *Modifiers* enable additional properties and values for *operators*, *operands* as well as for *modifiers* themselves.

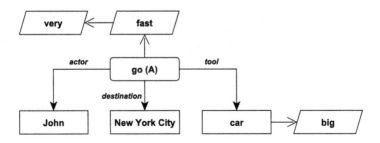

Fig. 3. Graph representing the sentence *"John goes very fast to New York City by big car."*

For example, the sentence shown in the previous section could be expanded to "*John goes very fast to New York City by big car.*" The graph representing it is shown in Fig. 3.

The operand "*car*" was supplemented by the modifier "*big*", while the operator (to) "*go*" was extended by the modifier "*fast*", and the modifier itself was extended by the modifier "*very*".

Quantifiers. *Quantifiers* significantly complement the information concerning operators and operands. The following types of information which can quantify these elements were selected: *certainty factor (cf)*, *time quantifier (tq)*, *space quantifier (sq)*, *intensity quantifier (iq)*. Certainty factor *(cf)* takes values between $[0, 1]$ and shows the degree of certainty about a described element of information. For example, the sentence from the previous section could be expanded to: "*Probably*$^{(cf=0.5)}$ *John goes very fast to New York City by big probably*$^{(cf=0.9)}$ *car.*" (Fig. 4).

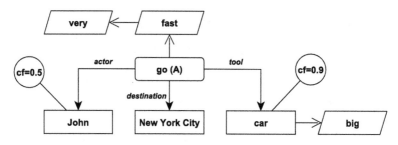

Fig. 4. Graph representing the sentence "*Probably*$^{(cf=0.5)}$ *John goes very fast to New York City by big probably*$^{(cf=0.9)}$ *car.*"

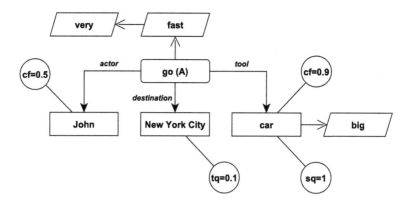

Fig. 5. Graph representing the sentence "*Probably*$^{(cf=0.5)}$ *John goes very fast rarely*$^{(tq=0.1)}$ *to New York City by big probably*$^{(cf=0.9)}$, *everywhere*$^{(sq=1)}$ *car.*"

It should be noted that the certainty factor (cf) does not pertain to an entire fact, but only to its selected element. Every quantifier can be completely independently defined for the operator and each operand.

The time quantifier (tq) and the space quantifier (sq) take values between $[0, 1]$, where 0 means never or nowhere respectively, while a value of 1 should be interpreted as always or everywhere. The graph presented in Fig. 5 contains both a time quantifier (tq) and a space quantifier (sq).

These quantifiers are linked to the operator or an operand; however, it is easier to interpret them in the context of the role they fulfill. An intensity quantifier (iq) defines the level of intensity with which a given element fulfills its role. If the role is e.g. an "*actor*", the intensity quantifier defines how deeply is he engaged in fulfilling the role of a given operator. An example of a graph representing information which takes the intensity quantifier (iq) into account is shown in Fig. 6.

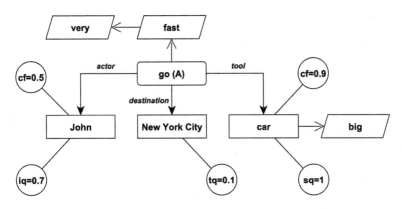

Fig. 6. Graph representing the sentence "*Probably*$^{(cf=0.5)}$ *John*$^{(iq=0.7)}$ *goes very fast rarely*$^{(tq=0.1)}$ *to New York City by big probably*$^{(cf=0.9)}$, *everywhere*$^{(sq=1)}$ *car.*"

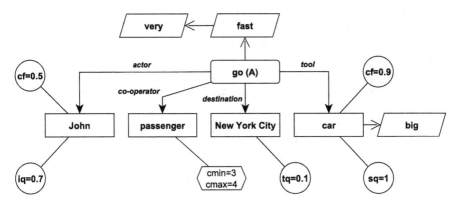

Fig. 7. Graph representing the sentence "*Probably*$^{(cf=0.5)}$ *John*$^{(iq=0.7)}$ *goes with 3 to 4 passengers very fast rarely*$^{(tq=0.1)}$ *to New York City by big probably*$^{(cf=0.9)}$, *everywhere*$^{(sq=1)}$ *car.*"

Multiplicity. The *multiplicity* of a relationship is defined by minimum (*cardinalitymin*) and maximum (*cardinalitymax*) number of elements constituting operands which participate in a given role. The default values are: $cardinalitymin = 1$, $cardinalitymax = 1$. Considering the sentence: "*Probably*$^{(cf=0.5)}$ *John goes with 3 to 4 passengers very fast rarely*$^{(tq=0.1)}$ *to New York City by big probably*$^{(cf=0.9)}$, *everywhere*$^{(sq=1)}$ *car.*" a graph shown in Fig. 7 can be created.

4   Association-Oriented Database Metamodel (AODB)

AODB [9,12] is complete and consistent database metamodel. All of its components have been elaborated exclusively for this metamodel and are dedicated to it, i.e. it does not use any language, data storage model or other element of known database systems.

Association-oriented metamodel is based on the following primitives:

- intensional (structures): *database* (*Db*), *association* (*Assoc*), *role* (*Role*), *collection* (*Coll*), *attribute* (*Attr*),
- extensional (data): *association object* (*AssocObj*), *object* (*Obj*), *role object* (*RoleObj*).

In the extensional matter the most important categories are the following: *association* (*Assoc*) and *collection* (*Coll*). *Association* ($\Diamond$) is primitive realizing conception of relationships between data and the function of *collection* ($\Box$) is to store data.

Association is the owner of *roles*. *Roles* are lists of references to linked elements, that can be either *objects* (instances of *collection*) or *association objects*(instances of *association*). *Roles* in given *association* can have any cardinality, which means unrestricted arity of relationships. Each role is defined with number of properties, such as: identifier (name) of role unique within association, multiplicity on the side of *association*, multiplicity on the side of linked element, lifetime dependency between linking and linked elements (both directions), furthermore navigability and restriction on number of reduplication of bound elements.

Apart from standard roles, in AODB one can specify *description* role. In certain sense, one can approximately treat them like specific and redefined kind of role, which features with having no identifier, can bind only association and collection, is unidirectional, i.e. objects bound with description do not store information about it, multiplicity constraint on the side of collection is 0..1 and there is no mutual constraint of lifetime of bound elements.

Collection corresponds to the concept of data storage. In the intensional sense, *collection* is well defined by set of attributes, which define types of values stored in objects.

Association does not have ability to store data, and *collection* does not have possibility to create relationships, because internal structure of those primitives of association-oriented metamodel forces completely distinct way of their usage.

Both categories independently are subject to the mechanisms such as inheritance. *Associations* can inherit from other *associations* and similarly trees of generalizations can be created for *collections*. Separation of data and relationships is complete, since in AODB each primitive performs only one, separate function. This means that if there are *collections* on database schema, one can conclude from the definition of association metamodel that they perform only one function strictly defined in terms of grammar. The situation known from relational or object metamodel does not occur, where *relation* or respectively *class* can perform function involving data storage (tuples of objects) and at the same time build *n*-ary relationship. AODB is completely unambiguous in terms of semantics of particular grammatical elements of metamodel. It is very significant not just whilst modeling, but also while attempting to analyze existing model or altering complex database schema.

Apart from definition of *association metamodel* ($\mathfrak{M}$), the descriptive part ($\mathfrak{D}$) and behavioral part ($\mathfrak{B}$) of AODB have been developed. The descriptive part comprise *Formal Notation* (AFN), which is strict, concise, formal and symbolical language of description of intensional and extensional part of metamodel. *Modeling Language* (AML) is graphical language of structures and data in AODB. It is fully consistent with AFN and metamodel definition. Both of the description languages namely (AFN and AML) are designed only for AODB and they are not any modification or subset of existing languages. The behavioral part of AODB contains *Query Language* (AQL) and *Data Language* (ADL). Both the languages fully correspond to metamodel and are completely original solutions for data selection or alteration problem, because they work directly on hypergraph structures, which represent the basis of AODB data model.

4.1   Modeling Language – AML

This section addresses the most important aspects of grammar and semantics of AML – Association-Oriented Modeling Language. It is a graphical language used to design database schemata in the *Association-Oriented Database Metamodel*.

Intensional Diagram. The graphical representation has been provided in Fig. 8. In AML, colors do not matter in syntactic terms, although using them may increase the diagram transparency.

(1) *Association* corresponds to a semantic category (*Assoc*).
(2) *Abstract Association* means that association cannot create its instances.
(3) *Collection* corresponds to a semantic category (*Coll*).
(4) *Abstract Collection* means that one collection cannot create its instances.
(5) *Role* corresponds to a semantic category (*Coll*). The graphical form in AML depends on navigability, directionality and composability.
(6) *Role Ownership,*
(7) *Navigability,*
(8) *Composition,*

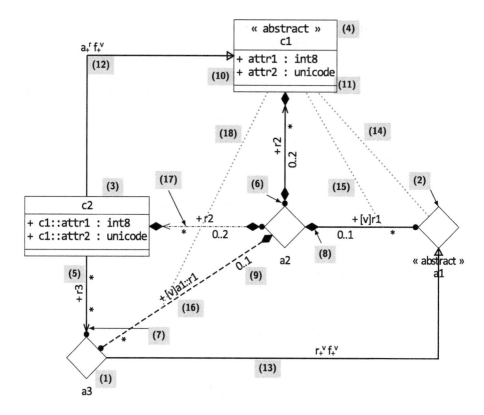

Fig. 8. Sample database schema diagram in AODB

(9) *Multiplicity* is a part of the graphical form of a role (*Role*).

(10) *Attribute* is a part of the graphical form o a collection (*Coll*). Attribute
     (*Attr*) has a name, scope of visibility, quantity, type and default value.

(11) *Attribute Type* is a part of the graphical form of an attribute (*Attr*).

(12) *Collection Generalization* is a relationship which may link two collections.

(13) *Association Generalization* is a relationship which may link two associa-
     tions. This relationship is described by an inheritance mode for roles.

(14) *Association Description* is a relationship which may link an association
     (*Assoc*) and a collection (*Coll*).

(15) *Role Description* is a relationship which may link a role and a collection.

(16) *Derived Role* is a relationship which may be represented in the diagram in
     a form similar to a role (*Role*).

(17) *Derived right to fulfill the Role* is a relationship which may be represented
     in the diagram analogically to the *Derived Role* case.

(18) *Derived Role Description* is a relationship which may be represented in the
     diagram, having an identical form as in the *Role Description*.

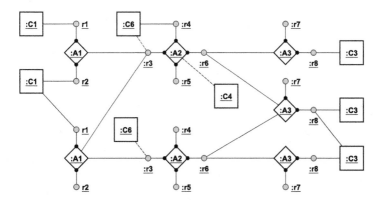

Fig. 9. An exemplary extensional AML diagram

Extensional Diagram. The Fig. 9 shows an exemplary extensional diagram for association-oriented metamodel. AML Extensional diagram may contain the items listed below.

- *Association objects* – diamonds containing names of *association objects*, colon and name of *association*, which they belong to. The name of *association object* may be omitted, if it is insignificant. The text containing name of *association object* and name of *association* is underlined.
- *Objects* – rectangles with names of the *objects*, colond and name of *collection*, which they belong to. The name of *object* may be omitted, if it is insignificant. The text containing name of *object* and name of *collection* is underlined.
- *Role objects* – small circles filled with grey or other collor that differs from white and black, with underlined *role* name.

  Moreover, it contains the following lines linking:

- *association objects* with role objects – solid lines with a small circle on the side of the association objects, indicating the role owner,
- *role objects* with *association objects* or *objects* – solid lines,
- *role objects* with *objects* describing them – dashed lines,
- *association objects* with *objects* describing them – dashed lines.

5   Extended Semantic Nets Module Implementation

A structure of SKB was modeled and realized in an *Association-Oriented Database Metamodel*. The AODB operates mainly on such primitive notions as: collection, attribute, association, role. Figure 10 shows a structure of ESNM^{SKB} in the AODB that has been slightly simplified for the purposes of this article. As ESNM is part of the SKB it fullfills all of its benefits and constrains, i.e. it is defined within AODB, stored with the use of AODB data storage model and can be queried by the use of AQL. SKB development involves the research over a specialised knowledge query language, that will also have dedicated solutions towards semantic networks. There is also a inference module being prepared. However, those topics are not a subjects of this paper and won't be evaluated here.

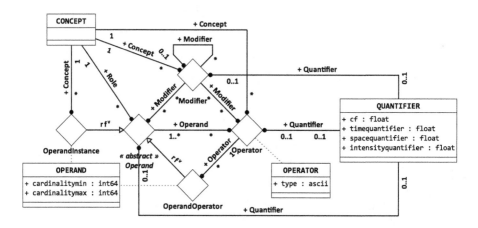

Fig. 10. Extended Semantic Nets Module (ESNM) diagram in SKB

The operator has a list of operands which has to contain at least a single operand. The operator is any CONCEPT. The association of the operands is abstract since it can take the form of either a particular CONCEPT or another operator. The latter option realizes the concepts of partitioned networks. Regardless, each operand fulfills a single Role defined by any CONCEPT and has a list of Modifiers. Modifiers can have their own modifiers, so a Modifier association has a role which links to itself. Furthermore, it refers to a CONCEPT and a QUANTIFIER which, however, is optional. The operator also has a list of Modifiers and an optional QUANTIFIER. Moreover, it is linked to the OPERATOR collection which constitutes its description, i.e. it stores the attribute defining the type of the operator. Similarly, the OPERAND collection describes the OperandInstance associations.

5.1   An Example Data Structure for ESNM in SKB

Figure 11 is a AODB data representation for the fact "*Probably*$^{(cf=0.5)}$ *John goes very fast rarely*$^{(tq=0.1)}$ *to New York City by big probably*$^{(cf=0.9)}$, *everywhere*$^{(sq=1)}$ *car.*". The semantics of the sentence and thus the illustration has been presented earlier, so it shall not be deeply elaborated here. However, it is important to observe that each and every element of the sentence that carries any direct and explicit meaning has its own representation as an object of the CONCEPT collection, i.e. John, go, fast, very, car, big, New York City. All of other words are represented by appropriate structures of ESNM, e.g. *probably*$^{(cf=0.5)}$ as the

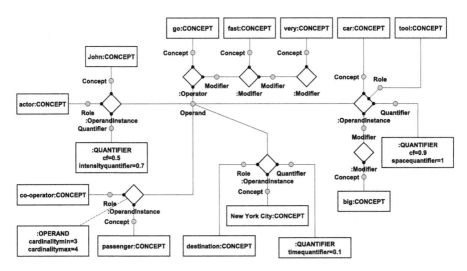

Fig. 11. A data diagram of the associative metamodel (AODB) for the conceptual graph in Fig. 10 expressed in the ESNM structure of an SKB system

object of QUANTIFIER collection with the *cf* attribute value set to 0.5, etc. Moreover, the are several objects of CONCEPT collection that represent the roles assigned to each of operands of the created semantic network, e.g. tool, actor, co-operator, etc. The diamond-shaped elements represent association objects of corresponding associations defined as elements of ESNM and identify, by the use of roles, the function of particular elements in the presented semantic network, e.g. the concept John fulfills the role of Operand being an actor.

6   Summary

1. An Extended Semantic Nets Module dedicated to storing complex facts and rules for the purposes of an SKB system was demonstrated.
2. The functional scope and a realization of an extension of the idea of semantic networks in the scope of implementing quantifiers and multiplicity of elements were presented.
3. The presented assumptions of extending the semantic networks have been verified by an implementation in a database structure developed in an advanced association-oriented database metamodel.

The authors see great strength in the Extended Semantic Nets Module, mainly stemming from the possibility of a flexible approach to the knowledge stored therein. It should be noted, however, that the strength of this module comes directly from the fact that every element used, whether as an operator, operand or even a modifier, has to be defined in advance in the SKB as a concept existing in the structural module. Such a solution is meant to efficiently manage polymorphism in the scope of stored and processed knowledge.

References

1. Collins, A.M., Loftus, E.F.: A spreading-activation theory of semantic processing. In: Readings in Cognitive Science: A Perspective from Psychology and Artificial Intelligence, pp. 126–136 (2013)
2. Collins, A.M., Quillian, M.R.: Retrieval time from semantic memory. J. Verbal Learn. Verbal Behav. **8**(2), 240–247 (1969)
3. Do, N.V.: Ontology COKB for knowledge representation and reasoning in designing knowledge-based systems. In: Fujita, H., Selamat, A. (eds.) SoMeT 2014. CCIS, vol. 513, pp. 101–118. Springer, Heidelberg (2015)
4. Do, N.V., Nguyen, H.D., Mai, T.T.: Designing an intelligent problems solving system based on knowledge about sample problems. In: Selamat, A., Nguyen, N.T., Haron, H. (eds.) ACIIDS 2013, Part I. LNCS, vol. 7802, pp. 465–475. Springer, Heidelberg (2013)
5. Dyachenko, O., Zagorulko, Y.: A collaborative development of ontology-based knowledge bases. In: Klinov, P., Mouromtsev, D. (eds.) KESW 2014. CCIS, vol. 468, pp. 219–228. Springer, Heidelberg (2014)
6. Gruber, T.R.: A translation approach to portable ontology specifications. Knowl. Acquisition **5**(2), 199–220 (1993)
7. Hendrix, G.G.: Encoding knowledge in partitioned semantic networks. In: Associative Networks: Representation and Use of Knowledge by Computers, pp. 51–92 (1979)
8. Iwanowski, M., Korzyńska, A.: Segmentation of moving cells in bright field and epi-fluorescent microscopic image sequences. In: Bolc, L., Tadeusiewicz, R., Chmielewski, L.J., Wojciechowski, K. (eds.) ICCVG 2010, Part I. LNCS, vol. 6374, pp. 401–410. Springer, Heidelberg (2010)
9. Jodłowiec, M., Krótkiewicz, M.: Semantics discovering in relational databases by pattern-based mapping to association-oriented metamodel—a biomedical case study. In: Piętka, E., Badura, P., Kawa, J., Wieclawek, W. (eds.) Information Technologies in Medicine. AISC, vol. 471, pp. 475–487. Springer, Switzerland (2016)
10. Kolonin, A.: Distributed knowledge engineering and evidence-based knowledge representation in multi-agent systems. In: Klinov, P., Mouromtsev, D. (eds.) KESW 2015. CCIS, vol. 518, pp. 291–300. Springer, Heidelberg (2015). doi:10.1007/978-3-319-24543-0_23
11. Koprowski, R., Wróbel, Z., Korzyńska, A., Chwiałkowska, K., Kwaśniewski, M.: Automatic analysis of 2D polyacrylamide gels in the diagnosis of DNA polymorphisms. Biomed. Eng. Online **12**(1), 1 (2013)
12. Krótkiewicz, M.: Association-oriented database model n-ary associations. Int. J. Softw. Eng. Knowl. Eng. (2016, accepted for publication)
13. Krótkiewicz, M., Wojtkiewicz, K.: An introduction to ontology based structured knowledge base system: knowledge acquisition module. In: Selamat, A., Nguyen, N.T., Haron, H. (eds.) ACIIDS 2013, Part I. LNCS, vol. 7802, pp. 497–506. Springer, Heidelberg (2013)
14. Krótkiewicz, M., Wojtkiewicz, K.: Functional and structural integration without competence overstepping in structured semantic knowledge base system. J. Logic, Lang. Inform. **23**(3), 331–345 (2014)
15. Quillian, M.R.: The teachable language comprehender: a simulation program and theory of language. Commun. ACM **12**(8), 459–476 (1969)

16. Tran, T.H., Nguyen, N.T.: Integration of knowledge in disjunctive structure on semantic level. In: Lovrek, I., Howlett, R.J., Jain, L.C. (eds.) KES 2008, Part I. LNCS (LNAI), vol. 5177, pp. 253–261. Springer, Heidelberg (2008)
17. Verhodubs, O., Grundspenkis, J.: Ontology merging in the context of a semantic web expert system. In: Klinov, P., Mouromtsev, D. (eds.) KESW 2013. CCIS, vol. 394, pp. 191–201. Springer, Heidelberg (2013)

Data Management

RDF Query and Inference in Prolog

M.B. Alves[1,2(✉)], C.V. Damásio[1], and N. Correia[3]

[1] CENTRIA, Universidade Nova de Lisboa, 2829-516 Caparica, Portugal
cd@fct.unl.pt
[2] ESTG, Instituto Politécnico de Viana do Castelo,
4900-348 Viana do Castelo, Portugal
mba@estg.ipvc.pt
[3] CITI, Universidade Nova de Lisboa, 2829-516 Caparica, Portugal
nmc@fct.unl.pt

Abstract. It is presented a system that allows the combination of a declarative language with a Semantic Web framework, namely, the Jena framework (https://jena.apache.org). and XSB Prolog using InterProlog [3], a library allowing the development of combined Java+Prolog applications. Our library allows RDF and SPARQL queries in Prolog predicates. In this way, we can develop Semantic Web applications that makes use of the power of declarative languages to construct sophisticated rule systems within Semantic Web environments. Benchmark results are presented, showing the practical impact of the use of the system.

1 Introduction

Most of the Web's content is designed for humans to read, not for computer programs to manipulate meaningfully. The Semantic Web brings structure to the meaningful content of Web pages, creating an environment where software agents roaming from page to page can readily carry out sophisticated tasks for users. Semantic Web provides enhanced information access based on the exploitation of machine-processable metadata. To achieve this purpose, the information is modelled in the Resource Description Framework (RDF) [9], a language that is a part of the W3C's Semantic Web Activity, which purpose is that Web information that has an exact meaning, can be understood and processed by computers and can integrate information from the web. With the aim of structuring RDF resources, RDF Schema [7] provides mechanisms for describing groups of related resources and the relationships between these resources. Since the expressivity of RDF and RDF Schema is very limited, OWL (Web Ontology Language) [2] extends RDFS with a means for defining description vocabularies for Web resources and its primary aim is to bring the expressive and reasoning power of description logic to the Semantic Web. The data described by an OWL ontology is interpreted as a set of "individuals" and a set of "terminological and property assertions" which define concepts, and relate these individuals to each other. SPARQL (Sparql Protocol And RDF Query Language) [8] is the language to query RDF. SPARQL can be used to express queries across diverse

© Springer International Publishing Switzerland 2016
A.-C. Ngonga Ngomo and P. Křemen (Eds.): KESW 2016, CCIS 649, pp. 191–201, 2016.
DOI: 10.1007/978-3-319-45880-9_15

data sources, when the data is stored as RDF. The SPARQL query language is based on matching graph patterns and its results can be result sets or RDF graphs.

The most well-known frameworks for the Semantic Web are provided in widely used computer programming languages, like Java. Our work fosters the development of Semantic Web applications where part of the knowledge can be defined with a logic program. This approach allows the use of a semantic web framework enriched with knowledge represented by Prolog rules. Our solution allows to overcome some limitations of Jena rules or SWRL rules as is stated in [6]. Therefore, to develop Semantic Web knowledge-rich applications, knowledge engineers need to integrate Prolog with a second language to combine the power of declarative languages with the main frameworks of the semantic web world. To achieve this purpose, we choose Jena as the semantic web framework and we use the InterProlog library to make the connection to Prolog. We use XSB Prolog as a state-of-the-art programming logic environment, and supported by the InterProlog library. Combining Prolog with a semantic web framework allows a focus on intrinsic issues of logic engines, delegating the Semantic Web issues to the proper frameworks, namely, RDF databases, RDFS and OWL reasoning and the SPARQL engine.

This document is structured as follows. We overview the capabilities of our system in Sect. 2, where we detail how we integrate Prolog and Jena, using the InterProlog API the Jena Framework. Section 3 presents the system architecture. In Sect. 4 we describe some benchmarks tests and we finish with our conclusions in Sect. 5.

2   Integrating Prolog and Jena, via InterProlog

The Jena framework is a free and open source Java framework for building Semantic Web applications. It provides a programmatic environment for RDF, RDFS, OWL, a query engine for SPARQL and it includes a rule-based inference engine. Jena is widely used in Semantic Web applications because it offers an "all-in-one" solution for Java, including a general purpose rule-based reasoner which is used to implement both the RDFS and OWL reasoners. Jena general reasoner supports rule-based inference over RDF graphs and provides forward chaining, backward chaining and a hybrid execution model.

InterProlog (http://www.declarativa.com/interprolog) is an open source library for developing Java + Prolog applications, providing a higher-level API directly mapping Java Objects to Prolog terms, inducing a more concise and declarative programming style [3]. Currently, InterProlog supports XSB Prolog.

To exemplify how Prolog integrates with Jena, consider the following OWL schema. To save space, the well know prefixes are not listed. Note that the examples that will be shown are with the purpose to exemplify how Prolog integrates with Jena. Some of them can be implemented with Jena rules. It is out of the scope of this work to demonstrate the limitations of Jena rules or SWRL rules and why a more declarative language is required. Some of these limitations are stated here [4] and here [1].

```
:Person                              :Father owl:intersectionOf
   owl:unionOf ( :Men :Women ) .        (:Men :Ascendent).

:Men                                 :Mother owl:intersectionOf
   owl:equivalentClass [                (:Women :Ascendent).
      rdf:type owl:Restriction ;
      owl:onProperty :hasSex ;       :hasParent
      owl:hasValue :Male                rdfs:domain :Person ;
   ].                                    rdfs:range :Person.

:Women                               :hasChild owl:inverseOf :hasParent .
   owl:equivalentClass [
      a owl:Restriction ;            :hasSex
      owl:onProperty :hasSex ;          rdfs:domain :Person ;
      owl:hasValue :Female              rdfs:range [ owl:oneOf (:Male
   ].                                                :Female)].

:Ascendent                           :hasFather rdfs:subPropertyOf
   owl:equivalentClass [                :hasParent ;
      rdf:type owl:Restriction ;        rdfs:range :Father .
      owl:minCardinality
           "1"^^xsd:nonNegativeInteger :hasMother rdfs:subPropertyOf
           ;                            :hasParent ;
      owl:onProperty :hasChild          rdfs:range :Mother .
   ] .
```

Next Prolog rule (Rule 1) defines when a X is sibling of a Y. For that, the rules resorts to the RDF database where is defined the family relationships. The query is done through invocation of the predicate jenaQueryModel, that allows queries to the knowledge base redirecting the question to Jena.

```
isSibling(X, Y) :-                   isSibling(X, Y) :-
   jenaQueryModel(triple(X,':hasParent',   jenaQueryModel([
      Z)),                                 triple(X, ':hasParent',
   jenaQueryModel(triple(Y,':hasParent',       Z),
      Z)),                                 triple(Y, ':hasParent',
   X \= Y.                                     Z)
           Rule 1                    ]), X \= Y.
                                             Rule 2
```

Notice in the previous example that **hasParent** instances are obtained by RDFS inferencing in the Jena Model. With this approach we leave to Jena the standard inferences related with the ontological model, and where we can configure the desired type of inference (RDFS, OWL, microOWL, miniOWL, etc.), and we use Prolog as a declarative environment.

In Prolog, we can register prefixes through predicate **registerPrefix/2** (e.g. registerPrefix('fam', 'http://www.example.org/family#')). The prefixes that are registered in Jena Model are also considered, and are automatically imported to the Prolog environment.

In the Prolog side, we have two predicates to make calls to the inference model on Jena, the predicate jenaQueryModel and the predicate jenaQuerySparql. The predicate jenaQueryModel/1 can receive two types of arguments. The argument may be a **triple** term whose arguments are the *Subject*, the *Predicate* and the *Object* of a RFD triple. The Rule 1 shows a call of the predicate jenaQueryModel/1 with a **triple** term. Alternatively, the argument of the predicate may be a list of **triple** terms, as is shown in Rule 2, where all the triples are sent to the Jena at a once. In each call to the Jena Model all triples that satisfy the condition are returned. Obtaining another solution doesn't imply another call to Jena, since InterProlog does not support non deterministic goals. In Rule 2, we can see that Z is a variable in the triples but is not used on the

Prolog side. In these situations, we can use SPARQL variables instead, by single quoting them (`triple(X, ':hasParent', '?z')`).

The predicate `jenaQuerySparql/3` allows retrieving data using a SPARQL command. The first argument of the predicate is the SPARQL command to be executed, the second argument is the functor that unifies the result and the third argument is a list with the results. Each tuple of the result is the functor declared in the second argument and all variables of the `Select` clause of the SPARQL command as arguments (see Rule 3).

```
isSibling(LResult) :-
  jenaQuerySparql('
  prefix :
       <http://www.example.org/family#>
  Select ?x ?y
  where { ?x :hasParent ?z . ?y :hasParent
       ?z .
  filter(?x != ?y) . } ', sibling,
     LResult).
```
Rule 3

The obtained result is:

```
[sibling(:Michael, :Carla),
   sibling(:Carla, :Michael),
   sibling(:Peter, :Martha),
   sibling(:Martha, :Peter)]
```

3   System Architecture

The architecture of the system is presented in Fig. 1. In Listing 1 is listed a set of coding excerpts that will support the description of the system. The numbers associated with arrows have correspondence in the excerpts from the listed programming code.

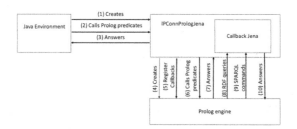

Fig. 1. System architecture

```
/* ---------- Coding excerpt 1 ---------- */
IPConnPrologJenaModel vIPConnPrologJenaModel = new IPConnPrologJenaModel();
    //(1)
vIPConnPrologJenaModel.init(getXSBloc(), getPrologFile(), infModel); //(1)

/* ---------- Coding excerpt 2 ---------- */
IP_callbacksModel vIP_callbacksModel = new IP_callbacksModel(infModel);

/* ---------- Coding excerpt 3 ---------- */
engine = new XSBSubprocessEngine(PATH_TO_XSB); // (4)
engine.consultAbsolute(new File(fprologfile));
int objID = engine.registerJavaObject(vIP_callbacks); // (5)
engine.command("retractall(ipObject(_))");
engine.command("assert(ipObject(" + objID +"))"); // (5)

/* ---------- Coding excerpt 4 ---------- */
String clause2 = "[TM]";
```

```
String clause1 = "findall(sibling(X, Y), isSibling(X, Y), L),
    buildTermModel(L,TM)";

Object [] bindings = vIPConnPrologJena.run(clause1, clause2); // (2) (3)
TermModel list = (TermModel) (TermModel)bindings[0]; // (3)
while(list.getChildCount()>0){
    System.out.println("result:"+list.getChild(0).toString());
    list = (TermModel) list.getChild(1);
}

/* ---------- Coding excerpt 5 ---------- */
public Object[] run(String clause1, String clause2) {
    Object[] bindings = engine.deterministicGoal(clause1, clause2); // (6)
        (7)
    return bindings;
}

/* ---------- Coding excerpt 6 ---------- */
isSibling(X, Y) :-
    jenaQueryModel(triple(X, 'fam:hasParent', Z)), % (8)
    jenaQueryModel(triple(Y, 'fam:hasParent', Z)), % (8)
    X \= Y.

/* ---------- Coding excerpt 7 ---------- */
public String[] queryRDF(String pSubject, String pProperty, String pObject) {
    String retV[] = null;

    if(getModel() != null) {
        retV = queryRDFModel(pSubject, pProperty, pObject);
    }
    return retV; //(10)
}
```

Listing 1

To connect a Prolog environment with Jena, we create an instance of the class IPConnPrologJenaModel, an extension of *IPConnPrologJena*, and initialise with the location of Prolog environment, the Prolog program, and the information model that represents our knowledge base (**Excerpt 1**). In the initialisation of the system, besides the creation of a *Prolog Engine* and consulting the Prolog source files (**Excerpt 3**), it is also created a Jena callback object (**Excerpt 2**) that will be responsible for receiving calls from Prolog to the Jena model. The Jena callback object makes the connection between Prolog and the Jena environment, answering to the queries of Prolog. The Jena callback object must be registered in the *Prolog Engine* to be able to redirect the RDF queries. We use this feature of InterProlog to allow calls to the knowledge base (predicated jenaQueryModel/1 and jenaQuerySparql/3), that is stored in Jena, from the Prolog environment (8, 9). The answers are redirected in the reverse way. The messages between Prolog and Jena callbacks are synchronous, when a query or a SPARQL command is done, the Prolog engine waits for the answer from the Jena callback (10).

In the Java side, we can make calls to the Prolog environment, via IPConnPrologJena that will redirect the query to the Prolog engine. In **Excerpt 4** is listed an excerpt of a program that makes a call to the Prolog environment and lists the answers. In **Excerpt 5**, the method run of the class IPConnPrologJena executes the call in the Prolog engine and returns the answers. If the Prolog rule needs to consult the knowledge base to perform reasoning in the Jena side, then makes a call to using the Jena callback. In the example **Excerpt 6** the predicate jenaQueryModel/1 sends a message to the callback object to query the knowledge base. The callback object receives the message, performs the task and returns the results **Excerpt 7**.

3.1   Jena RDF Query Engines

In Jena, there are two ways to query RDF data. One of them is using Jena as a Java API which can be used to create and manipulate RDF graphs. Jena has object classes to represent graphs, resources, properties and literals. The other way is using the ARQ engine, a query engine for Jena that supports the SPARQL RDF Query language.

In our system, when the predicate `jenaQueryModel` is invoked in a rule, and not in a list, is used the mechanisms of Jena API as RDF query engine. When is defined a list of conditions, the system uses the ARQ engine. This decision was supported by the benchmark tests performed and detailed in Sect. 4. However, we also allow the static definition of the type of the engine. We can also recommended that when we have a set of conditions is better to use a list of conditions (Rule 2) instead of declare all conditions individually (Rule 1). Furthermore, querying a single triple in a list is to be avoided.

3.2   Calling Prolog in Builtin Primitives of Jena

Jena allows procedural primitives which can be called by the Jena rules to process data beyond the triples inference. Some examples of Builtin functions are *less Than, equal, max*. These Builtin functions are each implemented by a Java object stored in a registry and additional primitives can be created and registered, called custom Builtin functions.

In our system we also allow the calling of Prolog predicates by custom Builtin functions. The purpose is allow to call Prolog predicates in functions that can be used, for instance, in SPARQL commands. The sole limitation is that these predicates cannot call SPARQL commands. In a custom Builtin functions, what is passed to the `IPConnPrologJena` is not a Jena model, but an instance of a class *RuleContext* that conveys context information from a rule engine to the stack of procedural builtins. This gives access to the triggering rule, the variable bindings and the set of currently known triples. However, the class *RuleContext* doesn't allow to retrieve data through SPARQL commands. Notice that Builtin functions returns *true* or *false* and can return single values as reference variables. To call a Prolog predicate, we developed the custom built-in *prologCall*. The constructor of this class receives the location of the Prolog executable and the location of the Prolog program. *RuleContext* assignment is done at runtime:

```
vPrologCall = new PrologCall(getXSBloc(), getPrologFile());
BuiltinRegistry.theRegistry.register(vPrologCall);
```

Now, let's consider the following Jena rule that makes a call to the custom builtin function *prologCall*. The first parameter is the predicate name to be called in Prolog, followed by the the parameters itself. This mechanism allows use to use Prolog reasoning, for instance, in SPARQL queries, as well as in Jena's rules.

```
(?x :isSibling ?y) <-                    Select ?x ?y
    (?x rdf:type :Person),               where { ?x :isSibling ?y .}
    (?y rdf:type :Person),
    prologCall(isSibling, ?x, ?y).
```

To develop custom Builtin functions based on *prologCall*, we implemented the class *IPBuiltin*, an extension of *prologCall*. A custom Builtin function based on *prologCall* can be found below, as well as how this rule can be used in a Jena Rule.

```
public class IsSibling extends IPBuiltin{       (?x fam:isSibling ?y) <-
    @Override                                       (?x rdf:type
    public String getName() {                           fam:Person),
        return "IsSibling";                         (?y rdf:type
    }                                                   fam:Person),
    @Override                                       IsSibling(?x, ?y).
    public int getArgLength() {
        return 2;
    }
    public IsSibling(String XSBloc, String
        fprologfile) {
        super(XSBloc, fprologfile,
            "isSibling");
    }
}
```

4   Benchmark Tests

We performed some benchmark tests to evaluate the performance of our RDF query and inference in Prolog, resorting to LUMB [5]. LUBM is a benchmark for OWL Knowledge Base Systems, generated by *the Lehigh University Benchmark*, consisting in 14 datasets whose number of triples ranged from about 16500 to 24500. The data generated is about an University system and the OWL schema can be consulted in http://swat.cse.lehigh.edu/onto/univ-bench.owl. To perform the benchmark tests, we use the tool JMH[1], a Java harness for building, running, and analysing nano/micro/milli/macro benchmarks. We also have a set of 14 SPARQL commands, also defined by *the Lehigh University Benchmark*. For each SPARQL command, we adapted it to the different encodings supported by our approach. For instance, consider the next SPARQL command:

```
SELECT ?X
WHERE { ?X rdf:type ub:GraduateStudent .
        ?X ub:takesCourse
            <http://www.Department1.University0.edu/GraduateCourse0> }
```

Based on this SPARQL command, we created the following 3 variants:

```
rule4(X) :-
    jenaQueryModel(triple(X, 'rdf:type', 'ub:GraduateStudent')),
    jenaQueryModel(triple(X, 'ub:takesCourse',
        "http://www.Department1.University0.edu/GraduateCourse0")).
rule5(X) :-
    jenaQueryModel([
        triple(X, 'rdf:type', 'ub:GraduateStudent'),
        triple(X, 'ub:takesCourse',
            "http://www.Department1.University0.edu/GraduateCourse0") ]).
rule6(LResult) :-
    jenaQuerySparql('
    SELECT ?X
```

---

[1] http://openjdk.java.net/projects/code-tools/jmh/.

```
WHERE { ?X rdf:type ub:GraduateStudent .
   ?X ub:takesCourse
         <http://www.Department1.University0.edu/GraduateCourse0>
}', solutions, LResult).
```

We performed 4 sets of benchmarks for each SPARQL command in each dataset:

1. Executing the SPARQL command in the ARQ engine.
2. Executing rules in the form of **Rule 4**, where triples are invoked individually.
3. Executing rules in the form of **Rule 5**, where are invoked as a set of triples.
4. Executing rules in the form of **Rule 6**, where a SPARQL command is called.

We compared item (2) and item (3) to figure out the best approach for RDF data retrieval in our system, considering the two engines provided by Jena, Jena API or ARQ engine. We performed two comparisons. In the first one, we ran rules where conditions were defined as a set of invocations of the predicate jenaQueryModel, as is exemplified in Rule1. For each rule, we tested the performance using the mechanisms of Jena API and using the ARQ engine. The same type of approach was adopted to evaluate the appropriate engine to use when querying with a list of triple patterns, as is exemplified in Rule 2. We performed the latter analysis for rules with two or more conditions because for one condition the behaviour would the same as the former. In summary, ARQ engine it was about 22 % faster. In the queries with individual triples the mechanisms of Jena API was about 73 % faster in average than the ARQ engine whereas for lists of triple patterns, the ARQ engine was about 25 % faster than the mechanisms of Jena API. However, we didn't had linear results with differences when we store the data in a TDB store or when we kept the model in memory. We can have a partial conclusion that item (3) is a better approach than item (2), except in the special situation where we query a single triple pattern. However, there are exceptions that we didn't figure out the pattern.

To assess the overhead of the system, we compared item (1) with item (4) returning the result of a *count* instead of the data itself. The rationale was testing the overhead only invoking the SPARQL command via Prolog. The operation of the item (4) takes in average about 15 times than the item (1). Therefore, we can conclude that our approach brings some overhead. The difference between item (3) and item (4) was not significant, almost the same timings. One factor which brings a significant overhead is the total amount of data returned, both the number records returned and the amount of bytes returned by a command. **Graphic 1** shows the number of records returned with the time taken (in milliseconds) by a given command to be executed. We can see, obviously, that the time increases with the number of records returned. However, this relation is not immediate, since there are some commands that take less time to obtain the same number of records. However, this is also obvious because two different commands can return a similar number of records but return a different total amount of bytes. **Graphic 2** relates the total amount of bytes returned with the time consumed by a given command to be executed. As expected, the time increases with the total amount of bytes returned but this relation is not linear. In this situation is not obvious and we argue that the number of records also contributes to the time spending by a command to be executed. Note that the data displayed is obtained by subtracting the time to execute a command in item (3) less the time to execute a command in item (1). The idea is to evaluate the commands putting emphasis on the total amount of data returned and abstracting out the complexity of their execution.

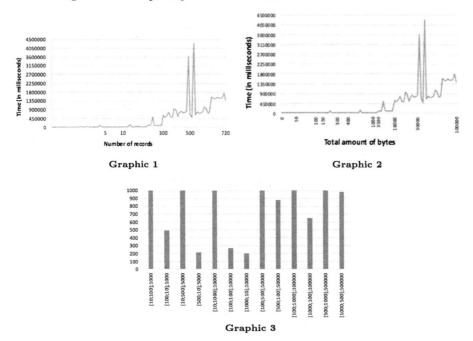

Graphic 1   Graphic 2

Graphic 3

We also tested the time spent in the data transfer by InterProlog. We created data structures with 10, 100, 500 and 1000 records combined with record size of 10, 100, 500 and 1000 bytes each one, considering the combinations of 1000, 5000, 10000, 50000, 100000, 100000 of total amount of data. The **Graphic 3** compares the different combinations. The label of each bar means *[number of records; size of each record in bytes]; total amount of data.* For purpose of legibility of the chart, all the data were converted to a same maximum. By the results of the **Graphic 3**, we can highlight that transfer a data structure with 100 records of 10 bytes each one is almost 2 times slower that 10 records of 100 bytes each one whereas in 500 records of 10 bytes each one is almost 3.5 times slower that 10 records of 500 bytes each one. For higher values of total amount of data transfer, the proportion reduces but keeps slower when the number of records is high. We can conclude that the number of records returned contributes more strongly to the overhead.

5 Conclusions and Discussion

Prolog is well–known for its qualities in the development of sophisticated rule systems. However, there are many tasks which requires logic programmers to integrate Prolog with a second language. This is the case of the Semantic Web, where the main frameworks are developed for the most widely used computer programming languages. In this work, we describe a system that allows to combine the power of declarative languages with a framework for the semantic web. In our work, we use InterProlog, an open source library for developing Java + Prolog applications, to connect Prolog with Jena framework. Our system allows Prolog predicates to query RDF databases. All inferences related with Semantic Web environments is done by the Jena framework, and supports both triples queries and SPARQL commands. Furthermore, we allow that Jena builtin functions can make calls to Prolog predicates to support Prolog reasoning in SPARQL commands, for instance. There are some systems that also integrate Prolog with RDF. For example, ClioPatria [11] is a Prolog framework builds on top of SWI-Prolog the to construct Semantic Web Applications. Thea OWL library [10] is a Prolog implementation of an OWL library where OWL parser uses SWI-Prolog Semantic Web library for parsing RDF/XML serialisations. Our approach has the advantage of allowing the use of Prolog in a complete Java semantic web framework as Jena, that allows different kinds of RDFS and OWL reasoning, supplies both a SPARQL engine and methods for RDF retrieval. For example, ClioPatria does not have SPARQL in the programming environment, recommends the translation of SPARQL commands into Prolog rules. For the client-side, ClioPatria has an incomplete implementation of SPARQL. Thea OWL library, as an OWL library, is not suitable for developing semantic web applications. We performed some benchmark tests that shows that our system can bring some overhead that must be carefully evaluated when the adoption of our system is to be considered. As a future work, we foresee the study the library InterProlog to look for improvements in performance. The Java library can be downloaded from https://github.com/mbentoalves/InterPrologJena and the example can be downloaded from https://github.com/mbentoalves/testPrologFamily. As a future work, we need to improve the benchmark tests to define when is better ARQ engine or when is better the mechanisms of Jena API.

References

1. Alves, M.B., Damásio, C.V., Correia, N.: SPARQL commands in Jena rules. In: Klinov, P., et al. (eds.) KESW 2015. CCIS, vol. 518, pp. 253–262. Springer, Heidelberg (2015). http://dx.doi.org/10.1007/978-3-319-24543-0_19
2. Bechhofer, S., van Harmelen, F., Hendler, J., Horrocks, I., McGuinness, D., Patel-Schneijder, P., Stein, L.A.: OWL web ontology language reference, 10 February 2004
3. Calejo, M.: InterProlog: towards a declarative embedding of logic programming in Java. In: Alferes, J.J., Leite, J. (eds.) JELIA 2004. LNCS (LNAI), vol. 3229, pp. 714–717. Springer, Heidelberg (2004)
4. Fortineau, V., Paviot, T., Louis-Sidney, L., Lamouri, S.: SWRL as a rule language for ontology-based models in power plant design. In: Rivest, L., Bouras, A., Louhichi, B. (eds.) Product Lifecycle Management. Towards Knowledge-Rich Enterprises, pp. 588–597. Springer, Heidelberg (2012). http://dx.doi.org/10.1007/978-3-642-35758-9_53

5. Guo, Y., Pan, Z., Heflin, J.: Lubm: a benchmark for owl knowledge base systems. Web Semant **3**(2–3), 158–182 (2005). http://dx.doi.org/10.1016/j.websem. 2005.06.005
6. Horrocks, I., Patel-Schneider, P., Bechhofer, S., Tsarkov, D.: Owl rules: a proposal and prototype implementation. Web Semantics: Science, Services and Agents on the World Wide Web **3**(1), 23–40 (2005). http://www.websemanticsjournal.org/ index.php/ps/article/view/62
7. Nejdl, W., Wolpers, M., Capelle, C.: The RDF schema specification revisited (2000)
8. Prud'hommeaux, E., Seaborne, A.: Sparql query language for RDF. Latest version available as http://www.w3.org/TR/rdf-sparql-query/, http://www.w3.org/TR/ 2008/REC-rdf-sparql-query-20080115/
9. Tauberer, J.: What is RDF (2006)
10. Vassiliadis, V., Wielemaker, J., Mungall, C.: Processing OWL2 ontologies using Thea: an application of logic programming. In: OWLED 2009 (2009)
11. Wielemaker, J., Hildebrand, M., van Ossenbruggen, J., Schreiber, G.: Thesaurus-based search in large heterogeneous collections. In: Sheth, A.P., Staab, S., Dean, M., Paolucci, M., Maynard, D., Finin, T., Thirunarayan, K. (eds.) ISWC 2008. LNCS, vol. 5318, pp. 695–708. Springer, Heidelberg (2008). http://dx.doi.org/10.1007/978-3-540-88564-1_44

An Approach for Structuring Sound Sample Libraries Using Ontology

Eugene Cherny[1,2], Johan Lilius[1], Johannes Brusila[1], Dmitry Mouromtsev[2], and Gleb Rogozinsky[3(✉)]

[1] Åbo Akademi University, Turku, Finland
{eugene.cherny,johan.lilius,jbrusila}@abo.fi
[2] ISST Laboratory, ITMO University, St. Petersburg, Russia
mouromtsev@mail.ifmo.ru
[3] The Bonch-Bruevich Saint-Petersburg University of Telecommunications,
St. Petersburg, Russia
gleb.rogozinsky@gmail.com

Abstract. Sound designers use big collections of sounds, recorded themselves or bought from commercial library providers. They have to navigate through thousands of sounds in order to find a sound pertinent for a task. Metadata management software is used, but all annotations are text-based and added by hand and there is still no widely accepted vocabulary of terms that can be used for annotations. This introduces several metadata issues that make the search process complex, such as ambiguity, synonymy and relativity. This paper addresses these problems with knowledge elicitation and sound design ontology engineering.

Keywords: Ontology engineering · Knowledge elicitation · Sound design · Metadata management · Web ontology language

1 Introduction

Sound design is an essential part of modern media production: documentary and fiction films, mobile applications, interactive installations, games and virtual reality—almost every part of digital content works with audio in order to bring in the top notch experience to the auditory. The sound is so tightly bound with other media that well-made sounds are no longer associated with multimedia in our everyday life. But without it almost every part of media becomes unfinished and not able to maintain one's attention, making digital content unattractive for the public. There are many types of sound design, depending on the application (games, films, theater, mobile applications, etc.), but for the sake of simplicity from now on we will be talking about gaming sound design.

There are two different approaches to sound design according to how a sound is generated and used [1]. A sample-based approach is based on the processing of already recorded sounds, which are triggered at specific game events. A procedural one implies building up signal synthesis and processing chains and making

© Springer International Publishing Switzerland 2016
A.-C. Ngonga Ngomo and P. Křemen (Eds.): KESW 2016, CCIS 649, pp. 202–214, 2016.
DOI: 10.1007/978-3-319-45880-9_16

them reactive to a in-game context in real-time. But the inherent complexity of implementing DSP algorithms for artistic tasks made this approach rarely used in the gaming industry, hence the sample-based one is used most to do sound design. However this approach has its own drawbacks, particularly in sound organization and search tasks. Sound sample management software does not completely solve the problem, because it is based on a per-sample hand-added metadata, which is often incomplete or even missing. This leads to the situations when a single sound query returns several dozens of sounds that are needed to be listened in order to find a relevant one that works for a task.

The purpose of this paper is to explore a *sound organization* problem in the context of sound sample libraries. The goal of the paper is to identify factors which cause difficulties with sample libraries usage and to propose a solution to overcome them. To accomplish this goal the interviews with professional sound designers will be conducted, and the ontology-based metadata integration solution will be proposed.

The rest of the paper is organized as follows. *Section* 2 describes the current state of sound libraries management solutions. In *Sect.* 3 an approach to a sound ontology engineering will be proposed. *Section* 4 will describe results of one iteration of the approach. *Section* 5 will conclude findings of the paper and will discuss further steps to develop the project.

2 Related Work

2.1 Timbre Descriptions and Ontologies

One of the first attempts to study and describe sound was conducted by the German scientist Von Helmholtz [2]. He studied the relation between verbally expressed qualities of tone and sound spectrum content using resonator objects. A number of other studies during the 20th century have been conducted to find out how the acoustic properties of the sound relate to the verbalization produced by humans. A review of such studies can be found in [3]. A lot of research focuses on how similar timbres are [4–6], how well they can be discriminated with verbalizations [7,8], and does not explore the sound domain terms and relationships that would help to structure the sound libraries.

We found very few papers describing projects that use ontologies to represent information about sound [9–11], but these does not provide ontologies themselves for study and have no pertinent resources available online.

W3C recommendation for "Ontology for Media Resources" [12] and the Music Ontology [13] which provide structure for describing and publishing media resources online with general metadata, like keywords, creator, distributor, etc.

2.2 Metadata Management

Nowadays professional sound designers have multi-terabyte sound sample collections[1], containing all kinds of recordings, from door locks to dolphins.

---

[1] Tim Prebble, Sound Library Storage Solutions, URL: http://www.musicofsound.co. nz/blog/sound-library-storage-solutions.

The metadata management software[2] is used to navigate through them, it is the description of the sound file and its contents filled in when the sound is added to library. Commercial libraries provide sample descriptions in different forms: proprietary metadata software formats, spreadsheets or at least a PDF file. The metadata is represented as the table data format with rows corresponding to sound files and columns to different information associated with them. Such information includes, for example, a filename, a creator's name, keywords, a description, and others. Metadata software does the text search in these fields.

3   Approach Overview

Our ontology engineering approach is based on the NeOn methodology [14]. This methodology provides guidelines for different scenarios of an ontology life cycle, covering specification, localization and other issues. For this project we employed several scenarios from the methodology: developing the specification, reusing and re-engineering non-ontological resources, and reusing and merging ontological resources. Before going into methodological details, we would like to give a brief overview of the approach.

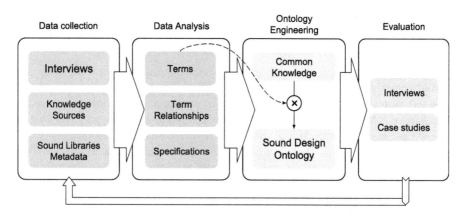

Fig. 1. A sound design ontology life cycle. The dashed shows the links between sound design ontology and the common knowledge ontology based on keywords.

A life cycle diagram (Fig. 1) depicts four major stages in the process. On the first stage we *collect* the data by interviewing professionals, selecting knowledge

---

[2] There are a number of metadata management software, but they essentially do the same. The differences are in the user experience and in the format of underlying metadata, often incompatible with each other. A comprehensive list of metadata software in the blog of the professional sound designer Tim Prebble: http://www.musicofsound.co.nz/blog/metadata-support-in-sound-library-apps.

sources (e.g. sound design related books or blog posts), and downloading metadata for the selected commercial sound sample libraries. This data is then *qualitatively analyzed* to find the important domain concepts, keywords, terms and relationships between them. The data is also used to define the domain problems and to write the ontology specifications. The *ontology engineering* stage includes creation of the sound design ontology, populating it with the data and interlinking this data with the common knowledge ontology. After this stage we *qualitatively evaluate* the results to find out how well they work for solving the defined problems. The ontology is then undergoes through all life cycle stages from the beginning in order to address issues found in the evaluation.

3.1   Data Collection

As was already mentioned, the approach involves three distinct data sources that can be used to create an ontology: interviews, knowledge sources (books and other resources on sound design) and metadata from commercial sample libraries. They provide a multifaceted view on the problem domain and naturally validate conclusions drawn from each of them separately.

The *purpose* of the data collection is to get a holistic view on the problem of sound organization by interpreting different sources of sound design related data. This data is then going to be analyzed in order to build a formal representation of the problem domain from the bottom up. The purpose statement is rather broad, but it allows to shift the research focus later on after we started to collect the data.

For the rest of this section we explain the role of each data source in the process.

Interviews. As was mentioned earlier, there is little research done on the sound design concepts' formalization. Books about using related methods and techniques concern mostly theoretical issues (basics of digital signal processing and psychoacoustics [1,15], recording techniques [16], etc.), and only rare blog posts shed some light on practical problems experienced by professionals when working with sound organization. Thus there is a gap between documented and practical knowledges which makes it impossible to move forward without cooperation with professionals.

The *purpose* of the interviews is to fill this gap by communicating on the existing issues of sound organization directly with the sound designers working in the industry. The interviews were designed to follow general recommendations for doing the qualitative inquiry [17] and the following two paragraphs describe their organization.

The interviews does not dictate a specific *setting*, hence the data can be collected from different sources: email, chat or phone conversation. All discussions will be documented in the researcher's notebook (verbal conversations are transcribed) and published online on the later stages of the project provided that participants agreed for the publication.

Interviews will be conducted in a *semi-structured* manner, using open- ended questions and following a loose structure to guide general direction of a discussion. The discussion topics include how participants organize their sound libraries, how they search for sounds, what they think about current sound organization solutions, and also more general questions on how they do their work.

Knowledge sources can include several books and other materials for the sound designers. Most of them regarded as trusted sources of structured knowledge that can be used to validate interview results.

The sample libraries metadata contain text annotations describing the sound file content. They have a special value as they provide professionally crafted metadata, designed to be practically useful; thus the terms used in this kind of metadata are of great value for designing the ontology. They also allow to connect the ontology to sound files, which may have possible applications in machine learning field (this aspect is beyond the scope of this paper).

The only issue with the metadata is licensing. A preliminary agreement from metadata owners should be received in order to use it for research.

3.2   Data Analysis

The *purpose* of the data analysis process is to conceptualize the information elicited on the previous stage. The conceptualization includes definition of important domain terms and their relationships, and also writing an ontology specifications in form of competency questions (CQs); together they are main components of the ontology design requirements in the NeOn methodology. CQs are the one sentence user stories[3], telling what question the ontology should answer or how it should be structured.

A typical qualitative data analysis process [17] is employed upon the interviews and knowledge sources[4]. The process can be summarized into the following steps:

1. Data preparation: transcribing, sorting and arranging.
2. Getting the general sense of data by reading through it.
3. Coding—labeling the data chunks, splitting them into categories.
4. Creating the list of terms and competency questions from the codes.

The text processing techniques are used to analyze sample libraries metadata. The main objectives here are to recognize the entities in the text annotations and to perform the exploration of used terms.

3.3   Ontology Engineering

The *purpose* of the ontology is to build a semantic layer on top of a text metadata to perform structural search in a sound database.

---

[3] They do not necessarily have to be questions; declarative sentences can also be used.
[4] Also the grounded theory or case study strategies can be used together with the described process.

This stage has three main objectives:

1. To create base classes and properties needed to represent the domain knowledge elicited from the previous steps.
2. To populate the ontology with sound instances collected from the commercial sample libraries using manual or automatic entity recognition methods.
3. To interlink the results with common knowledge ontology in order to do basic inference, for example, to find related concepts.

We added the third objective because sound metadata consists of common keywords (such as "car", "water", etc.) which are already structured in the common knowledge ontologies.

3.4 Evaluation

The *purpose* of the last stage is to evaluate how well the resulting ontology solves defined problems using qualitative procedures, such as interviews or case studies. Also quantitative metrics can be provided in order to assess such parameters as ontology size, number of interlinked entities, etc.

Qualitative evaluation (in the form of case studies or interviews) complements these metrics with a subjective evaluation of the ontology by assessing the domain knowledge representation and finding structural and terminological issues. This step also makes it easier to request the comments from professionals, because intermediate results are much easier to receive comments on than to abstract "how X should be done" questions.

New tasks for improving the ontology are defined after evaluation and the process starts over from the first stage with new or refined goals.

4   Implementation

This section describes the project findings up to the moment of finishing this paper, including data analysis and the ontology based on it.

4.1   Data Collection

The ontology engineering starts **interviews** with sound designers, who work professionally in the industry. At this point two professionals have agreed to participate in the project: the first (abbreviated as IO) works as a sound designer at the computer games development company[5], and the other one (abbreviated as AR) manages metadata for commercial sound sample libraries[6]. Thus we have representatives of the two different facets of the problem domain: the one who uses sample libraries and the one who creates them.

---

[5] Saber Interactive, URL: http://www.saber3d.com.
[6] Boom Library, URL: http://boomlibrary.com.

Three small interviews have been conducted up to this moment:

- A face-to-face interview with IO about basic sound design topics.
- An email interview with AR about metadata management issues.
- An internet chat interview with IO also about metadata management issues.

Although this paper focuses on the metadata issues, we would like to describe the sound design workflow in general. Sound design starts with a concept document describing an artificial setting. This document can contain textual descriptions of game elements, as well as visual references. The sound designer's goal is to create such a sound that would convey written and drawn concepts. The design process usually starts with looking for source sounds in a sound library. The new sound is then worked out of the found sounds being manipulated in different ways (cutting, slicing, rearranging, processing, etc.).

AR explained difficulties of metadata-based sound search in the interview:

When I would need the sound of a closing car door, this gives me a lot of good results. I could type in "car", "door" and would get a bunch of results. However, getting more into detail it gets a bit more tricky. When I would specifically search for a squeaking car door closing for example. Some manufacturers dont even include "open" or "close" or if a file consists of a recording opening and closing the door "open/close" or similar. Then others would note "opening" or "closing". "Opening" is not much of an issue, because "open" is in the word "opening" and it would be found. When typing "close" some metadata searches would not find "closing" though. Then even worse: squeaking might be described as "jarring", "squealing", "grating". Even though these words describe different things, it would be too much detail to work with for me personally. This leads to the most annoying part: there are tons of materials, objects, actions or feelings than can be described with a lot of different words. ... Soundminer[7] can do boolean search, but this is only half the deal, because then I got too many results.

One should also make a compromise between completeness and usefulness when adding metadata. Here is a thinking example for whoosh sound annotation:

Whoosh is used for many different things. But if I would work on a cartoonish thing typing in "whoosh" for a cartoon punch I would need a light, high whoosh sound. If only type in whoosh, a lot of things might be super heavy, trailer related things. So I need to add "light", "small", "high" or similar words and hope those are in the description. But then again, if there is a trailer library focussing on whooshes, there might be lighter, smaller, higher whooshes than others for trailers purposes, but still way too large for cartoonish usage. This specific example could be easily solved by adding "trailer" to the trailer whooshes, but this is only one example out of a million possible whoosh usages, so I can not fill in every possible usage/style of this specific whoosh sound without creating an overkill of description which is simply not readable in a nice way.

Metadata issues discussed in the interviews can be summarized as follows:

- Incompleteness: every sound description is always a trade-off between usefulness and completeness.

---

[7] A metadata management software.

- Ambiguity: sound libraries made by different companies can use different spellings of the same term or use synonyms. For example "GUI" and "UI"; "armor", "armour" and "chainmail".
- Relativity: annotations meanings highly depend on a context.
- Absence of any industrial standard or a guideline on adding metadata, which makes it hard to search in several libraries at the same time: sometimes users exclude libraries from search in order to reduce search results.
- The metadata is usable only when filled in thoroughly. Sound designers often does not have time to do this for their own sounds.

As we can see, most of the issues are caused by the textual format of the metadata. A well-made ontology can address these issues of sound organization, as the sound will be linked not simply to text keywords, but to concepts that may have different textual representation making the search more convenient and less dependable on spelling differences. The concepts interlinked with common knowledge can solve the problem of querying through using synonyms or closely related concepts. The ontology can also be used improve sound search by providing structure for common terms and adding query suggestions: for example suggesting the "car closing door" or "car engine" when putting in the "car" keyword into a search field. Besides search, similar mechanisms can be employed to improve sound annotation process.

In this project we also analyze the BOOM Library metadata, which the company has kindly granted permission to work with. The metadata was provided in the form of XLS files and available from the company's web site.

4.2   Data Analysis

We define the *ontology specification* as the following list of competency questions created based on the metadata management problems and elicited from the interviews and authors' understanding of the problem:

1. The basic concept is *sound file.*
2. A sound file has a *common sound metadata*: filename, designer, microphone[8].
3. A sound file belongs to a *sound library.*
4. A sound library has a textual *annotation* describing the contents.
5. A sound file contains *sound.*
6. A sound file has one or more terms associated with it.
7. A term can have one or more topics associated with it.

These CQs describe the ontology structure needed to represent the sound content in terms of keywords. The purpose of the ontology is to aid search and annotation tasks, hence we do not need to introduce linguistic variables and can limit ourselves to using only "crisp" formalisms. This structure can be revised when new topics are added to the ontology.

---

[8] The list was created based on the BOOM Library metadata files.

Existing metadata can be very useful for understanding the sound design concepts. To demonstrate this we had chosen a sound library containing user interface sounds[9] and elicited domain concepts from the supplied metadata.

The metadata is represented as 10-column table, including such columns as "filename", "license", "designer" and others. The most important one for our purposes is "description". It contains keywords written using uppercase letters followed by a concise and more detailed description of the sound, for example *"DIGITAL CLICK SYNTHETIC Short notes, clicks, high pitch"*. Merging these two data types in one text string was made in order to fit in the common metadata format.

Keywords Analysis. 23 keywords used in the sound library, which can be split into the following groups:

– *Sound mood*: "arcade", "digital", "generic", "orchestral", "organic". This group roughly represent the content of the sound, for example, "arcade" means that the sound is synthesized to resemble the arcade games sound, "digital" is for emotionless synthesized computer sounds, etc.
– *Sound form*: "jingle", "button", "click", "slide". This group represents the temporal evolution of the sound, for example, "jingles" are the little pieces of music, "clicks" are sounds with a short decay, "buttons" are recorded sounds of button pressing, etc.
– *Materials*: "human", "paper"—and also *material adjectives*: "synthetic", "plastic", "metallic", "woody", "glassy". This group provides the clues on what kind of sound sources were recorded. It can be confused with the first group, but there is a fundamental difference: the "sound mood" group represents an *intended* usage of the sound, but the "material" group represents the way how the sounds were made. For example, "digital" and "arcade" sounds are "synthetic"[10]. "Plastic", "metallic", "woody" and "glassy" materials are in the "generic" group[11].
– *Sound size*: "tiny", "small", "medium".
– *Interaction type*: "negative", "normal", "positive". For example, navigation sounds through the menu hierarchy (up, down, stay on the same level). Interaction sounds are labeled with the "feedback" keyword.

Descriptions Analysis. The second part of the textual annotations describing the sound content is descriptions. It consists of a short comma-separated statement, written in free form. They indicate, for example, *sound effects* applied, *instruments* or *objects* used in the record, or *action* performed on the

---

[9] BOOM Library – The Interface, URL: http://www.boomlibrary.com/boomlibrary/products/the-interface.
[10] According to metadata, i.e. synthetic sounds are also annotated either with "digital" or with "arcade" keyword.
[11] Although the terminology is questionable and can be improved, at we use it "as is" for now.

recorded object. There are 221 of these descriptions in the library, hence we do not list everything here and instead suggest the reader to have a look at the ontology [18].

4.3   Ontology Engineering

The ontology has 4 top-level classes: *SoundFile*, *SoundLibrary*, *SoundFileTerm*, and *Topic*. The first two represent basic metadata (file and library names, creator, licensing, etc.). *SoundFileTerm* class connects sound file to one or more abstract *topics* extracted from the library: it either represent a keyword or a comma-separated description. This structure reflects the keyword-based search procedures sound designers use today with additional categories introduced in the knowledge engineering process. Intermediate *SoundFileTerm* class connects several topics together but does not specify the connection type. It was introduced for the future development of enrichment methods using external ontologies to provide the connection between them (Fig. 2).

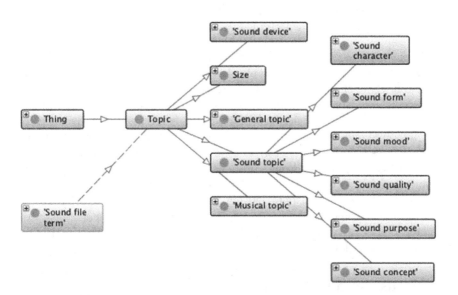

Fig. 2. "Topic" class and its subclasses. All subclasses except "SoundTopic" have their subclasses hidden. Dashed arrow is the ":hasTopic" property, all other—"rdfs:subClassOf".

Interlinking. We considered three common knowledge ontologies to interlink with: OpenCyc[12], DBpedia [19] and Wikidata [20]. We have done subjective evaluation of each of them and decided to go with OpenCyc, because it has better text annotations (including synonyms) and common concepts structure.

---

[12] OpenCyc for the Semantic Web, URL: http://sw.opencyc.org.

The interlinking process was done using OpenCyc's search and disambiguation facilities with some human aid when they did not work well. Suitable Open-Cyc concepts were linked with our ontology using *rdfs:seeAlso* property. We didn't use OWL axioms, because on the interlinking stage it is still unclear what keywords should be refactored from classes into properties, so *rdfs:seeAlso* is used as the marker, that a human attention will be needed later.

4.4   Evaluation

This section provides a subjective evaluation of the ontology.

A rather important issue is that the sound design field widely spreads between computer gaming and academic music avant-garde. Such a field determines a large list of sound-related specialties, i.e. composer, sound designer, sound engineer, sound programmer, arranger, etc. All of them operate a number of sound libraries, sometimes using absolutely different sets of notions. For instance, the situation when low-pass filter opens during trance sequence will be described differently according to an academic composer and to a DJ. The first thinks about harmonic contents and interprets the sound as an addition of higher notes (octaves and so on), while the latter thinks about filtering of sound complex without dissecting it into separate sounds. Meanwhile, somebody whos not familiar with sound technology thesauri can describe the same sound saying timbre is getting brighter. This simple example shows that more descriptors are needed, or probably there should be several sets of synonyms.

Another issue is the crisp nature of the descriptor set. For example the distinction between real-world and synthesized sounds might be obvious when comparing a violin playing legato notes (real-world sound) with Access Virus playing neurofunk bass (synthetic). But what if we take some samples of spectral music (i.e. Gèrard Grisey) or sound mass music (i.e. Iannis Xenakis or György Ligeti), it would be hard to describe some elements as real-world. At the same time, sounds of instruments with poor harmonic contents, i.e. bells, xylophones, jaw harps, kalimbas, could be synthesized from scratch very realistic.

These points reveal new epistemological depths of the problem and will be discussed in the following section.

5   Discussion and Future Work

The evaluation has shown that the structure elicited from the metadata should be revised in order to resolve terminological problems. From the other side, a great care should be taken when working out the professionals' comments, as some issues may be less important for the project's problem scope.

At the present state the ontology does not benefit from linking with the OpenCyc ontology, because the integration of these two is loose. In the future we plan to reflect many abstract OpenCyc concepts, for example, relative descriptions ("high", "low", etc.) to describe timbral similarity. After adding proper interlinking using OWL axioms the ontology can be validated using reasoner facilities.

The next steps of this project will be adding a number of other sample libraries to the ontology and implementing a sound search software to test it in the real situation. Once the demo software is ready we can involve more professionals into the project to build a better tool for the field.

6 Conclusions

In this paper we investigated key factors which cause difficulties with using sound libraries by interviewing working sound designers. Main difficulties are mostly caused by widely adopted text-based metadata format, which leads to such problems as synonymy, typos, misspellings and so on. To deal with this problem a knowledge-engineering approach was proposed. The approach was validated by demonstrating one iteration of the sound design ontology development. The outcome of this iteration was the ontology [18] with terminology elicited from "The Interface Library" by BOOM Library[13]. This terminology was manually structured and then linked with the OpenCyc ontology.

Acknowledgments. We would like to express our thanks to sound designers Ivan Osipenko (Saber Interactive, St. Petersburg, Russia) and Axel Rohrbach (BOOM Library, Mainz, Germany) for the provided interviews and discussions on the sound design and metadata topics. We also thank anonymous reviewers for providing valuable comments on our work.

This work has been partially financially supported by the Government of Russian Federation, Grant #074-U01.

This work was conducted using the Protégé resource, which is supported by grant GM10331601 from the National Institute of General Medical Sciences of the United States National Institutes of Health.

References

1. Farnell, A.: Designing Sound. MIT Press, Cambridge (2010)
2. Helmholtz, H.L., Ellis, A.J.: On the Sensations of Tone as a Physiological Basis for the Theory of Music. Cambridge University Press, Cambridge (2009)
3. Siedenburg, K., Fujinaga, I., McAdams, S.: A comparison of approaches to timbre descriptors in music information retrieval and music psychology. J. New Music Res. **45**(1), 27–41 (2016)
4. Torres, D., Turnbull, D., Barrington, L., Lanckriet, G.: Identifying words that are musically meaningful, pp. 1–6 (2007)
5. Bernays, M., Traube, C.: Verbal expression of piano timbre: multidimensional semantic space of adjectival descriptors. In: Williamon, A., Edwards, D., Bartel, L. (eds.) International Symposium on Performance Science. Association Europenne des Conservatoires, Acadmies de Musique et Musikhochschulen (AEC) (2011)
6. Zacharakis, A., Pastiadis, K., Reiss, J.D., Papadelis, G.: Analysis of musical timbre semantics through metric and non-metric data reduction techniques. In: Proceedings of the 12th International Conference on Music Perception and Cognition (ICMPC12) and the 8th Triennial Conference of the European Society for the Cognitive Sciences of Music (ESCOM 08), pp. 1177–1182 (2012)

---

[13] URL: http://www.boomlibrary.com/boomlibrary/products/the-interface.

7. Stepnek, J.: Musical sound timbre: verbal description and dimensions. In: Proceedings of the 9th International Conference on Digital Audio Effects (DAFx-06), pp. 121–126. Citeseer (2006)
8. Zacharakis, A., Pastiadis, K., Reiss, J.D.: An interlanguage unification of musical timbre: bridging semantic, perceptual, and acoustic dimensions. Music Percept.: Interdisc. J. **32**(4), 394–412 (2015)
9. Nakatani, T., Okuno, H.G.: Sound ontology for computational auditory scene analysis. In: Proceeding for the 1998 Conference of the American Association for Artificial Intelligence (1998)
10. Hatala, M., Kalantari, L., Wakkary, R., Newby, K.: Ontology and rule based retrieval of sound objects in augmented audio reality system for museum visitors. In: Proceedings of the 2004 ACM Symposium on Applied Computing, pp. 1045–1050. ACM (2004)
11. Hamadicharef, B., Ifeachor, E.C.: Intelligent and perceptual-based approach to musical instruments sound design. Expert Syst. Appl. **39**(7), 6476–6484 (2012)
12. Lee, W., Bailer, W., Bürger, T., Champin, P., Malaisé, V., Michel, T., Sasaki, F., Söderberg, J., Stegmaier, F., Strassner, J.:Ontology for media resources 1.0. W3C recommendation. World Wide Web Consortium (2012)
13. Raimond, Y., Abdallah, S.A., Sandler, M.B., Giasson, F.: The music ontology. In: ISMIR, vol. 422. Citeseer (2007)
14. Suárez-Figueroa, M.C., Gómez-Pérez, A., Fernández-López, M.: The NeOn methodology for ontology engineering. In: Suárez-Figueroa, M.C., Gómez-Pérez, A., Motta, E., Gangemi, A. (eds.) Ontology Engineering in a Networked World, pp. 9–34. Springer, Heidelberg (2012)
15. Cipriani, A., Giri, M.: Electronic Music and Sound Design - Theory and Practice with Max 7, vol. 1, 3rd edn. Contemponet, Rome (2016)
16. Viers, R.: The Sound Effects Bible: How to Create and Record Hollywood Style Sound Effects. Michael Wiese Productions, Studio City (2008)
17. Creswell, J.W.: Research Design: Qualitative, Quantitative, and Mixed Methods Approaches, 4th edn. Sage Publications Ltd., Los Angeles (2013)
18. Cherny, E., Lilius, J., Brusila, J., Mouromtsev, D., Rogozinsky, G.: A sound ontology for the paper "An approach for structuring sound sample libraries using ontology", July 2016. http://dx.doi.org/10.5281/zenodo.56833
19. Auer, S., Bizer, C., Kobilarov, G., Lehmann, J., Cyganiak, R., Ives, Z.G.: DBpedia: a nucleus for a web of open data. In: Aberer, K., et al. (eds.) ASWC 2007 and ISWC 2007. LNCS, vol. 4825, pp. 722–735. Springer, Heidelberg (2007)
20. Vrandečić, D., Krötzsch, M.: Wikidata: a free collaborative knowledgebase. Commun. ACM **57**(10), 78–85 (2014)

Efficient SPARQL to SQL Translation
with User Defined Mapping

Miloš Chaloupka$^{(\boxtimes)}$ and Martin Nečaský

Faculty of Mathematics and Physics, Charles University in Prague,
Malostranské nám. 25, 118 00 Prague, Czech Republic
{chaloupka,necasky}@ksi.mff.cuni.cz
http://www.ksi.mff.cuni.cz/

Abstract. The RDF framework is becoming popular for presenting data. It makes the data easily accessible and queryable. However, the most common way how to store structured data is to use a relational database system. It is essential to create a mapping between these two worlds, to publish the data stored in a relational database in the RDF format. That can be effectively achieved by a virtual SPARQL endpoint over relational data.

There are already existing tools providing virtual SPARQL endpoints, but as we will show in the paper there is still space for improvement. In this paper we propose an algorithm to query RDF data stored in a relational database with an user defined mapping. Our aim is to generate SQL queries which can be effectively executed on the relational engines. In comparison to existing approaches we do not rely only on the optimizations of the relational query, but the SPARQL query first.

Keywords: RDB2RDF · R2RML · SPARQL · Relational to RDF mapping

1 Introduction

The RDF framework is becoming a popular framework for presenting data as it is a part of the W3C standard - Linked Data. It makes data easily accessible and queryable without the need to publish information about data storage, etc. Users are able (using the SPARQL language [7]) to get all information without any further knowledge of a particular implementation.

On the other hand, the most common way how to store structured data is to use a relational database system. Relational databases benefit from their long theoretical and practical history. For most structured data, there is no intention to store them in some other way than in relational databases.

In this paper we present a mechanism which allows to create a virtual SPARQL endpoint over a relational database. There is no need to modify the relational database. It is only needed to specify the mapping between the relational and the RDF data representation. The mechanism presented in this paper

© Springer International Publishing Switzerland 2016
A.-C. Ngonga Ngomo and P. Křemen (Eds.): KESW 2016, CCIS 649, pp. 215–229, 2016.
DOI: 10.1007/978-3-319-45880-9_17

uses this mapping to transform the SPARQL queries over the final RDF form to the SQL queries over the source relational form. As we will show, our work outperforms the current approaches.

2   Related Work

In the recent years, the topic of the mapping from relational database to RDF representation seems to be getting more and more important. We do discuss only the approaches which provide virtual SPARQL endpoint over relational data using an user defined mapping.

Usually as the state-of-the-art tool for the mapping from relational databases to RDF is mentioned the D2R Platform [5]. It was the first working solution. The tool transforms the SPARQL query into multiple SQL queries and then processes the results in memory to create the result of the SPARQL query. However, it has a poor performance.

The Ultrawrap tool (see [13]) produces relational queries divided to two parts. The first part is an "R2RML view", single subquery which returns all triples according to the R2RML mapping file (see [6]). It is the union of selects (one per every triple mapping defined). The other part is the actual transformed SQL query over the "R2RML view". It is transformed using the Chebotko's approach [3] (the Chebotko's approach does not support any mapping it works only with a single table with three columns). The authors showed that the Ultrawrap is much more effective than the one used in D2R platform.

Another approach, the Morph translation algorithm [10], is also based on the Chebotko's work. They are redefining the Chebotko's algorithm, so it supports mapping and it is not needed to prepare data for it (as it is done in the Ultrawrap tool). The Morph generates a single unified query (the query is not split into "view" and "query" as it is done in the Ultrawrap tool). The main idea is that for every basic graph pattern it adds a subquery. These subqueries return all possible variable mappings. However, there is not a full union of all possibilities (as it is in the Ultrawrap's "R2RML view"), it selects only the candidates that may affect the query result. They do also support the self-join optimization.

Lately, there appeared a new tool - virtual SPARQL endpoint implemented in the -ontop- system [11]. This tool aims to generate optimized queries using well-known SQL optimization methods. The -ontop- seems to be actively used and from the found tools it has the most active community.

3   Motivation

We propose a new transformation algorithm to transform SPARQL queries to SQL queries. It uses R2RML files to define the mapping (see [6]). The tool aims to generate optimized queries. We use optimizations like candidate selection and self-join optimization (like the -ontop- system). The main difference is how we process the SPARQL algebra. We use a modified algebra which holds the mapping information. Therefore, we are able to perform some optimizations before a query is transformed to a relational form. Moreover, we are able to transform the SPARQL query in a form which is better suitable for the transformation. Thanks to it, the optimizations on the relational model are more effective.

Our aim is to provide an algorithm which translates the SPARQL as it is shown on the Fig. 1. This query is very similar to the one generated by the -ontop- system. In comparison to other approaches this query seems to be much more efficient. However, -ontop- transform SPARQL queries to datalog form and then performs all optimizations over the datalog form. We propose an algorithm which performs the optimizations over the SPARQL form of a query. Then the query is translated to the relational form and other optimizations are performed. As we will show in the evaluation section this approach is more efficient.

```
SELECT ?title, ?author
WHERE {
    ?p a x:paper;
        x:name ?title;
        x:author ?a.
    ?a x:name ?author
}
```

```
SELECT p.name, a.name
FROM papers AS p, authors AS a
WHERE
    p.id IS NOT NULL AND p.name IS NOT NULL AND
    p.author IS NOT NULL AND a.id IS NOT NULL AND
    a.name IS NOT NULL AND p.author = a.id
```

Fig. 1. Sample SPARQL query transformation

4   SPARQL Algebra

Our transformation works with the SPARQL algebra described in this section. Our algebra is based on the official W3C algebra [7]. In comparison to SPARQL$_C$ (proposed in [1,9]), we process the optional pattern differently. Our aim is to handle the optional pattern using the traditional LEFT JOIN clause from the relational algebra.

```
SELECT ?b ?c
WHERE {
    ?a :p1 ?b
    OPTIONAL {
        ?a :p2 ?c
        FILTER (?b > 3)
    }
}
```

Fig. 2. Scope of filter patterns

The reason why there are multiple approaches to OPTIONAL pattern is the fact that the filter pattern evaluation may depend on a variable defined somewhere else. More precisely, the scope of the filter pattern expression depends on the context. The filter pattern scope is defined (see [7]) as "a restriction on solutions over the whole group in which the filter appears" with an exception. If a filter pattern is a part of an optional pattern then the scope of the filter pattern is the same as the scope of the optional pattern. Without the mentioned exception, the filter expression from the Fig. 2 could not be ever evaluated to true (it will be always evaluated as error - see [7]). However, the scope of the filter pattern is the scope of the parent optional pattern so it is evaluated according to the value of ?b.

This problem complicates the transformation of the SPARQL pattern. Ideally, we want the algebra in a form, that it can be evaluated from bottom to up and every variable is in the scope when processing the operators recursively. Firstly, we show the algebra and then we will formally define the query without the mentioned problems, so called safe query, and we then will show how every query can be transformed to an equivalent safe query.

4.1   Syntax

For simplicity, we define the algebra only for the main language constructs without grouping and aggregation functions and solution modifiers (like DISTINCT etc., as defined in [7]). These constructs do not affect the validity of the proposed algebra and they will be defined in a separate paper. Moreover, we describe only SELECT query form. The other SPARQL query forms (CONSTRUCT, ASK and DESCRIBE) can be transformed to a corresponding SELECT query and then the solution mappings transformed to the expected result form.

Definition 1. *Let RDF-I be the set of all IRIs, RDF-L be the set of all literals and RDF-B be the set of all blank nodes. The set of* **RDF terms** *is defined as RDF-T = RDF-I ∪ RDF-L ∪ RDF-B.*

V denotes the set of all variables which is infinite and disjoint from RDF-T. A **variable** *is any $v \in V$. A* **solution mapping** *μ is a partial function, μ : $V \to RDF$-T. The domain of μ is denoted $dom(\mu)$. A* **solution sequence** *M is a list of solution mappings. A* **triple pattern** *is a tuple from $(RDF$-$T \cup V) \times (RDF$-$I \cup V) \times (RDF$-$T \cup V)$. A* **SPARQL query** *is a tuple (R, F, P) where R is a result query form, F is a set of dataset clauses (possibly empty) and P is graph pattern. A* **result query form** *is the expression SELECT W where $W \subset V$ is a set of variables. A* **dataset clause** *is in the form FROM g or FROM NAME g where $g \in RDF$-I.*

To define the SPARQL algebra, we firstly need to define the selection, which is then used as a parameter for the filter patterns.

Definition 2. *A* **selection** *is defined recursively as follows: A SPARQL expression[1] is a selection. If P is a graph pattern, then $EXISTS(P)$ is a selection. If σ_1 and σ_2 are selections, then $\sigma_1 \wedge \sigma_2$, $\sigma_1 \vee \sigma_2$ and $\neg\sigma_1$ are selections.*

Now, it is everything prepared to define the SPARQL algebra. We define the graph pattern, which represents the complete WHERE clause in the SPARQL query.

---

[1] For a complete description see [7].

Definition 3. *A **graph pattern** is defined recursively as follows:*

- *An $\emptyset$ is a graph pattern called **empty pattern**.*
- *An TP is a graph pattern called **triple pattern**.*
- *If P_1 and P_2 are graph patterns, then $P_1 \bowtie P_2$, $P_1 \cup P_2$ and $P_1 \setminus P_2$ are graph patterns called **join**, **union** and **minus** respectively.*
- *If P is a graph pattern and σ is a selection, then $\sigma(P)$ is a graph pattern called **filter**.*
- *If P_1 and P_2 are graph patterns and σ is a selection, then $P_1 \bowtie_\sigma P_2$ is a graph pattern called **left join**.*
- *If P is a graph pattern and $u \in RDF\text{-}I \cup V$ then $\Gamma_u(P)$ is graph pattern called **graph**.*
- *If P is a graph pattern, e is a SPARQL expression and v is a variable then $\varepsilon(P, v, e)$ is a graph pattern called **extend**.*
- *If M is a solution sequence, then Φ_M is a graph pattern called **values**.*

In other words, we can transform every SPARQL query to an **algebraic tree**, where the leafs are empty patterns, triple patterns and values. And the non-leaf nodes are joins, unions, minuses, filters, left joins, graphs and extends.

4.2 Safe Form

We have already discussed the problem with filter pattern scope. The query is in a safe form when there is no instance of this problem present in the query. Formally, the query is in a safe form when there is no instance of $P_1 \bowtie_{\sigma_1} \sigma_2(P_2)$ in the algebraic tree.

The solution is straightforward because the selection is part of the left join graph pattern. The algorithm goes recursively through the algebraic tree. Every instance of $P_1 \bowtie_{\sigma_1} \sigma_2(P_2)$ is replaced by $P_1 \bowtie_{(\sigma_1 \wedge \sigma_2)} P_2$ where $\sigma_1 \wedge \sigma_2$ is a conjunction of the selection in the original left join and the selection from the original filter pattern σ_2.

In this paper, we will not show the evaluation semantics. The evaluation is straightforward for queries transformed to their safe forms - the SPARQL and relational evaluation does not differ for them.

5 Transforming SPARQL Query to SQL Query

In this section, we will describe the process of the transformation of a SPARQL query to an SQL query and then the conversion of the results of the SQL query back to the solution mappings corresponding to the SPARQL query. The SPARQL algebra is defined in a recursive way, as a tree where every node is a subquery. We propose an algorithm that enhances this recursive nature and is able to generate a corresponding SQL query for every subtree.

To describe any SQL query in this paper we will use a very similar model to the domain relation calculus (DRC, described in [8]). An SQL query is described in the form $\{x_1, x_2, ..., x_n | p(x_1, x_2, ..., x_n)\}$ where $p(...)$ is a DRC formula.

We do not generate SQL queries which return an exact form of an SPARQL variable values. Instead of that, we are querying columns needed to generate these values. Therefore, our transformation algorithm has to provide not only the resulting SQL query for a given SPARQL query but also the information how to reconstruct the SPARQL variables from the SQL result. This information is needed also for the transformation algorithm itself - every time when we need to reference any SPARQL variable we need to have the information how is the value represented. To describe this information, we will use so called value binders. There will be exactly one value binder for every SPARQL variable in the query.

Definition 4. *A value binder is a structure* $vb(v, b, val)$ *where* $vb.v$ *is the corresponding SPARQL variable.* $vb.b$ *is the DRC formula that is true if and only if the variable is bound. And* $vb.val$ *is a DRC expression that calculates the value.*

For example, a sample value binder for a variable ?name that is generated using R2RML template from database columns [SurName] and [LastName] is $vb(name, b, val)$ where:

- $vb.b = \neg(NULL(SurName)) \wedge \neg(NULL(LastName))$ where $NULL(x)$ returns true if and only if the column x is null.
- $vb.val = (LastName + ',' + SurName)$.

The algorithm is transforming every graph pattern to the tuple (VB, X) where VB is the set of value binders and X is the domain relational calculus model of the SQL query which is the result of the transformation of the graph pattern. To get the SPARQL results, we execute the SQL query X to get the SQL result. Every returned row in the SQL query result represents one solution mapping in the SPARQL result set. The solution mapping is created from the value binders - every SPARQL variable is represented by exactly one value binder. For every variable c we find the value binder $vb(c, b, val)$. Evaluation of $vb.b$ gives us the information whether the variable is bound or not in the solution mapping. And if it is, we will evaluate $vb.val$ to get the value of the variable.

5.1   Adding the R2RML Mapping Information to the Algebra

To this point, the algebraic representation of the SPARQL query does not have any connection to the relational database. This connection is represented by the R2RML mapping. The operator that means the actual query to the dataset is the triple pattern. However, the triple pattern is not anyhow connected to the relational database. So, we will define a restricted triple pattern, the triple pattern that has a specific R2RML mapping assigned. A single valid combination of graph, subject, predicate, object mapping from the R2RML mapping [6]. The restricted triple pattern represents all triples generated only by the selected R2RML mapping.

Definition 5. *The **Restricted triple pattern** is the TP graph pattern restricted by a tuple $< g; s; p; o >$ of the graph, subject, predicate and object mapping. That means $eval(DS, TP_{<g;s;p;o>}) \subseteq eval(DS, TP)$ and $\forall \mu \in eval(DS, TP_{<g;s;p;o>})$ is true that μ is generated according to the R2RML mapping using the graph g, subject s, predicate p and object o mapping.*

Note 1. The notation $eval(DS, P)$ denotes the evaluation method of the graph pattern P over dataset DS.

The triple pattern needs to result in all triples in the dataset matching the patterns. According to the R2RML definition (as presented in [6]) that means to query all possible combinations of the graph, subject, predicate and object mappings. Therefore, we can replace all triple patterns TP in the query by the union of restricted triple patterns (using all possible combinations of the graph, subject, predicate and object mappings).

5.2   Creating the SQL Query

When the query is in the safe form and all triple patterns were replaced by the restricted triple patterns, we can finally transform the SPARQL algebra. The transformation is a recursive algorithm which processes the algebraic tree from bottom to up. The algorithm checks the graph pattern type and behaves according to it. We will describe the transformation of graph pattern types separately. In this paper we will cover only basic graph patterns to show the algorithm.

The Restricted Triple Pattern. Because our query is modified in a way that every triple pattern is restricted, we know the exact SQL query from R2RML mapping; it is part of the triples map assigned. The SQL query can be in two possible forms. A simple select clause from a table (or a statement) or select clause with inner join in the case when the object mapping is a referenced object map (see [6]).

So, we know the tuple $< g; s; p; o >$ that is in a triple map with defined logical table using the `rr:logicalTable` node. That node defines an SQL query that can be used to retrieve the data. If the object mapping contains a reference to another triple map, we know also the other SQL query and moreover, we know the join condition. In that case, the object mapping uses the columns from the other source to generate the value (using the subject mapping of the referenced triple map).

According to that, we start the transformation process with the SQL query $\{x_1, x_2, ..., x_n | \langle x_1, x_2, ..., x_n \rangle \in T\}$ where T is the table defined in the R2RML mapping (or with the query $\{x_1, x_2, ..., x_n, x_{n+1}, ...x_m | \langle x_1, x_2, ..., x_n \rangle \in T \wedge \langle x_{n+1}, ..., x_m \rangle \in T_2 \wedge jc\}$ where jc is a join condition from R2RML, when a reference to another triple map is defined). We start with $VB = \emptyset$.

The restricted triple pattern contains three patterns, for the subject, the predicate and the object. The pattern can contain a variable match pattern, a node match pattern and a blank node match pattern. We do the same for the subject, the predicate and the object:

The current transformation tuple $(VB, \{x_1, x_2, ..., x_n | p\})$ is transformed to another tuple according to the pattern type:

– **Variable match pattern** - We create a value binder vb using the corresponding R2RML mapping. Then the transformed tuple creation is split into two cases:
  - There is another value binder vb_2 that is for the same variable, then we replace the DRC formula p by $p \wedge vb.b \wedge vb.val = vb_2.val$.
  - Otherwise we add the value binder vb to VB and we replace the DRC formula p by $p \wedge vb.b$.
– **Node match pattern** - We create a value binder vb using the corresponding R2RML mapping. Then we add a condition to the SELECT statement to ensure that the value must be the same as the value in the node match pattern. The value binder is here used only to create the condition. So, we replace the DRC formula p by $p \wedge vb.b \wedge vb.val = val_p$ where val_p denotes is the value in the node match pattern.
– **Blank node match pattern** - The blank node pattern is handled like the variable pattern because in the SPARQL language the blank nodes in the where clause are very similar to standard variables.

After the subject, the predicate and the object is processed, we can reduce the variables on the left-side of the calculus. That means that we get the tuple in the form $(VB, \{x_1, x_2, ..., x_n | ...\})$ where $\forall x_i \in \{x_1, x_2, ..., x_n\}$ is true that x_i is needed for one of the value binders in VB.

Also, we have to think about the graph of the created triple. We have to add the condition, that the graph is one of the applicable (using value binders, the same mechanism as used for subject, predicate and object):

– When the triple pattern is under a graph operator $\Gamma_u(P)$:
  - $u \in RDF\text{-}I$: we add the condition, that the graph is u - same as node match pattern
  - $u \in V$: we add the condition that the graph is one of the graphs mentioned in the FROM NAME clause and to value binders we add the value binder for variable u with value of the graph IRI
– Otherwise we add the condition that the graph is in one of the graphs mentioned in the FROM clause

The Join Operator. Firstly, we run the algorithm on the child graph patterns. If the graph pattern P_1 is translated to $(VB_1, \{x_1, ... x_n | p_1\})$ and P_2 to $(VB_2, \{y_1, ... y_m | p_2\})$, then $P_1 \bowtie P_2$ is translated to $(VB, \{x_1, ... x_n, y_1, ... y_m | p_1 \wedge p_2 \wedge JC\})$ where JC is the join condition.

The join condition JC is defined as a conjunction of conditions created for every variables v present in both P_1 and P_2 (with value binders $vb_1 \in VB_1$ and $vb_2 \in VB_2$): $(\neg vb_1.b) \vee (\neg vb_2.b) \vee (vb_1.b \wedge vb_2.b \wedge vb_1.val = vb_2.val)$.

The resulting set of value binders VB is defined as follows:

- For all variables v present only in P_1 (with value binder $vb \in VB_1$), then $vb \in VB$.
- For all variables v present only in P_2 (with value binder $vb \in VB_2$), then $vb \in VB$.
- For all variables v present in both P_1 and P_2 (with value binders $vb_1 \in VB_1$ and $vb_2 \in VB_2$) we create a new value binder vb (this value binder works as the relational COALESCE operator):
  - $vb.v = v$
  - $vb.b = vb_1.b \vee vb_2.b$
  - $vb.val = (vb_1.b \rightarrow vb_1.val); (vb_2.b \rightarrow vb_2.val)$

The Selection Operator. The selection operator is only an added condition to the transformed inner graph pattern. So, if P is transformed to $(VB, \{x_1, x_2, ..., x_n | p\})$ then the selection $\sigma(P)$ is transformed to $(VB, \{x_1, x_2, ..., x_n | p \wedge p_\sigma\})$ where p_σ is the condition generated from σ. This transformation is straightforward. We only need to transform the operators and function in the SPARQL expression into the SQL form.

The Left Join Operator. Firstly, we run the algorithm on the child graph patterns. If the graph pattern P_1 is translated to $(VB_1, \{x_1, ...x_n | p_1\})$, P_2 to $(VB_2, \{y_1, ...y_m | p_2\})$ and the selection σ is translated to the condition p_σ, then the left join $P_1 \bowtie_\sigma P_2$ is translated to:

$$(VB, \{x_1, ...x_n, y_1, ...y_m | p_1 \wedge \langle y_1, ...y_m \rangle \in \bowtie ((x_1, ...x_n), \{y_1, ...y_m | p_2 \wedge JC \wedge p_\sigma\})\}).$$

Where the relational left join $\bowtie ((x_1, ...x_n), \{y_1, ...y_m | p\})$ is for every tuple $(x_1, ...x_n)$ defined as $M_j \cup M_n$ where:

- $M_j = \{y_1, ...y_m | p\}$
- $M_n = \{y_1, ...y_m | \neg \exists \langle y'_1, ...y'_m \rangle \in M_j \wedge (y_1 = \text{null}) \wedge ... \wedge (y_m = \text{null})\}$

The relational left join $\bowtie (\{y_1, ..:y_m | p\})$ works exactly as the LEFT OUTER JOIN relational clause, where $y_1, ...y_m$ are joined columns and p is the join condition in the clause ON.

The value binders VB and the join condition JC are created the exactly same way as for the join. The value binders are bound if and only if the source columns are not null. So, if the joined columns are null, the value binders are unbound. That is exactly the behavior we expect.

The Union Operator. The difference between the standard SQL union and the SPARQL union is that in SQL we do need that both operands have the same columns. Also, we need to handle the fact that the columns does not represent the SPARQL values.

Firstly, we run the algorithm on the child graph patterns. If the graph pattern P_1 is translated to $(VB_1, \{x_1, ...x_n | p_1\})$ and P_2 to $(VB_2, \{y_1, ...y_m | p_2\})$, then $P_1 \cup P_2$ is translated to $(VB, \{x_1, ...x_n, y_1, ...y_m, z | p_1 \wedge z = 1 \wedge y_1 = null \wedge ... \wedge y_m = null\} \cup \{x_1, ...x_n, y_1, ...y_m, z | p_2 \wedge z = 2 \wedge x_1 = null \wedge ... \wedge x_n = null\})$. As you can see, it is a standard union with added columns to match the columns in both sources and one extra added column (z) - this extra column is used in value binders to determine which value binder to use when creating the value.

The value binders are created from VB_1 and VB_2 in a way that for every variable v present in (at least) one of the operands we take its value binders $vb_1 \in VB_1$ and $vb_2 \in VB_2$. If there is one of them missing, we create an extra value binder that is always unbound. For example, if the variable is present only in P_1, then there is $vb_1 \in VB_1$ for variable v, but there is no $vb_2 \in VB_2$ for such variable. In that case we create the value binder with $vb_2.b = \mathtt{false}$. This value binder is always unbound - it does represent the variable in P_2 correctly.

Then, we create the value binder for variable v as follows (this value binder works as the relational `CASE` operator):

- $vb.v = v$
- $vb.b = ((z = 1) \wedge vb_1.b) \vee ((z = 2) \wedge vb_2.b)$
- $vb.val = (z = 1) \rightarrow vb_1.val(x); (z = 2) \rightarrow vb_2.val(x)$

6 Optimization

Although the algorithm we proposed already generates correct SQL query it is not yet efficient. We detect the cases when we are able statically decide that an operator will not return any result. Moreover, this observation we propagate through the algebraic tree to the ascendant operator. For example, if we found that in $(P_1 \bowtie P_2) \cup P_3$ the P_1 cannot return any result the whole fragment can be replaced by P_3.

In this paper, we will describe three optimization methods which seem to have the biggest impact on the effectiveness of the transformed queries.

6.1 Candidate Selection

We use a system how to replace the triple patterns by the union of the restricted triple patterns. If the triple pattern contains a node match in subject, predicate or object we can decide which of the restricted triple pattern can actually return some results.

For example, for a templated IRI mapping we can decide whether the template can match or not. We decide according to the possible values. Note that, if the template results in an IRI, the column values are converted into an IRI-safe version (see [6]). For example, the template s/{Code} can match the value s/12-45 but cannot match the value s/12/45. Moreover, we can load the column types and if we get that the `Code` column has integer type, then we know that the template cannot match the value s/12-45.

Thanks to this optimization, we will be able to keep the union of restricted triple patterns as small as possible. For standard scenarios we expect that even this optimization will be able to replace the union by a single restricted triple pattern (or only by an union of few triple patterns).

6.2   The Join of Unions Optimization

The other optimization method also uses the mapping information. In lots of cases, we can decide whether a join can return something.

For simplicity we will show the method on two joined triple patterns which share only one variable. The first triple pattern TP_1 is ?x rdf:type ?type. The other triple pattern TP_2 is ?x rdfs:label ?name.

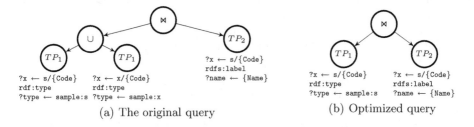

(a) The original query

(b) Optimized query

Fig. 3. The join of unions optimization

The join optimization works with the joins of restricted triple patterns. So firstly it is needed to transform triple patterns to restricted triple patterns. However, these restricted triple patterns are generated in a union. The join operator is usually the ascendant of these created unions, so we cannot apply the previous optimization immediately. However, we can use the distributivity of the union, that means that $Q_1 \bowtie (Q_2 \cup Q_3) = (Q_1 \bowtie Q_2) \cup (Q_1 \bowtie Q_3)$. Using this approach, we can enable the use of the join optimization so we can more precisely select the joins that can return some results.

To keep the example simple we will show it for the case when the triple pattern TP_1 is transformed (after candidate selection) to the union of restricted triple patterns $TP_{1<g;s_1;p_1;o_1>} \cup TP_{1<g;s_2;p_2;o_2>}$ and the triple pattern TP_2 is transformed to the restricted triple pattern $TP_{2<g;s_3;p_3;o_3>}$. After applying the distributivity of the union the whole join is transformed to $(TP_{1<g;s_1;p_1;o_1>} \bowtie TP_{2<g;s_3;p_3;o_3>}) \cup (TP_{1<g;s_2;p_2;o_2>} \bowtie TP_{2<g;s_3;p_3;o_3>})$. Now, for both joins we can apply the join of unions optimization. TP_1 and TP_2 shares the variable ?x in the subject. That means that the join $TP_{1<g;s_1;p_1;o_1>} \bowtie TP_{2<g;s_3;p_3;o_3>}$ can return any result if and only if mappings s_1 and s_3 can produce the same value.

A complete sample is shown on the Fig. 3. The restricted triple patterns TP_1 assigns to variables ?x and ?type. The restricted triple pattern TP_2 assigns to variables ?x and ?name. Firstly we move the union operator up to the root. Then we can decide, which join operators can return any result. The templates "x/{Code}" and "s/{Code}" cannot match so we can discard one of the joins and return only a single join.

In the common case, the method goes through all the shared variables. To improve the results it is better not to see the join as an operator with two operands, but to take nested joins as one large join and process all operands at once. That means that there may be some variable shared between more than two triple pattern operands, and therefore we will dismiss more operators that will not return any results.

However, it also has its drawbacks. We need to reckon with the fact that it can produce union of a large amount of joins (it is actually the Cartesian product of the unions). For example, when we have the join of ten triple patterns and every triple pattern will have five valid restricted triple patterns then we will create a union of $5^{10} = 9\,765\,625$ joins. To minimize this drawback, we filter the joins immediately during the creation of the Cartesian product. Therefore it will filter only the joins that can return some result. And because we do it on the fly we discard lots of joins before they are completed. Most of them will be discarded when it creates the join of two triple patterns with a shared variable because typically the shared variables are in the subject of the pattern and the IRIs are usually mapped using a uniquely identifying template. They will not match so often (commonly the subject IRI templates uniquely identify the R2RML triples map.

The merge with the join optimization can be even faster if we sort the joins (using the associativity and commutativity of the join operator) in a way that the triple patterns with shared variables will be as close as possible, so the join optimization will be able to discard the joins sooner (and therefore it will work faster).

6.3   The Self Join Elimination

The previous optimizations get lots of queries to the form containing joins unioned together. In many cases, the join will reference the same table multiple times - when the query refers to multiple properties of the same entity (usually that means stored in the same table). Using the database schema and the integrity constraints, we are able to detect the situations, when the tables are joined with themselves using an unique key (usually the primary key). In that case, the join is not necessary and the second table can be dropped, because it does not add any attributes to the result (see [2,4]). The only thing needed is to replace all references to columns from the dropped table, by the corresponding columns from the table that is kept in the query. However, the SPARQL algebra does not support these operations so this optimization is used over the relational model of queries.

7   Evaluation

For evaluation we used a part of the database of the State Institute for Drug Control of Czech republic[2]. As sample queries we have chosen representatives to cover standard usage of the dataset.

The first scenario demonstrate a sequence of SPARQL queries a UI client application needs to execute when displaying medicinal products to the end user. The query 1 retrieves a list of all medicinal products which contain ephedrine or more than 30 mg of pseudoephedrine. After that, one of the returned products is picked and using the query 2 the details of the selected medicinal product are retrieved. Then the query 3 returns the composition of the selected medicinal product. The second scenario demonstrates SPARQL queries an analytical application needs to execute to gather data. The query 4 retrieves a list of all medicinal products with the same addiction as before including all their details and their composition. The query 5 retrieves the same information but about medicinal products with any addiction level. The last scenario is the query 6 which gets the complete dataset dump.

The virtual dataset contains 2 987 422 triples. The queries returns 27, 1, 12, 585, 481, 26 180, 18 656, and 2 987 422 results respectively.

We evaluate our tool by comparison of the SPARQL endpoint created using our tool to the endpoint using -ontop- system. The -ontop- system was selected because it seems to be the most current and actively developed tool. For comparison, we include also the performance of a native solution - the Virtuoso Universal Server in version 07.10.3207[3]. The queries were executed using the Microsoft SQL Server 2014 Express running on a computer with the processor Intel Core i7-4600M and 4 GB of RAM. The machine runs Windows 10 64-bit (Fig. 4).

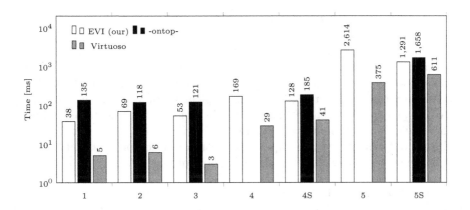

Fig. 4. Query execution times comparison

---

[2] See http://www.sukl.eu/.
[3] Available at http://virtuoso.openlinksw.com/.

The missing values indicates that there was a problem with a query execution. The query 4 and 5 were failing in the -ontop- system, it seems that they do have a defect in the CONSTRUCT queries. Therefore, there are queries 4S and 5S, which are exactly the same, but they do have SELECT instead of CONSTRUCT.

Our tool was the only one able to respond correctly on the query 6[4]. We were able to generate the complete dump (query 6) in 2 min and 1 s. The complete dump is a 381 MB file.

The complete query execution time can be divided into three parts:

1. Generating the SQL query
2. Query execution
3. Transforming the SQL results into RDF terms

Our tool produces very similar queries to the ones produced by ontop. So, the performance difference is in the first and last part. The reason why our tool is faster is that the query generation is faster. Moreover, the usage of value binders to transform the SQL results into RDF terms is more efficient than the ontop mechanism.

Virtuoso dominates the evaluation. It is caused by the fact that the dataset is not huge and that the Virtuoso does not need to translate the query and transform the results back to the RDF form (part one and three). As shown in the ontop evaluation (see [12]) if we stress more the second part (query execution) then the solutions based on a relational database starts to outperform Virtuoso. That is also the reason why we differ from the ontop evaluation. We are stressing first and second part and they were stressing the second part of the execution.

8   Conclusions

In this paper, we presented a formal model for a SPARQL algebra. The algebra was used to describe an algorithm transforming SPARQL queries to the SQL queries and to transform the SQL query result to the expected SPARQL form. The current implementation is only a proof of concept, but we have shown that it outperforms -ontop- (the current and actively developed tool for this task).

The next task will be to complete the implementation to support the whole SPARQL algebra. We believe that we will be able to keep the current performance while we will complete the SPARQL algebra support. Moreover, there are still some options how we can improve the transformation process, so the query will be generated faster and even slightly more optimized.

Acknowledgments. This work was supported in part by the Charles University in Prague, project GA UK No. #158215 and in part by the Czech Science Foundation (GACR), grant number 16-09713.

---

[4] Ontop failed to respond. Virtuoso returned a result but it was not complete.

References

1. Angles, R., Gutierrez, C.: The expressive power of SPARQL. In: Sheth, A.P., Staab, S., Dean, M., Paolucci, M., Maynard, D., Finin, T., Thirunarayan, K. (eds.) ISWC 2008. LNCS, vol. 5318, pp. 114–129. Springer, Heidelberg (2008)
2. Chakravarthy, U.S., Grant, J., Minker, J.: Logic-based approach to semantic query optimization. ACM Trans. Database Syst. **15**(2), 162–207 (1990)
3. Chebotko, A., Lu, S., Fotouhi, F.: Semantics preserving SPARQL-to-SQL translation. Data Knowl. Eng. **68**(10), 973–1000 (2009)
4. Cheng, Q., Gryz, J., Koo, F., Leung, T.Y.C., Liu, L., Qian, X., Schiefer, K.B.: Implementation of two semantic query optimization techniques in DB2 universal database. In: Proceedings of the 25th International Conference on Very Large Data Bases VLDB 1999, San Francisco, CA, USA, pp. 687–698. Morgan Kaufmann Publishers Inc (1999)
5. Cyganiak, R.: D2RQ: Accessing Relational Databases as Virtual RDF Graphs. http://d2rq.org/. Accessed 15 May 2015
6. Das, S., Cyganiak, R., Sundara, S.: R2RML: RDB to RDF Mapping Language. W3C Recommendation, W3C September 2012. http://www.w3.org/TR/2012/REC-r2rml-20120927/
7. Harris, S., Seaborne, A.: SPARQL 1.1 query language. W3C Recommendation W3C March 2013. http://www.w3.org/TR/2013/REC-sparql11-query-20130321/
8. Lacroix, M., Pirotte, A.: Domain-oriented relational languages. In: Proceedings of the Third International Conference on Very Large Data Bases, 6–8 October 1977, Tokyo, Japan, pp. 370–378. IEEE Computer Society (1977)
9. Pérez, J., Arenas, M., Gutierrez, C.: Semantics and complexity of SPARQL. ACM Trans. Database Syst. **34**(3), 16: 1–16: 45 (2009)
10. Priyatna, F., Corcho, O., Sequeda, J.: Formalisation and experiences of R2RML-based SPARQL to SQL query translation using morph. In: Proceedings of the 23rd International Conference on World Wide Web WWW 2014, pp. 479–490. ACM, New York (2014)
11. Rodríguez-Muro, M., Rezk, M.: Efficient SPARQL-to-SQL with R2RML mappings. Web Semant. Sci., Serv. Agents World Wide Web **33**(1), 141–169 (2015)
12. Rodriguez-Muro, M., Rezk, M., Hardi, J., Slusnys, M., Bagosi, T., Calvanese, D.: Evaluating SPARQL-to-SQL translation in ontop. In: Proceedings of the 2nd International Workshop on OWL Reasoner Evaluation (ORE). CEUR Workshop Proceedings, vol. 1015, pp. 94–100. CEUR-WS.org (2013). http://ceur-ws.org/Vol-1015/paper_16.pdf
13. Sequeda, J., Miranker, D.P.: Ultrawrap: SPARQL execution on relational data. Web Semant. Sci., Serv. Agents World Wide Web **22**, 19–39 (2013)

Comparison of Different Approaches for Hotels Deduplication

Ivan Kozhevnikov[1,3]([⊠]) and Vladimir Gorovoy[2,3]

[1] ITMO University, St. Petersburg, Russia
kozhevnikov@rain.ifmo.ru
[2] Graduate School of Management,
St. Petersburg State University, St. Petersburg, Russia
gorovoy@gsom.pu.ru
[3] Yandex, Moscow, Russia

Abstract. The present article addresses the problem of a hotel deduplication. Obvious approaches, such as name or location comparisons, fail, because hotel descriptions differ among different databases. The most accurate approach to solve this problem is to use the professionally trained content managers, but it is expensive, hence an automatic solution should be implemented. We propose a method to improve a hypothesis that a pair of hotels is identical, and compare its performance with alternative solutions. The proposed method satisfies business requirements set for the precision and recall of the hotel deduplication task. The method is based on machine learning approach with the use of some unique features, including those built with the help of computer vision algorithms.

Keywords: Deduplication · Entity resolution · Machine learning · Natural language processing

1 Introduction

Companies in the business of selling hotel inventory face a great challenge of matching hotel data, which are usually taken from quite a few sources with variable quality and completeness. Hotels vary in name, location, GPS coordinates, features, photos, e-mails, phone numbers, websites, etc. The sources of hotel data include B2C OTAs (Online Travel Agencies), such as Booking.com, Ostrovok.ru, Hotels.com, 101hotels.ru, tour operators (TUI, Thomas Cook, Pegas Touristik and others) and also the B2B players selling inventory to tour operators and OTAs (GTA, Hotelbeds and others). There are also a lot of meta-aggregators, which integrate hotel data from many sources. Among the most popular of them on the Russian market, we should mention travel.yandex.ru, Roomguru.com, tripadvisor.com, kayak.com, trivago.com and hotellook.com. Meta-aggregators and some OTAs, reselling hotel inventory from B2B partners, have to solve the hotel deduplication problem, i.e. they need to merge duplicate objects corresponding to the same entity in various sourses. The ideal solution should be

© Springer International Publishing Switzerland 2016
A.-C. Ngonga Ngomo and P. Křemen (Eds.): KESW 2016, CCIS 649, pp. 230–240, 2016.
DOI: 10.1007/978-3-319-45880-9_18

simultaneously cheap, accurate and fast, whereas the well-known solutions do not satisfy at least one of these requirements. For example, simple algorithms are cheap and fast, but inaccurate; to buy matching data from third-party companies, such as Tavisca and GIATA, can be fast, but costly and sometimes faulty; croudsourcing is expensive and again not accurate enough as we will show below. The most accurate solution, the manual matching by the professionally trained content managers, is costly and time-consuming.

For the first time we faced the challenging deduplication problem in travel.yandex.ru project. Naïve approaches, such as the hotel comparison solely by name, showed low performance on this task due to misspellings, the similarity of names used by diverse hotel data providers, and frequent changes of hotel names. Moreover, two different hotels can share the same name. Also, we could not make a decision based entirely on location. Hotels can cover a huge area of 5–10 sq. km, hence their provided coordinates may not be the same in different sourses. On top of that, there are problems with geocoding, owing to different sources use concurrent geocoders, such as Google maps and OpenStreetMap. This situation implies that geographic coordinates seldom match. Finally, sometimes coordinates are simply incorrect.

While solving the deduplication challenge, we had the following business goal: a solution should have a precision of more than 99 %, because it is really important to decrease the false positive error rate, i.e. the algorithm matches different hotels as the same entity. If a user book a wrong hotel due to our matching mistake, it could be a real problem for our customer support and reputation. The false negative errors, when we miss a duplicated hotel, are not so crucial, but it is still a big problem, because we might not be able to provide our customers with the best available price on the market. That is why the goal for the recall values was set to be greater than 95 %

We tried to apply different approaches to solve the deduplication task, including naïve algorithmic approaches, such as the mechanical turk approach and the mapping with the help of professional content managers, and machine learning approaches. We show below the results of different approaches and present a way to achieve the business goals stated above.

2   Related Work

The deduplication (also known as "Entity Resolution", or "Record Linkage") is a well-known problem which was studied by many researchers in the past [7], [9]. The common formulation of the problem is as follows: different data records refer to the same "real-world" entities and the goal is to "resolve" entities and to find the duplicated records.

One of the problems is the huge amount of data records. Almost always it is impossible to accomplish a full pairwise comparison between records. That is why researchers try to decrease the number of comparisons. For example, a method for general entity resolving was proposed in [2]. Authors suggest methods to minimize the number of comparisons. In this work the hotel deduplication was used as an example. Authors managed to achieve a 100 % precision, but the recall was not good enough.

Another approach is to use assessors to solve the problem [5, 11]. In [5] authors used human reviews for the hotel entity resolution. They managed to achieve an answer accuracy of 0.774. We have tried this approach and the results are described in our study.

There are many real world applications of the entity resolution. For example, the deduplication of users across social networks [3]. Authors had reached a good result with AUC 0.982 and a 95.9 % accuracy.

3   Dataset Description and Deduplication Challenges

Our hotel database contains more than 2.8 million hotels from 12 sources. Some sources provide us with a large number of hotels (0.8 million), some provide us with only 10 hotels. Almost each hotel is presented in 2 or more sources. We collected manually 9193 hotel pairs, for which we know for sure whether they refer to the same entity or not, for learning and validating purposes. Some of the mappings were provided by our partners, some of them were created by our content managers. Here is the description of our dataset, which tells how challenging our task is due to the data quality and incompleteness:

- 5755 pairs of identical hotels and 3438 of different hotels.
- 73 hotel pairs have a full name match, but are different in fact.
- 331 identical hotel pairs have a difference of 1 in stars and 360 identical hotels have a difference of 2 in stars. That means that a hotel from one source can have, for example, 5 stars having only 3 stars in another source.
- There are no hotel photos in 272 pairs.
- An average distance between identical hotels is 250 m. 10 % of identical hotels have a distance of more than 540 m.
- 722 hotel pairs are different, but a distance between them is less than 100 m, whereas for another 121 hotel pairs the distance is less than 10 m.

4   Methods

In this section we describe approaches which can be used to solve the hotel match problem. The main idea is to generate a possible pair of duplicates and then identify, whether they are identical or not. The number of all the possible hypotheses is $\frac{N(N-1)}{2}$, where N is the number of hotels. Thus, we need to split the hotels into buckets in order to reduce the number of hypotheses. The main idea of such a splitting is that the identical hotels are close to each other. As we have encountered only several cases when identical hotels had a distance between them of more than 10 km and in all of them coordinates of one of the hotels was in a sea we chose to split the Earth surface into squares with a side of 10 km. Then two similar hotels are expected be located in a single square, or in two adjacent squares. This helps us to reduce the number of hypotheses drastically. Having done that, we reduced the original problem to the task of the hotel pairs classification.

4.1   Expert Method

The first method, which can be applied, involves a professionally trained content manager being asked to assess whether two hotels are identical. We will refer to this method as "expert method".

Method Description. We developed an approach to check if two hotels are identical. It consists of the following steps:

- Look for photos, names, positions on the map, description, etc.
- Search for these hotels in different search engines.
- Search in different map services, such as Google maps.
- Read reviews for the considered hotels and compare them.
- Look for the official websites of those hotels and compare them.

Results. The decisions made by a trained content manager are very accurate, but the evaluation of a pair of hotels can take up to 20 min in peculiar cases. It means that only a small amount of work can be done by content managers. We estimated that provided with appropriate instruments a content manager could classify up to 150–200 hotel pairs per working day, what makes this method too slow and quite expensive.

4.2   Naïve Approach

A naive approach to match two hotels is to use solely their names and a distance between them. We tried this approach with our dataset with the following settings: hotels were considered identical if they had a full name match and a distance between hotels was less than 4 km. The results of this algorithm were the following: Precision 87.7 % and Recall 8.5 %. This result tells us, that there is a small number of duplicates, which have identical names. Also, the precision is rather small. It means that if two hotels are close to each other and their names match, it does not mean that the hotels are identical.

4.3   Crowdsourcing Approach

One of our hypotheses while solving the deduplication challenge was that duplicates could be easily found by an untrained human eye. Hotel pairs could be generated and then could be given to assessors to classify whether hotels matched. Owing to the assessors are not as well-trained as content managers, we expected that the majority of them would vote for the right answer. In order to check this crowdsourcing approach to our task Toloka[1] or Amazon Mechanical Turk could be used. We have used Toloka in our experiments.

---

[1] https://toloka.yandex.com/.

Method Description. As we had to pay for each classification task solved by assessors we were motivated to reduce the number of hotels pairs to classify. Thus we generated automatically hotel pairs in the following way: get the nearest hotel by distance and get the closest hotels in terms of name similarity within 10 km (and only those which had some intersection in names). We had understanding that some duplicates might be missed but we were interested to get a high precision firsthand. The tasks for assessors consisted of 5 hotel pairs and the assessors were paid two cents for doing it. To allow only good enough assessors to solve the task, the golden set of 2069 hotels was provided and the skill of each assessor was calculated as a percentage of golden set pairs classified correctly. All tasks were split into two task pools. In the first pool four assessors did the same classification task and only those who have not had a calculated skill or had it above 80 % were allowed to take a task. In the second pool one assessor did the task which was already taken in the first pool and only those with skill above 80 % were allowed to take it. So, finally, we had five classification results for each hotel pair and the final result for it was simply chosen by the majority of votes. The final price for each classification result was two cents.

Results. Overall, we have three cases for evaluation of crowdsourcing method: for pool with four tolokers, for pool with one skilled toloker and for the sum of both pools. For the case with four tolokers the results are following: precision 0.9702, recall 0.9377. For the case with one skilled toloker: precision 0.9956, recall 0.8744. And for the summed case: 0.9494 precision and 0.9904 recall. The evaluation was made with the golden set. None of the results satisfy our goals. We knew in advance that this method would not be a silver bullet and we would not be able to use it for the matching of the whole database, but, as a result of this experiment, we know, that assessors can be used for the evaluating of a gray zone (the cases where automatic methods are not able to make a decision with a high level of confidence).

4.4 Machine Learning-Based Approach

As we have shown in the previous subsections all the discussed approaches failed to achieve our business goal. The most accurate approach so far was the expert method so we came up with an idea to somehow simulate it automatically. We assume that there is a strong correlation between hotels similarity and similarity of photos, names, locations, features. Thus, we can invent some metrics based on hotel features, but it is still not clear, how to combine all these metrics manually in order to make a decision if the hotels match. That is why machine learning was chosen to help us to do it automatically.

Method Description. The first step is to extract a pair of hotels we want to compare. As we stated above, we can split the Earth surface into squares with side of 10 km. Identical hotels would be located in the same or adjacent squares. This approach not only reduces the number of hypotheses, but also gives us an opportunity to run algorithm concurrently (and even distributed on several machines), because each cell can be processed independently. The following step is preprocessing. As each hotel belongs to many pairs, we can prepare data for calculating features. It can save us a lot of processor time. The next step is to extract features for each pair. After features are extracted machine learning model should be trained and evaluated.

Features. *Name-Based Features.* Each hotel can have several names. They can be translations to other languages or just synonyms. Sometimes it can have its former name. So, each hotel have several strings describing it. Before calculating, features strings are preprocessed. A preprocessing includes the following steps:

– Removing any accents on symbols. For example, replace 'Å' with 'A'.
– Lowercasing. All characters become lowercase.
– All punctuation is converted to spaces.
– Normalizing white spaces. Any sequence of whitespace symbols (tabulation, spaces, new line) is replaced with one whitespace.
– Removing markers. Some words were considered as markers. For example: hotel, hostel, guest house, village. All these words are eliminated from the string.
– Transliteration. Some names are not in English. That is why we transliterate them to English words.

Henceforth, we are going to define several string distances. For any pair of hotels a pairwise name comparison is calculated and the minimal distance is chosen. In this study we extracted 5 name-based features:

– **3-gramm name similarity**. A name is splitted into words. Each word is splitted for 3-gramms. Then all the 3-gramms are gathered into a set. For example, "Best tar tar" is turned into ['bes', 'est', 'tar']. The sets are compared by the Jaccard index [1].
– **Name set similarity**. A name is splitted into a word set. For example, "beach and spa" is turned into ['beach', 'and', 'spa']. Again, the Jaccard index [1] is used for the set comparison.
– **Levenshtein distance** [4] — edit the distance to convert one string into another with the help of edit, delete and insert character operations.
– **TF-IDF** [13] **similarity**. TF-IDF stands for the term frequency–inverse document frequency. The method helps understand how important a word is to a document in a collection. Similarity of two hotels is measured via cosine distance between their names.
– **Local TF-IDF similarity**. This feature is like an ordinary TF-IDF except for the IDF part, which is calculated for all the hotels in the radius of 3 km from the current hotel pair. Using local context helps with cases, where hotels share some word describing local district (like 'Manhattan') where they are located, while being different. The idea of local context for names is also used in [12].

Location-Based Features. The location is one of the most important features. The closer hotels are the higher the probability that they are identical. But the concept of "closeness" differs in different regions. For example, if a distance between hotels is 1 km and they are in Moscow, probably they are different. But somewhere in Sahara Desert, where there are few hotels on a vast territory, the hotels probably are identical.

- **Local density** shows how many different hotels are in the area of the considered hotels.
- **Average distance** shows the average distance between hotels in the considered area. Some area near the hotels is chosen and the distance between all pairs of hotels is calculated. Then sum of distances is divided by the number of pairs.
- **Bounded distance** reflects the "closeness" of two hotels. If two hotels are in the same point, then value of this feature is 0. The further the hotels are located, the closer the value is to 1.
- **Address number similarity**. This features extracts all numbers from hotel address. Sets with numbers are compared by the Jaccard Index.
- **Address TF-IDF**. This feature is like **TF-IDF similarity**, but is calculated over hotel addresses.

Photo-Based Features. Most hotels have photos. Comparing hotel images is one of the easiest way for human to determine if two hotels are identical. But photos are rather hard for comparison using machines. One of the difficulties is that photos are taken at different times. Also, identical objects can be shown in the photos, but they were taken from different angles. We used 2 approaches for the photo comparison.

The first is the perceptual hashing [6] (phash). A perceptual hash is a fingerprint of a multimedia file derived from the various features of its content. Unlike cryptographic hash functions which rely on the avalanche effect of small changes in input leading to drastic changes in the output, perceptual hashes are "close" to one another if the features are similar. Another approach is to aggregate deep convolutional features from image with the help of a neural network. A neural network is trained for the classification on a large image dataset, for example, in Imagenet [10]. Then outputs of some hidden layer can be used as a feature vector [8]. The Euclidean distance between feature vectors of two images is lower for similar images and greater for different images.

- **Count of similar by phash photos** — compare photos of two hotels and count number of photos which a phash distance is lower than some threshold. Threshold is chosen by an expert.
- **Distance between nearest photos by phash** — assume we have N photos of one hotel and M photos of another. We have NM possible distances between photos. Minimal one is chosen.
- **Count of similar by neural network features photos** — the same as **Count of similar by phash photos** but the distance between neural network features is used

- **Distance between nearest photos by neural network features** — the same as **Distance between nearest photos by phash** but the distance between neural network features is used

Other Features

- **Same source similarity** is a binary feature. It equals to 0 if two hotels came from the same source, 1 otherwise. We have found a lot of duplicates in our partners' database, so we had to deduplicate even hotels coming from the same source, though the probability of finding duplicate in another source is higher than in the same source.
- **Same hotel type distance.** Hotels have types. Types examples are: hotel, hostel, apartment, villas, guest house, chalet etc. The types are extracted from a hotel name or they can come from the source as a separate feature. If types match then distance is 0, otherwise it is set to 1.
- **Phone similarity.** A phone number is just a string. All non-digit symbols are removed. If one cleaned string is a substring of another, then feature value is 0, otherwise it is 1.
- **Phone suffix similarity.** Phone numbers are converted to digit strings as in the previous feature. Then each phone is reversed and with the pairwise comparison the longest prefix is chosen. Then length of this suffix is divided by the length of minimum length of the considered phones.
- **Phone length** is just the length of the longest phone number.
- **Url similarity.** The feature value is 0 if one hotel's url substring of another, 1 otherwise.
- **Url host similarity.** A host is extracted from a hotel url. If they match then feature value is 0, otherwise it is 1.
- **Star similarity.** Most of the hotels have a star description. The higher the star value the higher is the comfort class of the hotel. If star values are identical in the considered pair of hotels then feature value is 0. If the difference between stars is 1 then feature value is set to 0.5. If the difference is more than 1 than feature value is set to 1.

Results. Python sklearn[2] was used as a machine learning framework. Our data set was splitted with stratification into training and holdout parts in a ratio of 7:3. The training part was used to train the model and the evaluation was made on the holdout part. We have tried several classifiers and used a grid search for the hyperparameters tuning. The best result was achieved the with the help of a random forest classifier with 30 trees, which showed the following results: Recall: 0.980; Precision: 0.991; ROC AUC score: 0.997. All features contributed to this great result above, except for 'Url similarity', 'Star similarity', 'Count of similar by neural network features photos' and 'Average distance'. Our hypotheses of why these factors did not contribute are the following:

---

[2] http://scikit-learn.org/.

- 'Url similarity' many times is worse in terms of predicting whether hotels refer to the same entity than just having the same url host
- 'Star similarity' as we stated above is not great in prediction because of the big differences in hotel stars in different sources for the same hotels
- 'Average distance' heuristic is better used with 'Local density' factor
- 'Count of similar by neural network features photos' heuristic probably works not so well because other factors based on computer vision predict similarity better.

Results Without Computer Vision Features. There are real-life scenarios when we do not have hotel image data from our partners. We tried our machine learning approach for this case, removing all factors generated from photos and got the following results: Recall: 0.975; Precision: 0.987; ROC AUC score: 0.997.

Results above were achieved using a random forest classifier with 30 trees with all non-image factors except for 'Url similarity', which actually made results a little bit worse and hence it was removed.

On our houldout set we got 22 false positives (compared to 15 with computer vision features) and 43 false negatives (compared to 34 with computer vision features). It is 1,47 as many critical errors, as when image data is available, but in some cases it might be applicable depending on the specific business requirements.

Results with Coordinates, Name and Stars Features. Some hotel data providers, like tour operators, might not have a lot of hotel features and often they can offer only hotel names, coordinates and stars for matching purposes. If we apply machine learning only with the factors based on this data on our dataset we get the following results: Recall: 0.962; Precision: 0.977; ROC AUC score: 0.991; False positive errors: 40; False negative errors: 66. Nine factors were used ('TF-IDF similarity', 'Local TF-IDF similarity', 'Same partner similarity', 'Local density', 'Bounded distance', 'Levenshtein disatnce', 'Name set similarity' '3-gramm name similarity', 'Star similarity'). 'Average distance' made results worse, and for this scenario 'Star similarity' brought a lot of value (without false positives rose from 40 to 59).

5   Results Comparison

Several methods were described in this paper. Each of them has it's own pros and cons. Expert method is most accurate but is not automatic and is expensive. Naive approach is cheap, automatic but is not accurate enough. Crowdsourcing is semiautomatic, solves task with rather high precision and recall, but also needs money to be used. The last and the best is machine learning approach. Precision and recall of different methods are compared in the Table 1. Expert method is not present in the table as it is not automatic.

Table 1. Different methods results

	Naive	Crowdsourcing	ML. All features	ML. W/o CV	ML. Names, coordinates and stars
Precision	87.7%	99.0%	99.1%	97.5%	96.2%
Recall	8.5%	94.9%	98.0%	98.7%	97.7%

6 Conclusions and Future Work

We have compared several approaches to solve the hotel deduplication problem, including the crowdsourcing, the naive algorithmic approach and the expert method. And found out that none of them can satisfy our requirements: they are either expensive or do not provide accurate results. So we proposed a machine learning approach using several invented factors capturing an intuition of how experts solve the hotel classification problem, which showed great results satisfying all the desired business goals. The developed method could be used for the fast, cheap and accurate hotel deduplication in production environment.

In our study, we implicitly required the presence of coordinates for each hotel to split data into buckets. But in the real world, not all sources can provide us with complete data. Other bucket splitting functions can be found. For example, name can be used for such purposes, as commonly sources can provide names for each hotel. It is a subject for a future research. Also, it is worth investigating how the proposed method could be applied for a more common task of the organizations deduplication.

Acknowledgement. Authors would like to thank Vladislav Dolbilov for his active involvement in hypotheses testing, features implementation and machine learning experiments; Margarita Pyartel for the help with the preparation of the final learning dataset and providing expert classification results in difficult cases; Andrey Filchenkov for valuable advice and reviewing this article; Andrey Tarkhov for proofreading; Yandex.Travel team for support and help; our partners for providing us with the hotel data and Yandex computer vision team for their expertise. This work was financially supported by the Government of Russian Federation, Grant 074-U01.

References

1. Jaccard, P.: Distribution de la flore alpine dans le Bassin des Dranses et dans quelques regions voisines. Bull. Soc. Vaudoise Sci. Natur. **37**(140), 241–272 (1901)
2. Benjelloun, O., et al.: Swoosh: a generic approach to entity resolution. VLDB J. Int. J. Very Large Data Bases **18**(1), 255–276 (2009)
3. Peled, O., et al.: Matching entities across online social networks (2014). arXiv preprint arXiv:1410.6717
4. Levenshtein, V.I.: Binary codes capable of correcting deletions, insertions, and reversals. Sov. Phys. Dokl. **10**(8), 707–710 (1966)
5. Su, Q., et al.: Internet-scale collection of human-reviewed data. In: Proceedings of the 16th International Conference on World Wide Web, pp. 231–240. ACM (2007)
6. Zauner, C.: Implementation and benchmarking of perceptual image hash functions (2010)

7. Brizan, D.G., Tansel, A.U.: A. survey of entity resolution and record linkage methodologies. Commun. IIMA **6**(3), 5 (2015)
8. Babenko, A., Lempitsky, V.: Aggregating deep convolutional features for image retrieval (2015). arXiv preprint arXiv:1510.07493
9. Getoor, L., Diehl, C.P.: Link mining: a survey. ACM SIGKDD Explor. Newsl. **7**(2), 3–12 (2005)
10. Image database organized according to the WordNet hierarchy. http://www.image-net.org/
11. Wang, J., et al.: Crowder: crowdsourcing entity resolution. Proc. VLDB Endow. **5**(11), 1483–1494 (2012)
12. Dalvi, N., et al.: Deduplicating a places database. In: Proceedings of the 23rd International Conference on World Wide Web, pp. 409–418. ACM (2014)
13. Salton, G., McGill, M.J.: Introduction to Modern Information Retrieval. McGraw-Hill Inc., New York (1986)

Fostering Accessibility of OpenCourseWare with Semantic Technologies – A Literature Review

Mirette Elias[1(✉)], Steffen Lohmann[2], and Sören Auer[1,2]

[1] University of Bonn, Bonn, Germany
melias@uni-bonn.de, auer@cs.uni-bonn.de
[2] Fraunhofer IAIS, Sankt Augustin, Germany
{steffen.lohmann,soeren.auer}@iais.fraunhofer.de

Abstract. Accessibility has become a fundamental requirement for web applications, especially when it comes to e-learning and educational websites for OpenCourseWare. There are various types of disabilities and numerous ways of addressing them. Using semantic technologies to structure and represent the available concepts and taxonomies enables sharing and reusing the knowledge in a variety of systems. This paper provides a literature review of standards and ontologies that were developed to address accessibility requirements. The findings and recommendations reveal missing and future needs for building accessible OpenCourseWare services.

Keywords: Accessibility · OpenCourseWare · OCW · Semantic Web · Disabilities · Ontologies · E-learning · Accessibility standards

1 Introduction

Accessibility has gained significant attention over the past decades due to the equality laws enforced by governments to ensure accessibility for systems and products. Also, the wide range of internet usage has urged web accessibility to address different user preferences and needs. By accessibility, we refer to designing systems, products, and services in a way that they are adaptable for people with disabilities [46], to help them interact with their environment equally as other users. In the literature, the topic of accessibility has also been addressed using other terms, such as Design-for-All, Universal Design, Access-for-All, etc. Most work performed in these contexts shares the objective to define and describe issues, guidelines, standards, and techniques to build accessible systems, including hardware and software devices, mobile, and web applications.

In our work, we are concerned with web accessibility of OpenCourseWare systems. OpenCourseWare (OCW) describes the idea of distributing educational material freely over the web. Accessibility is a key requirement when dealing with OCW systems which provide educational content to a wide range of learners, also taking into consideration people with disabilities. Designing a single OCW that meets the needs of all types of learners is usually not possible, especially

© Springer International Publishing Switzerland 2016
A.-C. Ngonga Ngomo and P. Křemen (Eds.): KESW 2016, CCIS 649, pp. 241–256, 2016.
DOI: 10.1007/978-3-319-45880-9_19

when we are addressing people with different types of disabilities at various levels of severity. For example, a blind person does not benefit from an image if it is not accompanied by a text description, whereas a person with dyslexia might instead prefer an image over a text description.

The goal of this paper is to review and compare guidelines and best practices for creating accessible OpenCourseWare using Semantic Web technologies. In particular, we propose the use of ontologies to represent knowledge required for developing accessible OCW systems, as this allows sharing, integrating, reusing, and extracting this knowledge as well as adapting the OCW accordingly. The remainder of this paper is organized as follows: Section 2 briefly summarizes relevant background knowledge on accessibility and types of disabilities. Section 3 provides a comprehensive review of the state of the art on standards and ontologies addressing accessibility on the web and in e-learning contexts. Section 4 presents findings and possible directions for future research derived from the review, before the paper is concluded in Sect. 5.

2   Background

As reported by the European Health and Social Integration Survey (EHSIS), there were more than 70 million people with disabilities aged 15 and older living in the European Union in 2012, which is equivalent to 17.6% of the EU population [13]. Approximately one out of four people with disabilities (25.6%) is reported to have accessibility problems in education and training contexts. Only 5% of public websites comply fully with web accessibility standards on average, while a larger number are partially accessible [41]. Consequently, the European Accessibility Act [39] proposed a new directive for establishing the laws and regulations needed for products and services to be accessible, including requirements for computers, mobile, and web applications.

In general, disabilities can be grouped into four categories [33,48]:

- *Visual impairments*, such as blindness, low vision, and color-blindness.
- *Hearing impairments*, such as deafness and hard-of-hearing.
- *Motor impairments*, such as the inability to use a pointing device (e.g. a mouse) due to limited movement and control of arms, hands, and fingers.
- *Cognitive impairments*, such as language and learning disabilities, distractibility, inability to remember or focus on large amounts of information (e.g. dyslexia, dementia, etc.).

Each of these impairments has different variations and severity levels that require different types of adaptations for web accessibility. For example, visually-impaired persons commonly use screen readers, while a video script should be provided for persons with hearing impairments. These accessibility requirements are at best taken into account from the beginning of a new web project (e.g., by adding descriptive information about each image to support the use with screen readers). A number of standards and guidelines are available to support developers designing accessible web applications. We will discuss the most relevant ones in the following section.

3   State of the Art

In this section, we review the available accessibility standards, guidelines, and ontologies. The section is divided into two parts: The first surveys available standards, guidelines, checklists, and techniques addressing accessibility in web and e-learning contexts. The second provides a review of ontologies available in the accessibility domain.

3.1   Accessibility Standards and Guidelines

A number of standards and guidelines have been developed to address various accessibility requirements for software, hardware, web, etc. A categorized list of accessibility standards is available at Cardiac-EU [25]. Since we are concerned with web accessibility in OpenCourseWare, we have investigated only those standards and guidelines that are related to accessibility for web and e-learning.

Web accessibility has got much significance recently due to the ubiquitous internet greatly facilitating information access. Accordingly, a number of regulations and standards have been created to assure that the web is accessible by a wider range of users, including users with disabilities and elderly people. Table 1 provides a list of relevant standards, guidelines, checklists, and techniques we considered in our review. The list can be organized into three categories: (i) standards and guidelines that make web applications more accessible, (ii) standards that focus on representing disabilities, and (iii) standards that address accessibility of e-learning websites and educational resources.

The World Wide Web Consortium (W3C) was among the first who developed web accessibility standards and guidelines with their Web Content Accessibility Guidelines (WCAG 1.0). WCAG 1.0 [31] is a technical standard composed of a list of general guidelines and checkpoints for designing accessible web content together with technical recommendations and examples using the Hypertext Markup Language (HTML), Cascading Style Sheets (CSS), Synchronized Multimedia Integration Language (SMIL), and the Mathematical Markup Language (MathML). With WCAG 2.0 [32], the W3C released a more mature and better structured version of WCAG 1.0. It is organized into four design principles of web accessibility (perceivable, operable, understandable, and robust) that are each composed of a list of guidelines. Every guideline has testable success criteria that have to meet one of three conformance levels (A, AA, and AAA). WCAG 2.0 is an approved ISO standard (ISO/IEC 40500:2012). Lately, the W3C published another standard and set of guidelines called Accessible Rich Internet Applications (WAI-ARIA) [29]. It provides a technical specification for presenting dynamic content and advanced user interface controls developed with client-side technologies, such as HTML, JavaScript, Ajax, and related technologies, to make web content more accessible to people with disabilities.

Table 1. Accessibility standards and guidelines

Name	Type	Creator	Released (last updated)	Focus
Barrier Walk-through	Checklist	Giorgio Brajnik	2009	Disabilities, incl. set of barriers and tips to address them
BBC Accessibility Standards and Guidelines	Organization standards and guidelines	BBC	2008 (2013)	Web and mobile accessibility
IBM Accessibility	Checklists and techniques	IBM	2008 (2011)	Web, software and hardware accessibility (based on Section 508 of the US Rehabilitation Act, W3C recommendations and IBM Research)
ICF	Standard	WHO	2001	Body functions and disabilities
IMS Access For All	Guidelines and metadata specifications	IMS Global Learning Consortium	2004 (2012)	Adaptation/personalization of learning resources and applications
ISO/IEC 24751	Standard	ISO	2008	E-learning, education and training accessibility
ISO/IEC TR 29138	Standard and guidelines	ISO	2009	User needs and their mappings to available standards
Section 508*	Standard and guidelines with checklist	US Government	2000	Electronic and information technology
WAI-ARIA	Technical specification	W3C	2014	Web accessibility guidelines
WCAG 1.0	Technical standard and guideline with checklist	W3C	1999	Web accessibility guidelines
WCAG 2.0	Technical standard and guideline with checklist and success criteria	W3C	2008	Web accessibility guidelines

*There are similar accessibility initiatives and standards in other countries (e.g., the British Standard 8878 (BS 8878) [24]).

Section 508 of the U.S. Rehabilitation Act[1] was the first accessibility initiative established in the context of the U.S. Standards for Electronic and Information Technology. Section 508 is a general-purpose standard, developed by the U.S. Access Board for application to electronic and information technology resources that are developed and used by US federal agencies. A checklist of Section 508 guidelines is provided by the Center for Persons with Disabilities at WebAim [21]. Recently, it has been proposed to update the accessibility requirements of Section 508 and to align them with WCAG 2.0.

Some organizations came up with own accessibility standards, such as IBM and BBC. IBM Accessibility [16] is a checklist with a number of guiding techniques for products including web, software, hardware, etc. The checklist of IBM incorporates guidelines from Section 508 of the U.S. Rehabilitation Act, recommendations of the W3C as well as experiences and findings from IBM Research.

---

[1] http://www.section508.gov.

The BBC also created accessibility standards and guidelines for web and mobile applications [10] on behalf of their experiences in making digital products accessible to the widest possible audience. Other standards and guidelines describe typical needs for different types of disabilities. For example, a person with low vision may have problems with color, moving contents, long lines of text, etc. ISO/IEC TR 29138-1:2009 provides a comprehensive summary of user needs to define accessibility barriers faced by people with disabilities when using information technology [28]. It defines relationships of these user needs with accessibility factors, in particular for the developers of standards and guidelines, such as the ISO/IEC Guide 71: "Guidelines to address the needs of older persons and people with disabilities".

A related guide is the Barrier Walkthrough [9], which is based on Section 508 and W3C initiatives. Disabilities of users are categorized into groups, and for each group, a list of barriers is created. For example, using a cascading menu is considered a barrier for people with motor disabilities, as it can be difficult for them to navigate to a second level menu. These barriers are then addressed with recommendations and guidelines from the available standards. Finally, a checklist is given for evaluating web accessibility with respect to each type of disability considered.

ISO/IEC 24751-1:2008 addresses accessibility for e-learning, education and training [27]. It provides a framework for describing and matching learner needs and preferences to digital learning interfaces and resources. The basic idea is that users specify their preferences and that the learning objects appear with respect to their needs. This requires that the learning objects consist of alternative resources that match various types of users (i.e., subtitles can be used with videos to address hearing problems).

IMS Access For All [19] is a related guideline and metadata specification, based on the standard ISO/IEC 24751-1:2008, for developing accessible learning applications and resources with respect to the user's preferences and needs. It links accessibility metadata and learning objects; for example, it defines whether the sensory mode of access is auditory, tactile, or textual, etc.

Overall, the reviewed standards and guidelines have several similarities in how they address accessibility issues. However, each standard addresses accessibility issues at a different level of granularity and from a different perspective. For example, Section 508 states on a general level that alternative descriptions for pictures are required, while WCAG is more precise by prescribing the use of the HTML "alt" attribute for images in websites. Further issues, checkpoints, tips, and technical aids are given in the other standards (e.g., IBM Web Accessibility Checklist, Section 508 standards, and WCAG 2.0 [17]). The W3C standards, i.e., the WCAG 2.0 and WAI-ARIA standards and guidelines for web accessibility, are most widely used and accepted. In particular, several governments, such as the Canadian and Australian ones, require conformance of all government-related websites to the WCAG 2.0 guidelines.

3.2   Accessibility Ontologies

In our literature review, we found a number of ontologies that were developed for the accessibility domain for different purposes. There are ontologies that focus on the description of disabilities or guidelines, while others define mappings of user preferences to assistive devices or are concerned with web content reformatting. Table 2 provides a list of the accessibility ontologies we found, including a brief description of their focus (user characteristics, assistive devices, guidelines, etc.), source (if available) and number of classes (where the ontology file is available for computation). We also included works that provide a classification or meta-modeling approach; although they are not proper ontologies, they can provide classifications that might help in developing new ontologies or merging concepts with existing ones.

The International Classification of Functioning, Disability and Health (ICF) and the Foundational Model of Anatomy (FMA) are two ontologies used for defining and describing disabilities. They contain body functions and disabilities from a medical perspective. These ontologies can be used to describe different types of disabilities and to specify user capabilities and needs. The ICF ontology [58], which is available in BioPortal [20], implements the corresponding standard of the World Health Organization (cf. Table 1) and provides a detailed classification of body functions and disabilities with a qualifier that describes their severity, using a special coding system [26]. For example, "b2-sensory functions and pain" includes "b210-seeing functions" and "b230-hearing functions", where each category contains a more detailed list of the functions. ICF is widely used for categorizing and describing various types of disabilities; its ontology provides good means of reusability and adaptability for different usage purposes. Much of the current literature uses the ICF standard to describe the characteristics of users with disabilities.

The Foundational Model of Anatomy (FMA) is a reference ontology for the domain of anatomy founded by the University of Washington [14]. It represents the anatomical entities, spatial structure, and relations that characterize the physical organization of the body at all notable levels of granularity [15]. FMA is a very large computer-based knowledge source, containing more than 100,000 classes, created for biomedical informatics. It can be used to represent disabilities but it is rather complex and difficult to apply and browse in those contexts.

A number of EU projects worked on accessibility and created ontologies for different purposes. We reviewed three of these ontologies: ASK-IT, ACCESSIBLE, and AEGIS. ASK-IT[2] was an EU project concerned with trip organization requirements for people with special needs, including accessibility of transportation, paths, and remotely accessing home appliances. An ontology was designed to commonly represent the information needs of mobility-impaired users in order to allow for easier search and information retrieval [51]. The ASK-IT ontology consists of user groups, supported services, transportation, and tourism-related content. It mainly focuses on mobility-impaired users, but it also includes

---

[2] http://www.ask-it.org.

Table 2. List of accessibility ontologies

Name	Year	Representation	Focus				Source	No. of Classes
			User Model	Assistive Technologies	Guidelines	Others		
Accessibility Metadata [2]	2012	Metadata tags	Sensory	-	WAI-ARIA WCAG	-	[1]	-
Accessibility Vocabularies in Multimodal UI [55]	2007	OWL	ICF, FMA, interaction effects	-	-	-	[3]	114
ACCESSIBLE [53]	2009	OWL	Based on ICF	Assistive devices	WAI-ARIA WCAG2	-	[4]	166
AccessOnto [52]	2008	UML	User profile	Interface objects	WAI, IBM, etc.	-	[5]	-
ADOLENA [49]	2008	OWL	Based on ICF	Device functionality	-	-	[23]	141
ADOOLES [54]	2012	OWL	Based on ADOLENA	Assistive mechanism	-	-	Not available	-
AEGIS [51]	2010	OWL	ACCESSIBLE ontology	Assistive devices	WAI-ARIA WCAG2	Personas	[6]	15
Affinto [38]	2010	OWL	Physical Cognitive Emotional	Software devices	-	Context model	[7]	86
ASK-IT [51]	2008	OWL	Mobility-impaired	Assistive devices	-	Agents and services	[8]	1,400
Classification for HCI [37]	2006	Structure	Extends ICF	Devices	-	-	-	-
Egonto [43]	2015	OWL	Cognitive Physical Sensory Affective	Hardware Software	-	Adaptation model	[12]	54
FMA [14]	2012	OWL	Body anatomy	-	-	-	[15]	104,145
ICF [26]	2012	OWL	Built for ICF standards	-	-	-	[20]	1,595
Impairment User Interface [47]	2007	OWL	Taxonomy [22]	-	-	Interface adaptation	[18]	114
INREDIS [44]	2010	OWL	Communication and modality	Software devices	-	Context model	Not available	-
Layered capability [35]	2012	OWL	Sample from ICF	-	-	-	Not available	-
OntoSAW [56]	2007	-	Visually-impaired	-	WAI	Web page components	Not available	-
SADIe [36]	2006	OWL	Visually-impaired	-	-	Web page components	Not available	-
WAFA [45]	2007	OWL	Visually-impaired	-	-	Web page components	[30]	152
WAI-ARIA Annotations and Ontologies [57]	2013	Conceptual model	Cognitive Physical Sensory	-	WAI-ARIA	Annotation, adaptation models	-	-

vocabularies for cognitive and sensory impairments. ACCESSIBLE[3] is another EU project that worked on developing an overall European Assessment Simulation Environment for aiding designers and developers to create accessible software applications and evaluate them [53]. The project had two main results: a number of accessibility assessment tools (i.e., for mobile apps, web pages, etc.)

---

[3] http://www.accessible-eu.org.

and an ontology describing different domains of accessibility and interaction between them. The ACCESSIBLE ontology contains characteristics for disabled users (based on ICF standards), descriptions of assistive devices and software applications, web accessibility standards and guidelines (based on WAI-ARIA and WCAG 2.0), and assessment rules for mapping user requirements and constraints [4]. The description of the user characteristics makes use of the ICF standard, by integrating the information required for illustrating the disabilities while neglecting medical and other biological details. Accordingly, ACCESSIBLE can be considered a general-purpose ontology for the domain of accessibility and a good reference to adapt to and extend from. Another EU project, AEGIS[4] developed an Open Accessibility Framework (OAF) which outlines the steps required to consider a computing platform accessible [40]. The AEGIS ontology adopted parts of the ACCESSIBLE ontology and extended it with the Persona concept used for mapping between accessibility concepts and accessibility scenarios.

INREDIS (INterfaces for RElations between Environment and people with DISabilities)[5] was a research and development project supported by the Spanish government in the context of the INGENIO 2010 initiative. The aim of the project was to develop an interoperable architecture that enables people with disabilities to interact with their environment via multiple devices (smartphones, tablets, etc.) in order to communicate with and control other devices (television, door locks, etc.). Three ontologies were developed and evaluated within the project: (i) The INREDIS ontology was developed to provide formal descriptions of disabilities and to check if a user can interact with a specific device. If a problem occurs, an assistive software was meant to provide alternative interaction means with respect to the user needs [44]. (ii) The Egonto ontology, an updated version of the INREDIS ontology, was created for the EGOKI system. EGOKI is a model-based generator for adaptive user interfaces developed in the INREDIS project. The Egonto ontology was used to support the automatic generation of user interfaces for ubiquitous services that are accessible by people with disabilities [43]. (iii) The Affinto ontology was designed as an extension for the EGOKI system, to provide information about sensory and perceptual capabilities of users in order to support the development of multimodal affective resources [38]. In addition, the Affinto ontology takes into account emotional and modality issues to provide knowledge about affective interactions, including the environment as well as social, task, and spatial-temporal perspectives [34].

SUS-IT (Sustaining IT use by older people to promote autonomy and independence)[6] was a project funded by the New Dynamics of Ageing (NDA) initiative in the UK. An ontology was developed in this project to assist a capability reasoning system in mapping users to appropriate assistive technologies [35]. It defines vocabularies for describing capabilities, their properties, and structural links between these capabilities. The user characteristics have been derived from

---

[4] http://www.aegis-project.eu.

[5] http://www.inredis.es.

[6] http://sus-it.lboro.ac.uk.

a subset of the ICF standards. The ADOLENA (Abilities and Disabilities Ontology for ENhancing Accessibility) ontology was developed for the South African National Accessibility Portal (NAP) [49]. It encompasses abilities, disabilities, devices, and functionalities. ADOLENA is considered an experimental ontology that has been created as a proof-of-concept for enhancing search capabilities by Ontology-Based Data Access (OBDA). ADOOLES (Ability and Disability Ontology for Online LEarning and Services) is an ontology based on ADOLENA to annotate learning resources. It represents knowledge in the domains of e-learning and disabilities [54].

Some ontologies have been developed for annotating and adapting user interface contents and components according to the preferences of impaired users. The Impairment User Interface ontology is such an ontology [47]. It describes both user impairments and user interface components. A taxonomy of user impairments was created with the help of a domain expert [22]. Rules were added to connect impairments with corresponding user interface components and to generate a list of suggestions. These suggestions were used afterwards to adapt the CSS styles of websites. Valencia et al. developed an adaptation system using a set of techniques based on the WAI-ARIA recommendations [57]. Their ontology models the user characteristics, adaptation techniques, annotation model, and the relations between them. The adaptation techniques are categorized into content, presentation, and navigation. Reasoning rules are used to infer the techniques that match the user needs. Obrenovic et al. created formalized vocabularies to describe human functionalities and anatomical structures needed for developing multimodal user interfaces [55]. The vocabularies are a combination of the ICF standards [26], FMA ontology [15], and additional concepts of interaction defined by the authors. The Accessibility Metadata Project[7] has been initiated to make educational resources more accessible by enriching their metadata. The developed metadata categories have partly been added to Schema.org [1]. The project is based on the Access for All (AfA) specification [19] and conforms to ISO/IEC 24751-1:2008: "Information technology - Individualized adaptability and accessibility in e-learning, education and training" [27].

AccessOnto provides a framework for integrating accessibility guidelines into requirement specification documents [52]. It is an ontology modeled in UML to describe existing guidelines (e.g., accessibility guidelines of WAI, IBM, Microsoft, Apple, and other companies and institutions), user characteristics, and objects in the interaction environment [5]. In another project, a classification supporting the modeling of human-computer interaction based on ICF was proposed to facilitate the matching of user capabilities (abilities/disabilities) to interaction capabilities of existing devices [37]. Strictly speaking, it is not an ontology but a categorization of the most frequently occurring disabilities and related assistive technologies.

Other ontologies focus specifically on the personalization of web pages for visually-impaired users. Xiong et al. presented an ontology-based approach for developing web user interfaces based on a structured representation of accessi-

---

[7] http://www.a11ymetadata.org/resources/.

bility guidelines [59]. The WAFA ontology is used as a controlled vocabulary to drive page transformations. It describes the structure, structural abstraction, meta-knowledge, spatial knowledge, and functionality of a web page [45]. WAFA is built upon the Travel Ontology of the Dante tool and contains structural and navigational knowledge about web pages [60]. OntoSAW represents the structural elements, attributes, and relationships between web page components, taking into account accessibility properties of the WAI guidelines [56]. It has been developed for the SAW tool, which uses the ontology to edit web page code in order to make it more accessible. SADIe [36] is a transcoding tool which generates semantic annotations for web documents. The contents of the web documents are structured in an ontology that is used in the transformation process.

In summary, we have reviewed 20 works on ontologies and related models designed to address accessibility. 17 of these works are ontologies and nearly all of them have been implemented in the OWL Web Ontology Language, but only 11 of those OWL ontologies are still available online. We examined those 11 ontologies in more detail and counted the number of classes they comprise of. As can be seen in Table 2, FMA is the largest ontology among them with more than 100,000 classes. However, it does not purely focus on accessibility but is a domain ontology representing detailed knowledge about human anatomy. Much smaller compared to FMA but still a large ontology is ICF, which comprises nearly 1,600 classes. ASK-IT is also a large ontology, as it contains many classes describing tourism, transportation, and travel services. Most of the other ontologies have between 50 and 170 classes, except for AEGIS that features only 15 classes as it is an extension of the ACCESSIBLE ontology.

From the above survey, we can summarize the main elements required for a comprehensive ontology supporting OCW accessibility as follows: disability types, assistive technologies, standards and guidelines, and learning objects. Disability types should be considered with details about their classification, specification, the level of severity together with a definition of their functional limitation. The functional limitation might differ from one person to another, even if they have the same disability. That is why some ontologies proposed using the term *capabilities* in order to specify what a user *can* do. Defining capabilities would make it easier to direct users to the most appropriate solution with respect to their abilities [35]. Assistive technologies should be clearly defined by their specification and capabilities in order to adapt content with their usage. Specifying the standards and guidelines (i.e. web and e-learning) which the system should comply to; either a specific country accessibility act and regulation or an accepted standards and guidelines. Finally, representing the learning objects and annotating them to facilitate their retrieval with respect to the user needs.

Among the reviewed ontologies, we find several available ones that satisfy these requirements. The ACCESSIBLE ontology [4], as explained above, is composed of a number of widely accepted standards and guidelines (ICF, WAI, etc.), which can be extended and reused for other purposes. Likewise, the AEGIS ontology [6] defines various personas with reference to the ACCESSIBLE ontology, which provide more insight into the needs and capabilities of people with specific disabilities. The Affinto ontology [7] focuses on extra elements, environmental factors (noise, light, etc.), and personal properties (emotion, mood, etc.) which may also be a useful information affecting a human behavior. From the perspective of learning objects, the Accessibility Metadata Project can be incorporated as it provides descriptive metadata for educational resources, which can help to map resources to user needs.

4   Findings and Research Directions

In our review, we found that most of the standards and guidelines for web accessibility are directed to sensory-impaired users (i.e., users with visual or hearing disabilities). The capabilities and needs of these users are described in detail, and many of them are addressed by web accessibility approaches and assistive technologies. Also, mobility impairments are addressed with some guidelines, in particular by describing the requirements of corresponding assistive technologies (i.e., creating web page content so that it can be accessed with the keyboard and assistive devices instead of requiring the use of a mouse).

However, we found little about standards and guidelines for cognitive impairments, except for a few research studies and recommendations [42]. Some of the existing guidelines can be considered useful to represent selected types of cognitive impairments, but not much of the needs of corresponding users are addressed so far. Recently, the task force "Cognitive A11Y TF" has been initiated by the W3C to focus on the web accessibility needs of cognitive-impaired users and people with learning disabilities [11].

Since our research focuses on OCW, cognitive impairments form an important category, as they encompass people with learning disabilities. There are multiple types and variations of learning disabilities; describing and structuring them in a formal ontology would make the knowledge available and reusable in OCW contexts.

From our review and findings, we derived three main aspects that must be addressed to develop accessible OCWs and that could be well supported by Semantic Web technologies. The aspects are described from a learner-centric perspective in the following:

4.1   Accessible Course Content

A *course* is a core entity of most OCW; representing the educational resources in an accessible form and managing them should be the first step (primary requirement). Three functional requirements are considered for course content:

1. *Course Content Representation.* ISO 24751 requires the course content to be represented in alternative forms (slides, audio, video,...) so that learners can select their preferred style of learning. This idea would be helpful if various types of educational resources could be created for a learning object that

can target certain types of learners with disabilities (e.g., a video would not be helpful for blind people). Creating a special type of educational resources for people with disabilities would also be an option (e.g., a plain document with notes for learners with learning disabilities). This would require designing course materials with respect to a list of accepted guidelines, including learning disabilities.

The process of generating different types of course materials could largely benefit from ontologies describing the required accessibility knowledge. For instance, authors could be supported in the creation process by ontology-based wizards and editing tools. In the best case, ontologies would even assist in partly automatizing the costly process of creating learning materials in different representations. Ontologies could also be effective when addressing persons with multiple disabilities. In those cases, a combination of the accessibility needs would be needed and could be derived from ontologies using reasoning and logics. Consequently, the most appropriate resources could be suggested, taking into consideration that the needs do not contradict each other.

2. *Course Content Validation.* For the validation of course content, the group of the targeted disabilities must first be specified together with the user needs. In a second step, a set of guidelines must be established with a checklist for the resource materials to conform to. Finally, validating the resources against the set of guidelines and highlighting accessibility issues is another requirement. Ontologies can help in meeting these requirements, for instance, by providing the conceptual basis for automatic validations and conformance checks. Furthermore, ontologies, such as the ACCESSIBLE ontology and the AEGIS persona approach, can be used to describe the characteristics and capabilities of the targeted impaired learners.

3. *Courses Management.* Managing courses with multiple educational resources coming from different sources is another challenge. Using an ontology for managing OCW courses and educational resources as well as annotating them with semantic data could make information retrieval easier. This would, for instance, allow adding a recommendation system suggesting appropriate materials to different groups of learners with disabilities.

4.2   Browsing and Navigation

Furthermore, it must be assured that the content and navigation of web pages conforms to general web accessibility standards. In particular, the W3C standards and guidelines for web accessibility (i.e., WCAG 2.0 and WAI-ARIA) can provide useful guidance in this regard; they are widely accepted and cover many accessibility issues. The browsing and navigation of OCW material must also conform to accessibility guidelines. Using an ontology for representing guidelines that can be used for adapting the design and content of web pages with respect to the user needs could be helpful for the customization and personalization of learning materials, like it is proposed by Valencia et al. [57].

4.3   Questions and Assessments

Designing and representing learning questions and assessments is another challenge, especially when dealing with disabilities. For example, using multiple choice can be preferable over essay questions for some disabilities. Using an ontology to describe the guidelines and list of targeted disabilities would aid in constructing appropriate questions and assessments for impaired learners. However, further research is needed for defining such guidelines and tailoring them to provide such a feature.

5   Conclusion

In this paper, we surveyed the state of the art and outlined requirements and challenges for developing accessible OCW platforms using the Semantic Web, in particular ontologies. We have reviewed existing standards and guidelines addressing web and e-learning accessibility, and we have summarized and categorized available ontologies with regard to different types of impairments. We also discussed and suggested some ontologies that can be reused and adapted for creating accessible OCW. We highlighted the need for additional accessibility standards, guidelines, and ontologies for special types of disabilities, such as cognitive impairments, and their importance for OCW. Finally, we defined a number of aspects that must be considered in the creation of accessible OCW and outlined how ontologies could help in addressing these aspects.

In our future work, we will apply selected ontologies to OCW in order to collect empirical data on how ontologies can support the creation and consumption of learning materials in such contexts. We are currently developing an OCW platform that will be used as a testbed for further investigating the topic of accessibility in such environments and how Semantic Web technologies can be used to foster accessible OCW.

Acknowledgment. This research was supported by the EU project SlideWiki (grant no. 688095).

References

1. Accessibility Metadata. http://schema.org/accessibilityFeature
2. Accessibility Metadata Project. http://www.a11ymetadata.org/
3. Accessibility Vocabularies in Multimodal User Interfaces. http://homepages.cwi.nl/~media/ontologies/multimodality.owl
4. ACCESSIBLE ontology. http://www.accessible-eu.org/index.php/ontology.html
5. AccessOnto ontology. http://shapevle.canterbury.ac.uk/repository/AccessOnto.xml
6. AEGIS ontology. http://160.40.50.89/AEGIS_Ontology/
7. Affinto ontology. http://sipt07.si.ehu.es/icearreta/Affinto.owl
8. ASK-IT ontology. http://www.hit-projects.gr/ask-it/ontology/

9. Barrier Walkthrough. https://users.dimi.uniud.it/~giorgio.brajnik/projects/bw/bw.html

10. BBC Accessibility Standards and Guidelines. http://www.bbc.co.uk/guidelines/futuremedia/accessibility/

11. Cognitive and learning disabilities accessibility task force (cognitive A11Y TF). https://www.w3.org/WAI/PF/cognitive-a11y-tf/

12. Egonto ontology. http://sipt07.si.ehu.es/icearreta/AdaptiveOntology.owl

13. EU Disability Statistics. http://ec.europa.eu/eurostat/statistics-explained/index.php/Disability_statistics_-_barriers_to_social_integration#Main_statistical_findings

14. Foundational Model of Anatomy (FMA). http://si.washington.edu/projects/fma

15. Foundational Model of Anatomy (FMA) ontology. http://www.bioontology.org/projects/ontologies/fma/fmaOwlFullComponent_2_0.owl. Also available at https://bioportal.bioontology.org/ontologies/FMA

16. IBM Accessibility Developer Guidelines. http://www-03.ibm.com/able/guidelines/index.html

17. IBM Web Accessibility Checklist 5.2, Section 508 standards, and WCAG 2.0. http://www-03.ibm.com/able/guidelines/web/ibm508wcag.html

18. Impairment User Interface ontology. http://www.ifs.tuwien.ac.at/~skarim/imp-v2.owl, http://www.ifs.tuwien.ac.at/~skarim/ui.owl

19. IMS Access for All. https://www.imsglobal.org/activity/accessibility

20. International Classification of Functioning, Disability and Health (ICF) ontology. https://bioportal.bioontology.org/ontologies/ICF

21. Section 508 checklist. http://webaim.org/standards/508/checklist

22. Taxonomy for user impairments. http://storm.ifs.tuwien.ac.at/publications/ImpTax.pdf

23. ADOLENA ontology. http://www.meteck.org/teaching/ontologies/adolena.owl

24. BS 8878:2010 web accessibility. Code of practice. http://shop.bsigroup.com/ProductDetail/?pid=000000000030180388

25. Cardiac-EU. http://www.cardiac-eu.org/standards/subjectlist.htm

26. ICF browser. http://apps.who.int/classifications/icfbrowser/

27. ISO, IEC 24751-1:2008 information technology - individualized adaptability and accessibility in e-learning, education and training. http://www.iso.org/iso/catalogue_detail?csnumber=41521

28. ISO, IEC TR 29138-1:2009 Information technology - accessibility considerations for people with disabilities. http://www.iso.org/iso/catalogue_detail?csnumber=45161

29. WAI-ARIA. https://www.w3.org/WAI/intro/aria.php

30. WAFA ontology. https://github.com/taurenshaman/semantic-web/blob/master/data/wafa.owl

31. Web Content Accessibility Guidelines 1.0. https://www.w3.org/TR/WCAG10/

32. Web Content Accessibility Guidelines 2.0. https://www.w3.org/TR/WCAG20/

33. Web AIM. http://webaim.org/intro/

34. Aizpurua, A., Cearreta, I., Gamecho, B., Miñón, R., Garay-Vitoria, N., Gardeazabal, L., Abascal, J.: Extending in-home user and context models to provide ubiquitous adaptive support outside the home. In: Martin, E., Haya, R.A., Carro, R.M. (eds.) User Modeling and Adaptation for Daily Routines: Providing Assistance to People with Special Needs. Human–Computer Interaction Series, pp. 25–59. Springer, London (2013)

35. Atkinson, M.T., Bell, M.J., Machin, C.H.C.: Towards ubiquitous accessibility: capability-based profiles and adaptations, delivered via the semantic web. In: Proceedings of the International Cross-Disciplinary Conference on Web Accessibility. W4A 2012. ACM, New York (2012)

36. Bechhofer, S., Harper, S., Lunn, D.: SADIe: semantic annotation for accessibility. In: Cruz, I., Decker, S., Allemang, D., Preist, C., Schwabe, D., Mika, P., Uschold, M., Aroyo, L.M. (eds.) ISWC 2006. LNCS, vol. 4273, pp. 101–115. Springer, Heidelberg (2006)

37. Billi, M., Burzagli, L., Emiliani, P.L., Gabbanini, F., Graziani, P.: A classification, based on ICF, for modelling human computer interaction. In: Miesenberger, K., Klaus, J., Zagler, W.L., Karshmer, A.I. (eds.) ICCHP 2006. LNCS, vol. 4061, pp. 407–414. Springer, Heidelberg (2006)

38. Cearreta, I., Garay-Vitoria, N.: Toward adapting interactions by considering user emotions and capabilities. In: Jacko, J.A. (ed.) Human-Computer Interaction, Part III, HCII 2011. LNCS, vol. 6763, pp. 525–534. Springer, Heidelberg (2011)

39. Commission, E.: European Accessibility Act. Technical report (2015). http://europa.eu/rapid/press-release_IP-15-6147_en.htm

40. Dionysia, K., Anastasios, D., Vasileios, D., Maria, G., Grammati-Eirini, K., Dimitrios, G., Athanasios, T., Dimitriou, J., Gianna, T., Aspiroz, J., Richards, J., Welche, P., Korn, P., Kalogirou, K.: Deliverable 1.2.2 - Common AEGIS context awareness ontologies, security, privacy, QoS and interoperability guidelines. Technical report, AEGIS (Grant Agreement No. 224348) (2011)

41. European Commission: European Disability Strategy 2010–2020: a renewed commitment to a barrier-free Europe. Technical report (2010). http://eur-lex.europa.eu/LexUriServ/LexUriServ.do?uri=CELEX:52010DC0636:en:NOT#top

42. Friedman, M.G., Bryen, D.N.: Web accessibility design recommendations for people with cognitive disabilities. Technol. Disabil. 19(4), 205–212 (2007)

43. Gamecho, B., Min, R., Aizpurua, A., Cearreta, I., Arrue, M., Garay-Vitoria, N., Abascal, J.: Automatic generation of tailored accessible user interfaces for ubiquitous services. IEEE Trans. Hum.-Mach. Syst. 45(5), 612–623 (2015)

44. González-Cabero, R.: A semantic matching process for detecting and reducing accessibility gaps in an ambient intelligence scenario. In: Proceedings 4th International Symposium of Ubiquitous Computing and Ambient Intelligence (UCAmI 2010), pp. 315–324 (2010)

45. Harper, S., Yesilada, Y.: Web authoring for accessibility (WAFA). Web Semant.: Sci. Serv. Agents World Wide Web 5(3), 175–179 (2007)

46. Henry, S.L., Abou-Zahra, S., Brewer, J.: The role of accessibility in a universal web. In: Proceedings of the 11th Web for All Conference, W4A 2014. ACM, New York (2014)

47. Karim, S., Tjoa, A.M.: Connecting user interfaces and user impairments for semantically optimized information flow in hospital information systems. J. Univ. Comput. Sci.: Proceedings of I-MEDIA'07 and I-SEMANTICS'07, KnowCenter, Austria 7, 372–379 (2007)

48. Kavčič, A.: Software accessibility: recommendations and guidelines. In: EUROCON 2005 - The International Conference on Computer as a Tool, vol. 2, pp. 1024–1027 (2005)

49. Keet, C.M., Alberts, R., Gerber, A., Chimamiwa, G.: Enhancing web portals with ontology-based data access: the case study of South Africa's accessibility portal for people with disabilities. In: OWLED, vol. 432 (2008)

50. Kehagias, D., Tsampoulatidis, I.: Deliverable 1.7.1 - ASK-IT Ontological framework. Technical report, ASK-IT (IST-2003-511298) (2006)

51. Kehagias, D.D., Tzovaras, D.: An ontology-based framework for web service integration and delivery to mobility impaired users. In: Lytras, M.D., Ordonez De Pablos, P., Ziderman, A., Roulstone, A., Maurer, H., Imber, J.B. (eds.) WSKS 2010. CCIS, vol. 111, pp. 555–563. Springer, Heidelberg (2010)

52. Masuwa-Morgan, K.R.: Introducing AccessOnto: ontology for accessibility require-
    ments specification. In: First International Workshop on Ontologies in Interactive
    Systems (ONTORACT 2008), pp. 33–38. IEEE (2008)
53. Mourouzis, A., Kastori, G.-E., Votis, K., Bekiaris, E., Tzovaras, D.: A harmonised
    methodology towards measuring accessibility. In: Stephanidis, C. (ed.) Univer-
    sal Access in HCI, Part I, HCII 2009. LNCS, vol. 5614, pp. 578–587. Springer,
    Heidelberg (2009)
54. Nganji, J.T., Brayshaw, M., Tompsett, B.: Ontology-driven disability-aware e-
    learning personalisation with ontodaps. Campus-Wide Inf. Syst. **30**(1), 17–34
    (2012)
55. Obrenovic, Z., Troncy, R., Hardman, L.: Vocabularies for description of accessibil-
    ity issues in multimodal user interfaces. In: MOG 2007 Workshop on Multimodal
    Output Generation, pp. 117–128 (2007)
56. Sánchez-Figueroa, F., Lozano-Tello, A., González-Rodríguez, J., Macías-García,
    M.: SAW: a set of integrated tools for making the web accessible to visually
    impaired users. UPGRADE Eur. J. Inf. Prof. **8**(2), 67–71 (2007)
57. Valencia, X., Arrue, M., Pérez, J.E., Abascal, J.: User individuality management
    in websites based on WAI-ARIA annotations and ontologies. In: Proceedings of
    the 10th International Cross-Disciplinary Conference on Web Accessibility, W4A
    2013. ACM, New York (2013)
58. World Health Organization: International classification of functioning, disability
    and health (ICF), Geneva (2001)
59. Xiong, J., Farenc, C., Winckler, M.: Towards an ontology-based approach for deal-
    ing with web guidelines. In: Hartmann, S., Zhou, X., Kirchberg, M. (eds.) WISE
    2008. LNCS, vol. 5176, pp. 132–141. Springer, Heidelberg (2008)
60. Yesilada, Y., Harper, S., Goble, C., Stevens, R.: Ontology based semantic anno-
    tation for enhancing mobility support for visually impaired web users. In: K-CAP
    2003 Workshop on Knowledge Markup and Semantic Annotation (2003)

Nearest Query on Distributed Binary Trees Starting from a Random Node

Francesco Gargiulo$^{(\boxtimes)}$, Flora Amato, Vincenzo Moscato, Antonio Picariello, and Giancarlo Sperli'

Italian Research Aerospace Centre, University of Naples Federico II, Naples, Italy
f.gargiulo@cira.it,
{flora.amato,amoscato,picus,giancarlo.sperli}@unina.it
http://www.cira.it
http://www.unina.it

Abstract. This paper proposes a new distributed data structure based on binary trees to support k-nearest neighbor queries over very large databases. The indexing structure is distributed across a network of "peers", where each one hosts a part of the tree and communication among nodes is realized by message passing. The advantages of this kind of approach are mainly two: it is possible to (i) handle a larger number of nodes and points than a single peer based architecture and (ii) to manage in an efficient way computation of multiple queries. In particular, we propose a novel version of the k-nearest neighbor algorithm that is able to start the query in a randomly chosen peer. Preliminary experiments have demonstrated that in about 65 % of cases a query, which starts in random node, does not involve the peer containing the root of the tree.

Keywords: Large databases · Distributed index · Multiple queries · k-nearest neighbor query algorithm

1 Introduction

A Distributed Binary Tree (DBT) is a data structure that maintains the links and the nodes of a Binary Tree (BT) but nodes are distributed over more than one peer. A DBT has mainly two advantages over a BT:

1. It can handle more nodes/points than a sequential BT.
2. It allows the elaboration of multiple queries simultaneously.

   Many strategies can be used to achieve a distributed version of a BT over a network of logical or physical peers. In general, a mapping between the set of nodes of the tree and the set of peers is required. In particular, if $N = \{n_1, ..., n_h\}$ is the set of the nodes of the BT and $PR = \{pr_1, ...pr_t\}$ is the set of the available peers then a function $MAP : N \rightarrow PR$ must be defined, where $MAP(n_i) = pr_j$ means that the peer pr_j hosts the node n_i. In order to reuse the well-know searching algorithms all the links of the BT must be preserved in the DBT.

© Springer International Publishing Switzerland 2016
A.-C. Ngonga Ngomo and P. Křemen (Eds.): KESW 2016, CCIS 649, pp. 257–271, 2016.
DOI: 10.1007/978-3-319-45880-9_20

This fact ensures that the elaboration of search algorithms with a DBT will visit the same nodes, in the same order with respect the elaboration with a BT. During the elaboration of these algorithms with a DBT, if n_i is current node and n_j is the next visited node, the main problem is that n_i and n_j may be on different peers, that is, $MAP(n_i) = pr_i$ and $MAP(n_j) = pr_j$ with $pr_i \neq pr_j$. To cope with this problem an update to search algorithms needs. In particular, the peer pr_i will delegate to pr_j the remaining part of search. To do this, pr_i will send a message containing all necessary information to pr_j. If the original search algorithm requires a feedback from the node n_j to the node n_i, after its own execution pr_j will send a message to pr_i with the result. The amount of information needed to implement this message passing mechanism is $O(1)$ in space, in fact, each node n of a BT has at most three neighbors (the parent, the left child and the right child) and for each of them n must store their peer.

The *Distributed k-Nearest Neighbor* algorithm (DKNN) quickly described is very similar to well-known *k-Nearest Neighbor algorithm* (KNN) and it is listed in Sect. 2.

The time complexity of KNN and DKNN algorithms is $O(\log N)$, where N is the number of nodes of the tree, both of them visit the same nodes in the same order. In addition, the DKNN algorithm must exchange a number of messages (hop) between peers. The number of hops is less than the number of nodes visited during the execution of the KNN algorithm because a message can be sent only when the KNN algorithm moves from one node to another one. Therefore, the number of hops is $O(\log N)$ and the DKNN is an efficient search algorithm. Suppose you have a distributed-BT obtained by (a) distributing the nodes using a map between tree nodes and peers and (b) changing the search algorithm as described. Such a distributed-BT achieves the objectives (1) and (2) listed in the preceding numbered list but this approach has some limitations.

Example. The following example illustrates the elaboration of a query with DBT and shows the limits of this approach. Suppose T is a DBT and q_1, q_2, q_3 and q_4 are queries. If *root* is the node root of T then the peer $pr_1 = MAP(root)$ at the beginning has four message in its queue (one for each query). The peer pr_1 starts the elaboration of the message for q_1 and cannot elaborate the message for q_2 until (i) the query q_1 ends or (ii) the execution of q_1 requires a node belonging to another peer pr_2. In the second case, pr_1 sends a message containing q_1 and the current result (possibly empty) to pr_2. From now on, pr_2 carries out q_1 and pr_1 starts the execution of q_2. When pr_2 ends its elaboration it sends a message, named for example $q1\_Result\_Msg$, to pr_1 containing the final result of q_1. The message $q1\_Result\_Msg$ enters the message queue of pr_1 and without any priority associated to $q1\_Result\_Msg$ there is no guarantee that it will be processed before the execution of the remaining query q_3 and q_4.

From the previous example it is possible to make some general remarks: the peer pr_1 is the bottleneck of the entire system and without an accurate message priority management the throughput of a DBT can be worse than a BT. This behavior does not depend on the distribution strategy (i.e. the mapping function MAP cannot solve this problem). This issues represent a substantial limit for DBT.

Hence, there is the need for a new distributed search algorithm that can start a query from any randomly chosen node/peer.

Binary trees, kd-trees and other tree based data structures can support knowledge management in many different ways. A rather simple and effective approach is the following. Suppose there is a set of object $C = \{c_1, ..., c_n\}$, for instance C can be the English words or the concepts of a domain ontology. With a little effort a metric M (or distance function) can be defined or adopted. A distance function is a function that defines a distance between each pair of elements of C, it measures how two objects are similar to each other.

For instance, for English words there are a set of well-known similarity measures (based on the lexical database Wordnet) that are also a measure of semantic relatedness: Pedersen et al. [1]. Instead, for the concepts of a domain ontology the distance function can be a custom function that reflects the idea of similarity of ontology developers. The set C with the metric M is a *metric space* and through the well-known mapping algorithms (MDS [2], FastMap [3]) a metric space can be easily mapped on a *vector space*. Now, a tree based data structure and be used to index the set C and their search algorithms can be used to retrieve its elements. For example, given a word or a concept, the k-nearest neighbor (i.e. the k-similar) or the nearest elements within a predefined range can be retrieved in very efficient way.

2   Distributed k-Nearest Neighbor with DBT

Let $q = (p, k)$ be a query on a BT with a query point p and an integer k and $N = \{n_1, ..., n_h\}$ the set of nodes.

Suppose $PR = \{pr_1, ..., pr_t\}$ is the set of peers and each peer:

- executes the same algorithm;
- can send a message M to any other peer pr calling $SendMessage(pr, M)$;
- has a priority queue containing the messages;
- after the startup waits for a message and on receiving a message it calls the function $onReceivingMessage$.

Also, suppose a mapping function $MAP : N \to PR$ between nodes and peer is defined. That is, $MAP(n) = pr$ means that the node n is allocated on peer pr.

Furthermore each node $n \in N$:

- has a structure *Status* containing the fields $\langle q, value \rangle$, where $q = (p, k)$ is a running query and *statusValue* is the status of q in n, $statusValue \in S = \{none, allVisited, rightVisited, leftVisited\}$. *Status* contains a distinct pair $\langle q, statusValue \rangle$ for each query q that traversed n. The methods $n.getStatus(q)$ and $n.setStatus(q, statusValue)$ respectively gets and sets the status value of q;
- knows the peer pr that hosts it;
- if n is a leaf, has a bucket of points and $n.points$ gets the points stored in it;

– has a value $n.SplitValue$ that determines which subtree of n must be visited.

Define *Results* as a data structure type that implements a priority queue having at top the farthest point from the query point p and it contains at most k points. In order to create a new instance of a *Results* the query point p and the size k of the queue must be specified. On an instance of a *Results* the *add* point operation can be performed. After the new point has been added to queue if the size of the queue exceeds k then the point at the top of the queue is deleted. *Results* has a flag that indicates if it is full and the operation *getFarthest* returns the top of the queue.

Each message carries the fields $\langle q, n, results, statusValue, priority \rangle$, where $q = (p, k)$ is the current query, $n \in N$, *results* is an instance of *Results*, $statusValue \in S$ and *priority* is the priority value of the message.

Assume that it is possible to access to the information in the message M with dot notation. For example, the query point p of the current query $q = (p, k)$ contained in M is $M.q.p$.

On receiving the message M, the peer containing the root of the tree sets the priority initial value to 1, creates a new instance *results* of *Results* for query point $q.p$ with size $q.k$ and starts the distributed search calling:

$$DKNN(M.q, M.n, M.results, M.status, 1) \tag{1}$$

The Distributed K-Nearest Neighbor algorithm DKNN checks if the current peer contains the node n. If the current peer contains n then set the status value of q in n and continues the elaboration calling DNN otherwise the current peer sends a message to the peer containing n and delegates to it the rest of the elaboration. Essentially the algorithm DNN alternates descending and ascending steps. During a descending step, if n is visited for the first time (i.e. the status value of q is *none*) then the algorithm compare the query point with the *splitValue* of n in order to decide which subtree of n must be visited. Otherwise, if n is visited for the second time (i.e. the status value of q is *leftVisited* or *rightVisited*) then the algorithm checks if the other subtree of n must be visited. When DNN reaches a leaf n it performs an ascending step returning to the parent of it. If the parent of n is NIL then n is the root of the tree end the elaboration ends.

Each node n has a distinct *statusValue* for each q because more than one query could be executed simultaneously. In fact, a new q_1 can start even if the previous query q is still running. Even if a single peer can execute one nearest neighbors at time, it can be idle because it is waiting for the elaboration of q in its subtree contained in another peer. In the meantime, the same peer can receive a new message for q_1 and then it starts the elaboration of q_1.

Please, note that the older messages the higher the priority is. Let r the root of the BT and suppose that $pr = MAP(r)$. The k-nearest neighbor search for query point p starts with the message $M \leftarrow \langle q, n, NIL,' none' \rangle$:

$$SendMessage(pr, M) \tag{2}$$

Algorithm 1. $DKNN\,(q, n, results, statusValue, priority)$

Require: q is query, n is a node of T, $results$ is an instance of $Results$, $statusValue \in$
S and $priority$ is an integer value
Ensure: return the k-nearest neighbor points of $q.p$ (the query point)
1. $pr \leftarrow MAP\,(n)$
2. **if** pr is the current peer **then**
3.   **if** $statusValue \neq NIL$ **then**
4. $n.setStatus\,(q, statusValue)$
5.   **end if**
6. $DNN\,(q, n, results, priority)$ {same peer}
7. **else**
8. $M \leftarrow \langle q, n, results, status, priority \rangle$
9. $SendMessage\,(pr, M)$ {another peer}
10. **end if**

Algorithm 2. $DNN\,(q, n, results, priority)$

Require: q is query, n is a node of T, $results$ is an instance of $Results$, $statusValue \in$
S and $priority$ is an integer value
1. **if** $n.isLeaf$ **then**
2. $results.add\,(q, n.points)$
3.   **if** $n.parent \neq NIL$ **then**
4. $DKNN\,(q, n.parent, results, NIL, priority + 1)$ {Ascend to parent}
5.   **else**
6. $G \leftarrow \langle q, NIL, results, NIL, priority \rangle$
7. $SendMessage\,(S, G)$ {Search completed. Returns $results$ of q to sender S}
8.   **end if**
9. **else**
10.   **if** $n.getStatus\,(q) =' none'$ **then**
11.   **if** $q.p < n.splitValue$ **then**
12. $n.setStatus(q, n,' leftVisited')$ {Descend to left}
13. $DKNN(q, n.left, results,' none', priority + 1)$
14.   **else**
15. $n.setStatus(q, n,' rightVisited')$ {Descend to right}
16. $DKNN(q, n.right, results,' none', priority + 1)$
17.   **end if**
18.   **end if**
19.   **if** $n.getStatus\,(q) =' leftVisited'$ **then**
20.   **if** $n.right \neq NIL$ **and** $mustBeVisited\,(n.right)$ **then**
21. $DKNN(q, n.right, results,' none', priority + 1)$
22.   **end if**
23.   **end if**
24.   **if** $n.getStatus\,(q) =' rightVisited'$ **then**
25.   **if** $n.left \neq NIL$ **and** $mustBeVisited\,(n.left)$ **then**
26. $DKNN\,(q, n.left, results,' none', priority + 1)$
27.   **end if**
28.   **end if**
29. **end if**

At the end of elaboration the data structure *results* in *pr* contains the results of the query.

The inputs of algorithm *mustBeVisited* are: a node n, an integer p (the query point) and an instance *results* of *Results*. It returns true if the subtree rooted in n must be visited, false otherwise. In particular, it returns true if:

$$distance(p, n.splitValue) < distance(p, results.getFarthest) \qquad (3)$$

as the well-known kd-tree k-nearest neighbor search algorithm [4].

3 R-DKNN: Starting a Query from Any Node Visited by DKNN Algorithm

This section introduces a new algorithm for k-nearest neighbor query, named $Random - DKNN$ (R-DKNN), that starts the search from any node visited by DKNN algorithm. For brevity, in the following description the message system is not specified but it is the same of the previous section. Assume that T is a BT and $q = (p, k)$ is a query and n is a node visited by DKNN algorithm during its elaboration. In particular, the node n is defined as *starting node* of the query. The instance *results* will contains the result of the query q after the call:

$$R - DKNN(n, q, results, 'none') \qquad (4)$$

Algorithm 3. R-DKNN(n, q, results, statusValue)

Require: n is a node of T, q is a query, *results* is an instance of *Results* and $statusValue \in S$

1. **if** $status \neq NIL$ **then**
2. $n.setStatus(q, statusValue)$
3. **end if**
4. randomDNN(n, q, results)

The differences between the algorithms 4 (randomDNN) and 2 (DNN) are the calls to the *mustBeSetParentStatus* procedure. In particular, at the end of each recursive call, the Algorithm 4 (randomDNN) checks the status of the parent node m of the current node n and if m has never been visited it sets its status calling *mustBeSetParentStatus* (Algorithm 5).

The algorithm *mustBeSetParentStatus* sets the *statusValue* of q in node m to *leftVisited* if n is the left child of m and *rightVisted* otherwise. At this point, the node m is elaborated by a call to R-DKNN. Without this change the Algorithm 4 would stop its elaboration because there is not a pending recursive call to R-DKNN and it would return an incorrect result.

Algorithm 4. randomDNN(n, q, results)

Require: n is a node of T, q is a query and *results* is an instance of *Results*
1. **if** $n.isLeaf$ **then**
2. $results.add(n.points)$
3.   **if** mustBeSetParentStatus(n, q) **then**
4.   R-DKNN(n.parent, q, results, NIL)
5.   **end if**
6. **else**
7.   **if** $n.getStatus(q) =' none'$ **and not** $n.isLeaf$ **then**
8.   **if** $p < n.SplitValue$ **then**
9. $n.setStatus(q,' leftVisited')$
10.   R-DKNN(n.left, q, results, 'none')
11.   **else**
12. $n.setStatus(q,' rightVisited')$
13.   R-DKNN(n.right, q, results, 'none')
14.   **end if**
15.   **end if**
16.   **if** $n.getStatus(q) =' rightVisited'$ **then**
17.   **if** $n.left \neq NIL$ **and** mustBeVisited(n, p, results) **then**
18.   R-DKNN(n.left, q, results, 'none')
19.   **end if**
20.   **if** mustBeSetParentStatus(n, q) **then**
21.   R-DKNN(n.parent, q, results, NIL)
22.   **end if**
23.   **end if**
24.   **if** $n.getStatus(q) =' leftVisited'$ **then**
25.   **if** $n.right \neq NIL$ **and** mustBeVisited(n, p, results) **then**
26.   R-DKNN(n.right, q, results, 'none')
27.   **end if**
28.   **if** mustBeSetParentStatus(n, q) **then**
29.   R-DKNN(n.parent, q, results, NIL)
30.   **end if**
31.   **end if**
32. **end if**

---

4 Finding a Starting Node

The Algorithm 3 (R-DKNN) works only if it starts from a node that is visited by the Algorithm 1 (DKNN) in Sect. 2. The following property helps to characterize this kind of nodes:

Theorem 1. *Starting Node Property (SNP).Let T be a BT and $M = \{m_1, ..., m_j\}$ the set of nodes visited at least once by the Algorithm 1 (DKNN) in Sect. 2 during the of elaboration of the query $q = (p, k)$. If for a node m holds:*

$$\begin{cases} m.minValue \leq p \leq m.maxValue & if\ m\ is\ a\ sleaf \\ m.left.splitValue \leq p \leq m.right.splitValue & otherwise \end{cases} \quad (5)$$

Algorithm 5. mustBeSetParentStatus(n, q)

Require: n is a node of T and q is a query
Ensure: Checks if the *status* of q in the parent of n must be set and it sets the correct value.

1. **if** $n.parent \neq NIL$ **then**
2.   **if** $n.parent.getStatus(q) = NIL$ **then**
3.   **if** $n.splitValue < n.getParent.splitValue$ **then**
4. $n.parent.setStatus(q, 'leftVisited')$ $\{n$ is a left child of $n.parent\}$
5.   **return true**
6.   **else**
7. $n.parent.setStatus(q, 'rightVisited')$ $\{n$ is a right child of $n.parent\}$
8.   **return true**
9.   **end if**
10.   **else**
11.   **return false**$\{ n.parent.getStatus(q) \neq NIL$ then do nothing$\}$
12.   **end if**
13. **else**
14.   **return false**$\{n$ is the root of the tree then do nothing$\}$
15. **end if**

Then $m \in M$ (*m.minValue* and *m.maxValue* are respectively the minimum and maximum values contained in the bucket of the leaf m).

Proof. It is a proof by contradiction. Let x be an internal node of the tree for which the (5) holds but such that $x \notin M$. Because (5) holds then in the bucket of the leaves of the subtree rooted in x there might be at least a point t that may be returned in the result of the query q. The value k determines whether the point t will be part of the result. If the algorithm does not visit the node x would not have the opportunity to evaluate whether to add t to the query result and then the search result may be incorrect. This is a contradiction of the correctness of standard search algorithm then $x \in M$. The proof in the case x is a leaf is the same.

The SNP in (5) can be used to build a recursive algorithm that, given a query, point p try to reach a node $m \in M$ starting from a random node n of T.

In particular, the algorithm $findStartingNode(p, n)$ (Algorithm 6) returns the node n if the SNP holds for it otherwise it moves to the father of n. It moves from bottom to top in the tree therefore its time complexity is $O(\log N)$. Please, note that for given the query $q = (p, k)$ the SNP depends only on the query point p and it do not depend on the value of k. Instead, the value k affects the probability that the root will be visited during the elaboration of the query: the larger k the larger the probability the root will involved. Furthermore, note also that the (5) is a sufficient but not necessary condition.

Algorithm 6. findStartingNode(p, n)

Require: p is the query point (it is an integer value) and n is a node of T
Ensure: return a starting node $m \in M$ for query point p
1. **if** $n.isRoot$ **then**
2.   **return** n
3. **else**
4.   **if** $n.isLeaf$ **then**
5.   **if** $n.minValue \leq p \leq n.maxValue$ **then**
6.   **return** n
7.   **else**
8.   **return** $findStartingNode(p, n.getParent)$
9.   **end if**
10.   **else**
11.   **if** $n.left.splitValue \leq p \leq n.right.splitValue$ **then**
12.   **return** n
13.   **else**
14.   **return** $findStartingNode(p, n.getParent)$
15.   **end if**
16.   **end if**
17. **end if**

---

5   Analysis of the *findStartingNode* Algorithm

The *findStartingNode* algorithm moves to the parent of the current node if the SNP does not hold and, of course, there is no guarantee that it has not reached the root of the tree. It is not very simple to determine if the algorithm will return the root, it depends on the value of p, the random node and the size of the tree. Please note that the higher the size of the bucket the lower the number of the leaves of the tree is. On the other hand, the higher the size of the bucket the worse the global performance of the search algorithms is because all points in the bucket will be inserted in the temporary results. For these reasons, an average on all possible choices of p over a set of trees with increasing number of nodes and increasing size of bucket was made.

Let T be a tree and p the query point, in order to estimate how many times in average the *findStartingNode* returns the root of T the test listed in Algorithm 7 (*testFindStartingNode*) can be executed.

In other words, the algorithm elaborates all nodes in the tree but the root. At the end, *percRoot* is the percentage in average of how many times the algorithm returns the root node. Of course, $percNoRoot = 100 - percRoot$. Now, suppose that $treeSet = \{T_{256}, T_{512}, T_{1.024}, T_{2.048}, T_{4.096}, T_{8.192}, T_{16.384}, T_{32.768}\}$ is a set of binary trees where the tree T_i contains points from 0 to $i - 1$. A test that calculates the same percentage of previous test over all the trees in *treeSet* varying the query point p is the Algorithm 8 (*testAverageFindStartingNode*).

At the end, the *avgPercRoot* variable will contains percentage of how many times on average the root node on each tree in *treeSet* is returned by the algorithm *findStartingNode* regardless of the query point.

Algorithm 7. testFindStartingNode(T, p, percRoot, percNoRoot)

Require: T is BT, p is an integer value, $percRoot$ and $percNoRoot$ are double values
Ensure: Calculate how many times in percentage the $findStartingNode$ returns the root node with the tree T and query point p. It returns $percRoot$ and $percNoRoot$
1. $nrRoot \leftarrow 0$
2. $nrNoRoot \leftarrow 0$
3. **for all** $n \in T.allNodes$ **do**
4.   **if** $n \neq T.root$ **then**
5. $resultNode \leftarrow findStartingNode\,(p, n)$
6.   **if** $resultNode = T.root$ **then**
7. $nrRoot \leftarrow nrRoot + 1$
8.   **else**
9. $nrNoRoot \leftarrow nrNoRoot + 1$
10.   **end if**
11.   **end if**
12. **end for**
13. $percRoot \leftarrow 100 * nrRoot/(T.allNodes.size - 1)$ {do not count the root}
14. $percNoRoot \leftarrow 100 * nrNoRoot/(T.allNodes.size - 1)$

Algorithm 8. $testAverageFindStartingNode\,()$

Ensure: Calculate how many times in percentage the $findStartingNode$ returns the root node with the trees in $treeSet$
1. $avgPercRoot \leftarrow 0$
2. $avgPercNoRoot \leftarrow 0$
3. **for all** $T \in treeSet$ **do**
4. $queryPoint \leftarrow 0$
5.   **for** $queryPoint < T.numPoints$ **do**
6.   testFindStartingNode(T, queryPoint, percRoot, percNoRoot)
7. $sumPercRoot \leftarrow sumPercRoot + percRoot$
8. $sumPercNoRoot \leftarrow sumPercNoRoot + percNoRoot$
9. $queryPoint \leftarrow queryPoint + 1$
10.   **end for**
11. $avgPercRoot \leftarrow sumPercRoot/T.numPoints$
12. $avgPercNoRoot \leftarrow sumPercNoRoot/T.numPoints$
13. **end for**

Of course, $avgPercNoRoot = 100 - avgPercRoot$. The results of tests with bucket size of 5, 10, 20, 30 and 40 points shows that if the bucket size is much smaller than the number of points the results do not depend on either the bucket size nor the number of points and in about 65.5 % of runs it returns the root node. When the bucket size approaches the number of points, the results should not be considered significant because the trees that result would have very few nodes, e.g. for 64 points and a bucket dimension of 40 points it results a tree with only 3 nodes.

6 Improving the $findStartingNode$ Algorithm

Intuitively, the algorithm $findStartingNode$ can be improved choosing a random node in the left subtree of the root r of T if $p \leq r.splitValue$
Algorithm 9 lists the $findStartingNodeSide$ algorithm.

Algorithm 9. $findStartingNodeSide\,(queryPoint)$

Require: $queryPoint$ is an integer value
Ensure: return a starting node for query point p
1. **if** $queryPoint < root.splitValue$ **then**
2.   {let $randomNode$ be a randomly chosen node in left subtree of the root of T}
3. **else**
4.   {let $randomNode$ be a randomly chosen node in right subtree of the root of T}
5. **end if**
6. $startNode \leftarrow findStartingNode\,(queryPoint, randomNode)$
7. **return** $startNode$

Each node of the DBT can easily be labeled with $left$ if it belongs to the left subtree of T and $right$ otherwise. During construction of the DBT, every new node inherits the label of the father.

Results shows that in about 65.3 % of runs the $findStartingNodeSide$ algorithm does not returns the root node of T. Therefore, about 65 % of the queries does not start in the root of the tree.

7 The MAP Function

This section report some observation about the role of the MAP function. The MAP function does not say enough about the way of distributing the BT on the network peers. In fact, the MAP function may be available only after the BT was distributed on the peers and for this reason it provides no indication on the distribution strategy.

Conversely, the MAP function might also be defined in advance and then it shall be the rule with which to move the nodes to peers: MAP itself is the distribution strategy.

This work does not deal with the issues relating to distribution strategy of BT and the definition of MAP function. Anyway, the MAP is used in the algorithm in Sect. 6 ($findStartingNodeSide$) after that a left (or right) node $randomNode$ is randomly chosen to obtain the peer that hosts $randomNode$ in order to send to that peer the message that starts the query.

Please, note that the lists of left and right nodes must be maintained. If the MAP function assures that each peer hosts only left nodes (or right nodes) then the respective lists of left and right peers can be maintained. Of course, the peer list has less items than the node list. Instead, if the MAP function assures that each peer has at least one left node and at least one right node than each peer can starts a query. In this case, the only information that should be passed to the peer is whether to start from a right or a left node and the lists are no longer needed.

8   Equivalence of R-DKNN and DKNN Algorithms

Suppose $q = (p, k)$ is a query and define $Leaves(A, q)$, $Nodes(A, q)$ and $Results$ (A, q) as the unordered sets of leaves, nodes elaborated by the algorithm A with query q and the result (the k-nearest points) of A with query q. Because are unordered sets assume that two sets are equivalent if and only if they contain the same nodes regardless the order the nodes are listed, e.g. $Leaves(DKNN, q) = Leaves(R - DKNN, q)$ if and only if they contain the same nodes regardless the order. The objective is to demonstrate that:

Theorem 2. *if R-DKNN starts in* $n \in Nodes(DKNN, q) \Rightarrow$ *Results(DKNN, q) = Results(R-DKNN, q).*

The following theorems will help to demonstrate Theorem 2.

Theorem 3. *If Leaves(DKNN, q) = Leaves(R-DKNN, q)* $\Rightarrow$ *Results(DKNN, q) = Results(R-DKNN, q).*

Proof. The *Results* data structure described in Sect. 2 is a queue that stores the k-nearest neighbor points of the point p regardless the order the leaves are processed.

Theorem 4. *If R-DKNN starts in a node* $n \in Nodes(DKNN, q) \Rightarrow$ $Nodes(DKNN, q) = Nodes(R - DKNN, q).$

Before the proof, observe that, during the elaboration of the DKNN algorithm, a node of the tree can be visited zero, one, two or three times. Only leaves are visited exactly zero or one times. Root node is visited one time if it has not children. Internal nodes are visited exactly two times if only one child is visited. Also, the root is visited two times if it has at least a child. Finally, internal nodes visited three times are those ones for which DKNN visit both children.

Proof. Suppose that n is an internal node, i.e. $n \notin Leaves(DKNN, q)$ and n is not the root (if R-DKNN starts in the root then the proof is trivial). If DKNN visit only one or both of the children of n then R-DKNN do the same because DKNN and R-DKNN checks the same conditions in node n (as stated in Sect. 3). The difference between the elaboration of DKNN and R-DKNN could be the order they visit the children of n but this do not affect the result. After, both DKNN and R-DKNN move its own elaboration to the parent node m of n. Now, the difference between DKNN and R-DKNN could be the status the of the node m (remember that R-DKNN sets the status of m calling $mustBeSetParentStatus$). If R-DKNN has never visited m then it will check if the other child of m must be visited. If DKNN already visited m and the other child of m also, then DKNN will move to the parent of m. R-DKNN will visit the other child of m because DKNN did it and because R-DKNN will check the same condition of DKNN. Therefore, R-DKNN will visit the children of m in reverse order with respect to DKNN but this not affect the result. At this point R-DKNN will move to the parent of m as DKNN. The same reasoning can be applied to the parent of m and so on up to the root.

Therefore, if n is an internal node then DKNN and R-DKNN visit the same nodes.

If n is a leaf then both DKNN and R-DKNN in the next step will move the elaboration to the parent node of n and the proof will proceeds as in the previous case.

Finally, observing that if $Node(DKNN, q) = Nodes(R - DKNN, q)$ then also $Leaves(DKNN, q) = Leaves(R-DKNN, q)$ the proof of Theorem 2 follows the Theorems 3 and 4.

Please note that if R-DKNN starts in a node $n \notin Nodes(DKNN, q)$ then there is no guarantee that during the elaboration any node in $Nodes(DKNN, q)$ will be processed. In this case, the elaboration of R-DKNN can reach the root with the incorrect result.

9 Decentralized Data Structures

In the last decade multi-dimensional and high-dimensional indexing in decentralized peer-to-peer (P2P) networks, received extensive research attention. Naturally, most such methods are tree-based and the data space is hierarchically divided into smaller subspaces (regions), such that the higher level data subspace contains the lower level subspaces and acts a guide in searching. These methods can be data-partitioning based, where data subspaces are allowed to overlap (eg. R-tree) or space-partitioning based, where data subspaces are disjoint (eg. kd-tree) and they can be classified into two broad categories: *tree-based* and *DHTs-based.*

TerraDir [5] is a tree-based structured P2P system. It organizes nodes in a hierarchical fashion according to the underlying data hierarchy. Each query request will be forwarded upwards repeatedly until reaching the node with the longest matching prefix of the query. Then the query is forward to the destination downwards the tree. In TerraDir, each node maintains constant number of neighbors and routing hops are bounded in $O(h)$, where h is the height of the tree. In [6] Mohamed et al. proposed a distributed kd-tree based on MapReduce framework [7]. In such index structures queries are processed similar to the centralized approach, i.e., the query starts in root node and traverse the tree. These methods exhibit logarithmic search cost, but face a serious limitation. Peers that correspond to nodes high in the tree can quickly become overloaded as query processing must pass through them. In centralized indices this was a desirable property because maintaining these nodes in main memory allow the minimization of the number of I/O operations. In distributed indices it is a limiting factor leading to bottlenecks. Moreover, this causes an imbalance in fault tolerance: if a peer high in the tree fails than the system requires a significant amount of effort to recover. The VBI-tree proposed in [8] provided a solution to the bottlenecks and imbalance problems introducing a distributed framework (inspired to BATON – Balanced Tree Overlay Network [9]) based on multidimensional tree structured overlays, e.g., R-tree. It provides an abstract tree structure on top of an overlay network that supports any kind of hierarchical tree indexing structures. Furthermore, it can cause unfairness as peers corresponding to nodes high in the tree are heavily hit.

Approaches based on distributed hash tables (DHTs) employ a globally consistent protocol to ensure that any peer can efficiently route a search to the peer that has the desired content, regardless of how rare it is or where it is located. A DHT system provides a lookup service similar to a hash table; (key, value) pairs are stored in a DHT, and any participating node can efficiently retrieve the value associated with a given key. Responsibility for maintaining the mapping from keys to values is distributed among the nodes, in such a way that a change in the set of participants causes a minimal amount of disruption. This allows a DHT to scale to extremely large numbers of nodes and to handle continual node arrivals, departures, and failures. DHT systems include Chord [10], Tapestry [11], Pastry [12], CAN [13] and Koorde [14]. The routing algorithms used in Tapestry and Pastry are both inspired by Plaxton [15]. The idea of the Plaxton algorithm is to find a neighboring node that shares the longest prefix with the key in lookup message, repeat this operation until find a destination node that shares the longest possible prefix with the key. In Tapestry and Pastry, each node has $O(logN)$ neighbors and the routing path takes at most $O(logN)$ hops. MIDAS [16] is similar to these works and in particular, MIDAS implements a distributed kd-tree, where leaves correspond to peers, and internal nodes dictate message routing. The proposed algorithms process point and range queries over the multidimensional indexed space in $O(logn)$ hops in expectance. Two algorithms for Nearest Neighbor Queries are described: the first (expected $O(logn)$) has low latency and involve a large number of peers; the second (expected $O(log^2n)$) has higher latency but involves far fewer peers.

References

1. Pedersen, T., Patwardhan, S., Michelizzi, J.: WordNet: similarity: measuring the relatedness of concepts. In: Demonstration Papers at HLT-NAACL 2004, pp. 38–41. Association for Computational Linguistics, May 2004
2. Cox, T.F., Cox, M.A.: Multidimensional Scaling. CRC Press, Boca Raton (2000)
3. Faloutsos, C., Lin, K.I.: FastMap: a fast algorithm for indexing, data-mining and visualization of traditional and multimedia datasets. ACM SIGMOD Rec. **24**(2), 163–174 (1995)
4. Samet, H.: Foundations of Multidimensional and Metric Data Structures. Morgan Kaufmann, San Francisco (2006)

5. Silaghi, B., Bhattacharjee, S., Keleher, P.J.: Query routing in the TerraDir distributed directory. In: ITCom 2002: The Convergence of Information Technologies and Communications, pp. 299–309. International Society for Optics and Photonics, July 2002
6. Aly, M., Munich, M., Perona, P.: Distributed kd-trees for retrieval from very large image collections. In: Proceedings of the British Machine Vision Conference (BMVC), August 2011
7. Dean, J., Ghemawat, S.: MapReduce: simplified data processing on large clusters. Commun. ACM **51**(1), 107–113 (2008)
8. Jagadish, H.V., Ooi, B.C., Vu, Q.H., Zhang, R., Zhou, A.: Vbi-tree: a peer-to-peer framework for supporting multi-dimensional indexing schemes. In: 2006 Proceedings of the 22nd International Conference on Data Engineering ICDE 2006, pp. 34–34. IEEE, April 2006
9. Jagadish, H.V., Ooi, B.C., Vu, Q.H.: Baton: a balanced tree structure for peer-to-peer networks. In: Proceedings of the 31st International Conference on Very Large Data Bases, pp. 661–672. VLDB Endowment, August 2005
10. Balakrishnan, H., Kaashoek, M.F., Karger, D., Morris, R., Stoica, I.: Looking up data in P2P systems. Commun. ACM **46**(2), 43–48 (2003)
11. Zhao, B.Y., Huang, L., Stribling, J., Rhea, S.C., Joseph, A.D., Kubiatowicz, J.D.: Tapestry: a resilient global-scale overlay for service deployment. IEEE J. Sel. Areas Commun. **22**(1), 41–53 (2004)
12. Rowstron, A., Druschel, P.: Pastry: scalable, decentralized object location, and routing for large-scale peer-to-peer systems. In: Guerraoui, R. (ed.) Middleware 2001. LNCS, vol. 2218, pp. 329–350. Springer, Heidelberg (2001)
13. Ratnasamy, S., Francis, P., Handley, M., Karp, R., Shenker, S.: A scalable content-addressable network. ACM SIGCOMM Comput. Commun. Rev. **31**(4), 161–172 (2001)
14. Kaashoek, M.F., Karger, D.R.: Koorde: a simple degree-optimal distributed hash table. In: Kaashoek, M.F., Stoica, I. (eds.) IPTPS 2003. LNCS, vol. 2735, pp. 98–107. Springer, Heidelberg (2003)
15. Plaxton, C.G., Rajaraman, R., Richa, A.W.: Accessing nearby copies of replicated objects in a distributed environment. Theor. Comput. Syst. **32**(3), 241–280 (1999)
16. Tsatsanifos, G., Sacharidis, D., Sellis, T.: Index-based query processing on distributed multidimensional data. GeoInformatica **17**(3), 489–519 (2013)

Ethnicity Sensitive Author Disambiguation Using Semi-supervised Learning

Gilles Louppe[1], Hussein T. Al-Natsheh[2], Mateusz Susik[3(✉)],
and Eamonn James Maguire[1]

[1] CERN, Geneva, Switzerland
{g.louppe,eamonn.james.maguire}@cern.ch
[2] Univ Lyon, ISH, USR 3385, CNRS, and ERIC Lab, EA 3083, Lyon, France
hussein.al-natsheh@ish-lyon.cnrs.fr
[3] University of Warsaw, Warszawa, Poland
msusik@student.uw.edu.pl

Abstract. Building on more than one million crowdsourced annotations that we publicly release, we propose a new automated disambiguation solution exploiting this data (i) to learn an accurate classifier for identifying coreferring authors and (ii) to guide the clustering of scientific publications by distinct authors in a semi-supervised way. To the best of our knowledge, our analysis is the first to be carried out on data of this size and coverage. With respect to the state of the art, we validate the general pipeline used in most existing solutions, and improve by: (i) proposing new phonetic-based blocking strategies, thereby increasing recall; (ii) adding strong ethnicity-sensitive features for learning a linkage function, thereby tailoring disambiguation to non-Western author names whenever necessary; and (iii) showing the importance of balancing negative and positive examples when learning the linkage function.

1 Introduction

In academic digital libraries, author name disambiguation is the problem of grouping together publications written by the same person. It is often difficult because an author may use different spellings or name variants across their career (synonymy) and/or distinct authors may share the same name (polysemy). Most notably, author disambiguation is often more troublesome for researchers from non-Western cultures, where personal names may be traditionally less diverse (leading to homonym issues) or for which transliteration to Latin characters may not be unique (leading to synonym issues). With the fast growth of the scientific literature, author disambiguation has become a pressing issue since the accuracy of information managed at the level of individuals directly affects: the relevance search of results (e.g., when querying for all publications written by a given author); the reliability of bibliometrics and author rankings (e.g., citation counts or other impact metrics, as studied in [28]); and/or the relevance of scientific network analysis [21]. Thus, even small improvements in the field significantly improve the usability of the digital libraries to some users.

© Springer International Publishing Switzerland 2016
A.-C. Ngonga Ngomo and P. Křemen (Eds.): KESW 2016, CCIS 649, pp. 272–287, 2016.
DOI: 10.1007/978-3-319-45880-9_21

Solutions to author disambiguation have been proposed from various communities [18]. On the one hand, libraries have maintained authorship control through manual curation, either in a centralized way by hiring professional collaborators or through developing services that invite authors to register their publications themselves (e.g., Google Scholar or Inspire-HEP). Recent efforts to create persistent digital identifiers assigned to researchers (e.g., ORCID or ResearcherID), with the objective to embed these identifiers in the submission workflow of publishers or repositories (e.g., Elsevier or arXiv), would univocally solve any disambiguation issue. As the centralized manual authorship control is expensive and the success of persistent digital identifiers requires large and ubiquitous adoption by both researchers and publishers, fully automated machine learning-based methods have been proposed to provide immediate, less costly, and satisfactory solutions to author disambiguation. In this work, we study how labeled data obtained through manual curation (either centralized or crowdsourced) can be exploited (i) to learn an accurate classifier for identifying coreferring authors, and (ii) to guide the clustering of scientific publications by distinct authors in a semi-supervised way. Our analysis of parameters and features of this large dataset reveal that the general pipeline commonly used in existing solutions is an effective approach for author disambiguation. Moreover, we propose better strategies for blocking (i.e., partitioning) based on the phonetization of author names to increase recall and ethnicity-sensitive features for learning a linkage function which tailor our author disambiguation to non-Western author names.

The remainder of this report is structured as follows. In Sect. 2, we briefly review machine learning solutions for author disambiguation. The components of our method are then defined in Sect. 3 and its implementation described in Sect. 4. Experiments are carried out in Sect. 5, where we compare approaches to the problem and explore feature choice. Finally, conclusions and future works are discussed in Sect. 6.

2 Related Work

As reviewed in [7,17,26], author disambiguation algorithms are usually composed of two main components: (i) a linkage function determining whether two publications have been written by the same author; and (ii) a clustering algorithm producing clusters of publications assumed to be written by the same author. Approaches can be classified along several axes, depending on the type and amount of data available, the way the linkage function is learned or defined, or the clustering procedure used to group publications. Methods relying on supervised learning usually make use of a small set of hand-labeled pairs of publications identified as being either from the same or different authors to automatically learn a linkage function between publications [4,12,13,32,33].

Training data is usually not easily available, therefore unsupervised approaches propose the use of domain-specific, manually designed, linkage functions tailored towards author disambiguation [14,20,25,27]. These approaches have the advantage of not requiring hand-labeled data, but generally do not perform as well as supervised approaches. To reconcile both worlds, semi-supervised methods make use of small, manually verified clusters of publications and/or high-precision domain-specific rules to build a training set of pairs of publications, from which a linkage function is then built using supervised learning [8,17,31]. Semi-supervised approaches also allow for the tuning of the clustering algorithm when the latter is applied to a mixed set of labeled and unlabeled publications, e.g., by maximizing some clustering performance metric on the known clusters [17].

In this context, we position this work as a semi-supervised solution for author disambiguation, with the significant advantage of having a very large collection of more than 1 million crowdsourced annotations of publications whose true authors are identified. The extent and coverage of this data allows us to revisit, validate and nuance previous findings regarding supervised learning of linkage functions, and to better explore strategies for semi-supervised clustering. Furthermore, by releasing our data in the public domain, we provide a benchmark on which further research on author disambiguation and related topics can be evaluated.

3   Semi-supervised Author Disambiguation

Formally, let us assume a set of publications $\mathcal{P} = \{p_0, ..., p_{N-1}\}$ along with the set of unique individuals $\mathcal{A} = \{a_0, ..., a_{M-1}\}$ having together authored all publications in $\mathcal{P}$. Let us define a signature $s \in p$ from a publication as a unique piece of information identifying one of the authors of p (e.g., the author name, his affiliation, along with any other metadata that can be derived from p, as illustrated in Fig. 1). Let us denote by $\mathcal{S} = \{s|s \in p, p \in \mathcal{P}\}$ the set of all signatures that can be extracted from all publications in $\mathcal{P}$.

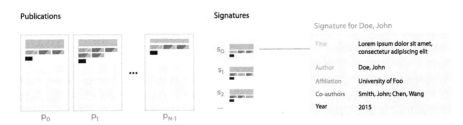

Fig. 1. An example signature s for "Doe, John". A *signature* is defined as unique piece of information identifying an author on a publication, along with any other metadata that can be derived from it, such as publication title, co-authors or date of publication.

Author disambiguation can be stated as the problem of finding a partition $\mathcal{C} = \{c_0, ..., c_{M-1}\}$ of $\mathcal{S}$ such that $\mathcal{S} = \cup_{i=0}^{M-1} c_i$, $c_i \cap c_j = \phi$ for all $i \neq j$, and where subsets c_i, or clusters, each corresponds to the set of all signatures belonging to the same individual a_i. Alternatively, the set $\mathcal{A}$ may remain (possibly partially) unknown, such that author disambiguation boils down to finding a partition

$\mathcal{C}$ where subsets c_i each correspond to the set of all signatures from the same individual (without knowing who). Finally, in the case of partially annotated databases as studied in this work, the set extends with the partial knowledge $\mathcal{C}' = \{c'_0, ..., c'_{M-1}\}$ of $\mathcal{C}$, such that $c'_i \subseteq c_i$, where c'_i may be empty.

The distinctive aspect of our work is the knowledge of more than 1 million crowdsourced annotations, indicating together that all signature $s \in c'_i$ are known to correspond to the same individual a_i.

Our algorithm is composed of three parts (Fig. 2): (i) a blocking scheme whose goal is to pre-cluster signatures $\mathcal{S}$ into smaller groups; (ii) the construction of a linkage function d between signatures using supervised learning; and (iii) the semi-supervised clustering of all signatures within the same block, using d as a pseudo distance metric.

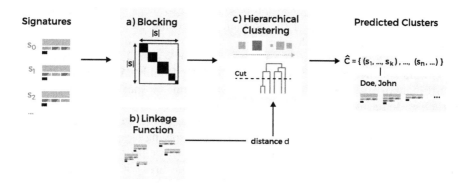

Fig. 2. Pipeline for author disambiguation: (a) signatures are *blocked* to reduce computational complexity, (b) a linkage function is built with supervised learning, (c) independently within each block, signatures are grouped using hierarchical agglomerative clustering.

3.1 Blocking

As in previous works, the first part of our algorithm consists of dividing signatures $\mathcal{S}$ into disjoint subsets $\mathcal{S}_{b_0}, ..., \mathcal{S}_{b_{K-1}}$, or *blocks* (i.e. partitions) [6], followed by carrying out author disambiguation on each one of these blocks independently. By doing so, the computational complexity of clustering (see Sect. 3.3) typically reduces from $O(|\mathcal{S}|^2)$ to $O(\sum_b |\mathcal{S}_b|^2)$. Since disambiguation is performed independently per block, a good blocking strategy should be designed such that signatures from the same author are all mapped to the same block, otherwise their correct clustering would not be possible in later stages of the workflow. As a result, blocking should be a balance between reduced complexity and maximum recall.

The simplest and most common strategy for blocking, referred to hereon in as *Surname and First Initial (SFI)*, groups signatures together if they share the same surname(s) and the same first given name initial. Despite satisfactory performance, there are several cases where this simple strategy fails to cluster related pairs of signatures together, including:

1. There are different ways of writing an author name, or signatures contain a typo (e.g., "Mueller, R." and "Muller, R.").
2. An author has multiple surnames (or a patronymic) and some signatures place the first part of the surname within the given names (e.g., "Martinez Torres, A." and "Torres, A. Martinez").
3. An author has multiple surnames and, on some signatures, only the first surname is present (e.g., "Smith-Jones, A." and "Smith, A.")
4. An author has multiple given names and they are not always all recorded (e.g., "Smith, Jack" and "Smith, A. J.")
5. An authors surname changed (e.g., due to marriage).

To account for these issues we propose instead to block signatures based on the phonetic representation of the normalized surname. Normalization involves stripping accents (e.g., "Jabłoński, L" → "Jablonski, L") and name affixes that inconsistently appear in signatures (e.g., "van der Waals, J. D." → "Waals, J. D."), while phonetization is based either on the Double Metaphone [23], the NYSIIS [29] or the Soundex [30] phonetic algorithms for mapping author names to their pronunciations. Together, these processing steps allow for grouping of most name variants of the same person in the same block with a small increase in the overall computational complexity, thereby solving case 1.

In the case of multiple surnames (cases 2 and 3), we propose to block signatures in two phases. In the first phase, all the signatures with a single surname are clustered together. Every different surname token creates a new block. In the second phase, the signatures with multiple surnames are compared with the blocks for the first and last surname. If the first surnames of an author were already used as the last given names on some of the signatures, the new signature is assigned to the block of the last surname (case 2). Otherwise, the signature is assigned to the block of the first surname (case 3). Finally, to prevent the creation of too large blocks, signatures are further divided along their first given name initial. The biggest limitation of this method is leaving the cases 4 and 5 unhandled. This method might result in blocking together signatures from many different authors with similar names.

3.2 Linkage Function

Supervised Classification. The second part of the algorithm is the automatic construction of a pair-wise linkage function between signatures for use during the clustering step which groups all signatures from the same author.

Formally, the goal is to build a function $d : \mathcal{S} \times \mathcal{S} \mapsto [0, 1]$, such that $d(s_1, s_2)$ approaches 0 if both signatures s_1 and s_2 belong to the same author, and 1 otherwise. This problem can be cast as a supervised classification task, where inputs are pairs of signatures and outputs are classes 0 (same authors), and 1 (distinct authors). In this work, we evaluate Random Forests (RF, [1]), Gradient Boosted Regression Trees (GBRT, [9]), and Logistic Regression [5] as classifiers.

Input Features. Following previous works, pairs of signatures (s_1, s_2) are first transformed to vectors $v \in \mathbb{R}^p$ by building so-called similarity profiles [33] on which supervised learning is carried out. In this work, we design and evaluate fifteen standard input features [7,17] based on the comparison of signature fields, as reported in the first half of Table 1. Noteworthy, the author metadata, both provided or derived, are far more important than the publication content itself. As an illustrative example, the *Full name* feature corresponds to the similarity between the (full) author name fields of the two signatures, as measured using as combination operator the cosine similarity between their respective (n, m)-*TF-IDF* vector representations[1].

Authors from different origins or ethnic groups are likely to be disambiguated using different strategies (e.g., pairs of signatures with French author names versus pairs of signatures with Chinese author names) [3,34]. For example, scientist coming from China might more/less often change affiliations, and this dependency, if learn't by the classifier, should improve the fit. To support our disambiguation algorithm, we added seven features to our feature set, with each evaluating the degree of belonging of both signatures to an ethnic group.

More specifically, using census data extracted from [24], we build a support vector machine classifier (using a linear kernel and one-versus-all classification scheme) for mapping the $(1, 5)$-TF-IDF representation of an author name to one of the ethnic groups, as defined in United States federal censuses. These groups are: *White, Black of African American, American Indian and Alaska Native, Asian, Native Hawaiian and Other Pacific Islander, Japanese, Chinese, Others*. Given a pair of signatures (s_1, s_2), the proposed ethnicity features are each computed as the estimated probability of s_1 belonging to the corresponding ethnic group, multiplied by the estimated probability of s_2 belonging to the same group. Each of the seven races is used to create a single new feature. In doing so, the expectation is for the linkage function to become sensitive to the actual origin of the authors depending on the values of these features. Indirectly, these features also hold discriminative power since if author names are predicted to belong to different ethnic groups, then they are also likely to correspond to distinct people.

Building a Training Set. 1 million of crowdsourced annotations (see Sect. 3) can be used to generate positive pairs $(x = (s_1, s_2), y = 0)$ for all $s_1, s_2 \in c'_i$, for all i. Similarly, negative pairs $(x = (s_1, s_2), y = 1)$ can be extracted for all $s_1 \in c'_i, s_2 \in c'_j$, for all $i \neq j$.

The most straightforward approach for building a training set on which to learn a linkage function is to sample an equal number of positive and negative pairs, as suggested above. By observing that the linkage function d will eventually be used only on pairs of signatures from the same block S_b, a further refinement for building a training set is to restrict positive and negative pairs (s_1, s_2) to only those for which s_1 and s_2 belong to the same block. In doing so, the trained classifier is forced to learn intra-block discriminative patterns rather than inter-block differences. Furthermore, as noted in [16], most signature pairs

---

[1] $(n, m) - TF\text{-}IDF$ vectors are *TF-IDF* vectors computed from $n, n + 1, ..., m$-grams.

Table 1. Input features for learning a linkage function

Feature	Combination operator
Full name	Cosine similarity of $(2,4)$-TF-IDF
Given names	Cosine similarity of $(2,4)$-TF-IDF
First given name	Jaro-Winkler distance
Second given name	Jaro-Winkler distance
Given name initial	Equality
Affiliation	Cosine similarity of $(2,4)$-TF-IDF
Co-authors	Cosine similarity of TF-IDF
Title	Cosine similarity of $(2,4)$-TF-IDF
Journal	Cosine similarity of $(2,4)$-TF-IDF
Abstract	Cosine similarity of TF-IDF
Keywords	Cosine similarity of TF-IDF
Collaborations	Cosine similarity of TF-IDF
References	Cosine similarity of TF-IDF
Subject	Cosine similarity of TF-IDF
Year difference	Absolute difference
Any ethnicity feature	Product of probabilities estimated by SVM

are non-ambiguous: if both signatures share the same author names, then they correspond to the same individual, otherwise they do not. Rather than sampling pairs uniformly at random, we propose to oversample difficult cases when building the training set (i.e., pairs of signatures with different author names corresponding to same individual, and pairs of signatures with identical author names but corresponding to distinct individuals) in order to improve the overall accuracy of the linkage function.

3.3 Semi-supervised Clustering

The last component of our author disambiguation pipeline is clustering - the process of grouping together, within a block, all signatures from the same individual (and only those). As for many other works on author disambiguation, we make use of hierarchical clustering [35] for building clusters of signatures in a bottom-up fashion. The method involves iteratively merging together the two most similar clusters until all clusters are merged together at the top of the hierarchy. Similarity between clusters is evaluated using either complete, single or average linkage, using as a pseudo-distance metric the probability that s_1 and s_2 correspond to distinct authors, as calculated from the custom linkage function d from Sect. 3.2.

To form flat clusters from the hierarchy, one must decide on a maximum distance threshold above which clusters are considered to correspond to distinct authors. Let us denote by $\mathcal{S}' = \{s | s \in c', c' \in \mathcal{C}'\}$ the set of all signatures for which partial clusters are known. Let us also denote by $\widehat{\mathcal{C}}$ the predicted clusters for all signatures in $\mathcal{S}$, and by $\widehat{\mathcal{C}}' = \{\widehat{c} \cap \mathcal{S}' | \widehat{c} \in \widehat{\mathcal{C}}\}$ the predicted clusters restricted to signatures for which partial clusters are known. From these, we evaluate the following semi-supervised cut-off strategies, as illustrated in Fig. 3:

- *No cut:* all signatures from the same block are assumed to be from the same author.
- *Global cut:* the threshold is chosen globally over all blocks, as the one maximizing some score $f(\mathcal{C}', \widehat{\mathcal{C}}')$.
- *Block cut:* the threshold is chosen locally at each block b, as the one maximizing some score $f(\mathcal{C}'_b, \widehat{\mathcal{C}}'_b)$. In case $\mathcal{C}'_b$ is empty, then all signatures from b are clustered together.

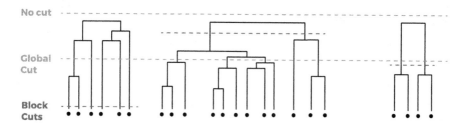

Fig. 3. Semi-supervised cut-off strategies to form flat clusters of signatures. Every dendogram represents a single block.

4   Implementation

As part of this work, we developed a stand-alone application for author disambiguation, publicly available online[2] for free reuse or study. Our implementation builds upon the Python scientific stack, making use of the Scikit-Learn library [22] for the supervised learning of a linkage function and of SciPy for clustering. All components of the disambiguation pipeline have been designed to follow the Scikit-Learn API [2], making them easy to maintain, understand and reuse. Our implementation is made to be efficient, exploiting parallelization when available, and ready for production environments. It is also designed to be runnable in an incremental fashion in which our approach is considered to be scalable. We adopt the blocking phase in order to reduce the computational complexity from $O(N^2)$ to $O(\sum N_i^2)$, which in practice tends to $O(N)$ when $N_i \ll N$. This also means that instead of having to run the disambiguation process on the whole signature set, the process could be run only on specified blocks if desired.

---

[2] https://github.com/inspirehep/beard

5   Experiments

All the solutions proposed in this work are evaluated on data extracted from the *INSPIRE* portal [10], a digital library for scientific literature in high-energy physics. Overall, the portal holds more than 1 million publications $\mathcal{P}$, forming in total a set $\mathcal{S}$ of more than 10 million signatures. Out of these, around 13 % have been *claimed* by their original authors, marked as such by professional curators or automatically assigned to their true authors thanks to persistent identifiers provided by publishers or other sources. Together, they constitute a trusted set $(\mathcal{S}', \mathcal{C}')$ of 15,388 distinct individuals sharing 36,340 unique author names spread within 1,201,763 signatures on 360,066 publications. This data covers several decades in time and dozens of author nationalities worldwide.

Following the *INSPIRE* terms of use, the signatures $\mathcal{S}'$ and their corresponding clusters $\mathcal{C}'$ are released online[3] under the CC0 license. To the best of our knowledge, data of this size and coverage is the first to be publicly released in the scope of author disambiguation research.

5.1   Evaluation Protocol

Experiments carried out to study the impact of the proposed algorithmic components and refinements, follow a standard 3-fold cross-validation protocol, using $(\mathcal{S}, \mathcal{C}')$ as ground-truth dataset. To replicate the $|\mathcal{S}'|/|\mathcal{S}| \approx 13\,\%$ ratio of claimed signatures with respect to the total set of signatures, as on the INSPIRE platform, cross-validation folds are constructed by sampling 13 % of claimed signatures to form a training set $\mathcal{S}'_{\text{train}} \subseteq \mathcal{S}'$. The remaining signatures $\mathcal{S}'_{\text{test}} = \mathcal{S}' \setminus \mathcal{S}'_{\text{train}}$ are used for testing. Therefore, $\mathcal{C}'_{\text{train}} = \{c' \cap \mathcal{S}'_{\text{train}} | c' \in \mathcal{C}'\}$ represents the partial known clusters on the training fold, while $\mathcal{C}'_{\text{test}}$ are those used for testing.

As commonly performed in author disambiguation research, we evaluate the predicted clusters over testing data $\mathcal{C}'_{\text{test}}$, using both B3 and pairwise precision, recall and F-measure, as defined below:

$$P_{\text{B3}}(\mathcal{C}, \widehat{\mathcal{C}}, \mathcal{S}) = \frac{1}{|\mathcal{S}|} \sum_{s \in \mathcal{S}} \frac{|c(s) \cap \widehat{c}(s)|}{|\widehat{c}(s)|} \qquad R_{\text{B3}}(\mathcal{C}, \widehat{\mathcal{C}}, \mathcal{S}) = \frac{1}{|\mathcal{S}|} \sum_{s \in \mathcal{S}} \frac{|c(s) \cap \widehat{c}(s)|}{|c(s)|} \qquad (1)$$

$$F_{\text{B3}}(\mathcal{C}, \widehat{\mathcal{C}}, \mathcal{S}) = \frac{2P_{\text{B3}}(\mathcal{C}, \widehat{\mathcal{C}}, \mathcal{S}) R_{\text{B3}}(\mathcal{C}, \widehat{\mathcal{C}}, \mathcal{S})}{P_{\text{B3}}(\mathcal{C}, \widehat{\mathcal{C}}, \mathcal{S}) + P_{\text{B3}}(\mathcal{C}, \widehat{\mathcal{C}}, \mathcal{S})} \qquad (2)$$

$$P_{\text{pairwise}}(\mathcal{C}, \widehat{\mathcal{C}}) = \frac{|p(\mathcal{C}) \cap p(\widehat{\mathcal{C}})|}{|p(\widehat{\mathcal{C}})|} \qquad R_{\text{pairwise}}(\mathcal{C}, \widehat{\mathcal{C}}) = \frac{|p(\mathcal{C}) \cap p(\widehat{\mathcal{C}})|}{|p(\mathcal{C})|} \qquad (3)$$

$$F_{\text{pairwise}}(\mathcal{C}, \widehat{\mathcal{C}}) = \frac{2P_{\text{pairwise}}(\mathcal{C}, \widehat{\mathcal{C}}) R_{\text{pairwise}}(\mathcal{C}, \widehat{\mathcal{C}})}{P_{\text{pairwise}}(\mathcal{C}, \widehat{\mathcal{C}}) + R_{\text{pairwise}}(\mathcal{C}, \widehat{\mathcal{C}})} \qquad (4)$$

and where $c(s)$ (resp. $\widehat{c}(s)$) is the cluster $c \in \mathcal{C}$ such that $s \in c$ (resp. the cluster $\widehat{c} \in \widehat{\mathcal{C}}$ such that $s \in \widehat{c}$), and where $p(\mathcal{C}) = \cup_{c \in \mathcal{C}} \{(s_1, s_2) | s_1, s_2 \in c, s_1 \neq s_2\}$ is

---

[3] https://github.com/glouppe/paper-author-disambiguation.

the set of all pairs of signatures from the same clusters in $\mathcal{C}$. The F-measure is the harmonic mean between these two quantities. In the analysis below, we rely primarily on the B3 F-measure for discussing results, as the pairwise variant tends to favor large clusters (because the number of pairs is quadratic with the cluster size), hence unfairly giving preference to authors with many publications. By contrast, the B3 F-measure weights clusters linearly with respect to their size. General conclusions drawn below remain however consistent for pairwise F.

5.2   Results and Discussion

Baseline. The simplest baseline against which we compare our results consists in grouping all signatures sharing the same (normalized) surname(s) and the same (normalized) first given name initial. It provides a simple and fast solution yielding decent results, as reported at the top of Table 2.

State-of-the-Art. Most methods proposed in related works have released neither their software, nor their data, making a fair comparison very difficult. Yet, we believe solutions reported in the literature can be closely matched to our generic pipeline, provided the blocking strategy, the linkage function and the clustering algorithm are properly aligned. In particular, we consider hereon as the *state-of-the-art* solution the following combination of components:

- Blocking: same surname and the same first given name initial strategy (SFI);
- Linkage function: all 22 features defined in Table 1, gradient boosted regression trees as supervised learning algorithm and a training set of pairs built from $(S'_{\text{train}}, C'_{\text{train}})$, by balancing easy and difficult cases.
- Clustering: agglomerative clustering using average linkage and block cuts found to maximize $F_{\text{B3}}(C'_{\text{train}}, \widehat{C}'_{\text{train}}, S'_{\text{train}})$.

Below we study each component individually and discuss results with respect to the underlined state-of-the-art solution.

Blocking Choices. The good precision of the state-of-the-art (0.9901), but its lower recall (0.9760) suggest that the blocking strategy might be the limiting factor to further overall improvements. Our experiments showed the maximum $B3$ recall (i.e., if within a block, all signatures were clustered optimally) for SFI is 0.9828, which corroborates the estimation of this technique on real data by [31]. At the price of fewer and therefore slightly larger blocks, the proposed phonetic-based blocking strategies show better maximum recall (all around 0.9905). Better recall pushes further the upper bound on the maximum performance of author disambiguation, as the signatures that belong to the same author and different groups can not be clustered together by our algorithm. Let us remind that the reported maximum recalls for the blocking strategies using phonetization are also raised due to the better handling of multiple surnames, as described in Sect. 3.1.

As Table 2 shows, switching to either Double metaphone or NYSIIS phonetic-based blocking allows to improve the overall F-measure score. In particular, the

Table 2. Average precision, recall and F-measure scores on test folds. Components correspond to the state-of-the-art choices.

Description	B3			Pairwise		
	P	R	F	P	R	F
Baseline	0.9024	0.9828	0.9409	0.8298	0.9776	0.8977
Blocking = SFI	0.9901	0.9760	0.9830	0.9948	0.9738	0.9842
Blocking = Double metaphone	0.9856	0.9827	0.9841	0.9927	0.9817	0.9871
Blocking = NYSIIS	0.9875	0.9826	**0.9850**	0.9936	0.9814	**0.9875**
Blocking = Soundex	0.9886	0.9745	0.9815	0.9935	0.9725	0.9828
Classifier = GBRT	0.9901	0.9760	0.9830	0.9948	0.9738	0.9842
Classifier = Random Forests	0.9909	0.9783	**0.9846**	0.9957	0.9752	**0.9854**
Classifier = Linear Regression	0.9749	0.9584	0.9666	0.9717	0.9569	0.9643
Training pairs = Non-blocked, uniform	0.9793	0.9630	0.9711	0.9756	0.9629	0.9692
Training pairs = Blocked, uniform	0.9854	0.9720	0.9786	0.9850	0.9707	0.9778
Training pairs = Blocked, balanced	0.9901	0.9760	**0.9830**	0.9948	0.9738	**0.9842**
Clustering = Average linkage	0.9901	0.9760	**0.9830**	0.9948	0.9738	**0.9842**
Clustering = Single linkage	0.9741	0.9603	0.9671	0.9543	0.9626	0.9584
Clustering = Complete linkage	0.9862	0.9709	0.9785	0.9920	0.9688	0.9803
No cut (baseline)	0.9024	0.9828	0.9409	0.8298	0.9776	0.8977
Global cut	0.9892	0.9737	0.9814	0.9940	0.9727	0.9832
Block cut	0.9901	0.9760	**0.9830**	0.9948	0.9738	**0.9842**
Combined best settings	0.9888	0.9848	**0.9868**	0.9951	0.9831	**0.9890**
Best settings without ethnicity features	0.9862	0.9819	0.9841	0.9937	0.9815	0.9876

NYSIIS-based phonetic blocking shows to be the most effective when applied to the state-of-the-art (with an F-measure of 0.9850) while also being the most efficient computationally (with 10,857 blocks versus 12,978 for the baseline).

Linkage Function Choices. Let us first comment on the results regarding the supervised algorithm used to learn the linkage function. As Table 2 indicates, both tree-based algorithms appear to be significantly better fit than Linear Regression (0.9830 and 0.9846 for GBRT and Random Forests versus 0.9666 for Linear Regression). This result is consistent with [33] which evaluated the use of Random Forests for author disambiguation, but contradicts results of [17] for which Logistic Regression appeared to be the best classifier. Provided hyper-parameters are properly tuned, the superiority of tree-based methods is in our opinion not surprising. Indeed, given the fact that the optimal linkage function is likely to be non-linear, non-parametric methods are expected to yield better results, as the experiments here confirm.

Second, properly constructing a training set of positive and negative pairs of signatures from which to learn a linkage function yields a significant improvement. A random sampling of positive and negative pairs, without taking blocking into account, significantly impacts the overall performance (0.9711). When pairs are drawn only from blocks, performance increases (0.9786), which confirms our intuition that d should be built only from pairs it will be used to eventually cluster. Finally, making the classification problem more difficult by oversampling complex cases (see Sect. 3.2) proves to be relevant, by further improving the disambiguation results (0.9830).

Moreover, we observed that the ethnicity features serve a purpose. When these features were not included in the features set, using the best combined settings, the algorithm yields worse performance (0.9841).

Using Recursive Feature Elimination [11], we next evaluate the usefulness of all fifteen standard and seven additional ethnicity features for learning the linkage function. The analysis consists in using the state-of-the-art algorithm first using all twenty two features, to determine the least discriminative from feature importances [19], and then re-learn the state-of-the-art algorithm using all but that one feature. That process is repeated recursively until eventually only one feature remains. Results are presented in Fig. 4 for one of the three folds with the state-of-the-art, starting from the far right, *Second given name* being the least important feature, and ending on the left with all features eliminated but *Chinese*. As the figure illustrates, the most important features are ethnic-based features (*Chinese, Other Asian, Black*) along with *Co-authors, Affiliation* and *Full name*. Adding the remaining other features only brings marginal improvements. Overall, these results highlight the added value of the proposed ethnicity features. Their duality in modeling both the similarity between author names and their origins make them very strong predictors for author disambiguation. The results also corroborate those from [14] or [8], who found that the similarity between co-authors was a highly discriminative feature. If computational complexity is a concern, this analysis also shows how decent performance can be achieved using only a very small set of features, as also observed in [33] or [17].

Semi-supervised Clustering Choices. The last part of our experiment concerns the study of agglomerative clustering and the best way to find a cut-off threshold to form clusters. Results from Table 2 first clearly indicate that average linkage is significantly better than both single and complete linkage.

Clustering together all signatures from the same block (i.e., baseline) is the least effective strategy (0.9409), but yields anyhow surprisingly decent accuracy, given the fact it requires no linkage function and no agglomerative clustering – only the blocking function is needed to group signatures. In particular, this

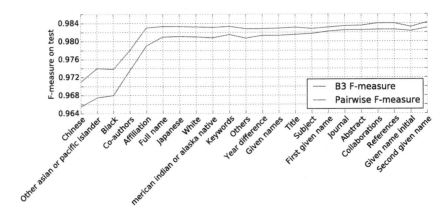

Fig. 4. Recursive feature elimination analysis.

result reveals that author names are not ambiguous in most cases[4] and that only a small fraction of them requires advanced disambiguation procedures. On the other hand, both global and block cut thresholding strategies give better results, with a slight advantage for the block cuts (0.9814 versus 0.9830), as expected. In case S'_b is empty (i.e. partial clusters are not known for any of the signatures from the block), this therefore suggests that either using a cut-off threshold learned globally from the known data would in general give results only marginally worse than if the claimed signatures had been known.

Combined Best Settings. When all best settings are combined (i.e., Blocking = NYSIIS, Classifier = Random Forests, Training pairs = blocked and balanced, Clustering = Average linkage, Block cuts), performance reaches 0.9862, i.e., the best of all reported results. In particular, this combination exhibits both the high recall of phonetic blocking based on the NYSIIS algorithm and the high precision of Random Forests.

Execution Time. Our implementation takes around 20 h to process the complete set of the data (for 10M signatures, on a 16 cores machine with 32GB of RAM). Related work [15] reports execution times around 24 h to cluster 4M signatures. Note also that shorter execution times can be achieved, at the expense of worse results, by reducing the set of the features used.

6 Conclusions

In this work, we have revisited and validated the general author disambiguation pipeline introduced in previous independent research work. The generic approach is composed of three components, whose design and tuning are all critical

---

[4] This holds for the data we extracted, but may in the future, with the rise of non-Western researchers, be an underestimate of the ambiguous cases.

to good performance: (i) a blocking function for pre-clustering signatures and reducing computational complexity, (ii) a linkage function for identifying signatures with coreferring authors and (iii) the agglomerative clustering of signatures. Making use of a distinctively large dataset of more than 1 million crowdsourced annotations, we experimentally study all three components and propose further improvements. With regards to blocking, we suggest to use phonetization of author names to increase recall while maintaining low computational complexity. For the linkage function, we introduce ethnicity-sensitive features for the automatic tailoring of disambiguation to non-Western author names whenever necessary. Finally, we explore semi-supervised cut-off threshold strategies for agglomerative clustering. For all three components, experiments show that our refinements all yield significantly better author disambiguation accuracy. In general, the results encourage further improvements and research. For blocking, one of the challenges is to manage signatures with inconsistent surnames or first given names (cases 4 and 5, as described in Sect. 3.1) while maintaining blocks to a tractable size. As phonetic algorithms are not yet perfect, another direction for further work is the design of better phonetization functions, tailored for author disambiguation. For the linkage function, the good results of the proposed features pave the way for further research in ethnicity-sensitivity. The automatic fitting of the pipeline to cultures and ethnic groups for which standard author disambiguation is known to be less efficient (e.g., Chinese authors with many homonyms) indeed constitutes a direction of research with great potential benefits for the concerned scientific communities. Exploring other name-to-ethnicity datasets with deeper coverage of names is another future work worth considering.

Moreover, the techniques presented in this work can be easily adapted to the broader problem of named entity disambiguation and thus might significantly improve accuracy of semantic search algorithms.

As part of this study, we also publicly release the annotated data extracted from the *INSPIRE* platform, on which our experiments are based. To the best of our knowledge, data of this size and coverage is the first to be available in author disambiguation research. By releasing the data publicly, we hope to provide the basis for further research on author disambiguation and related topics.

References

1. Breiman, L.: Random forests. Mach. Learn. **45**(1), 5–32 (2001)
2. Buitinck, L., Louppe, G., Blondel, M., Pedregosa, F., Mueller, A., Grisel, O., Niculae, V., Prettenhofer, P., Gramfort, A., Grobler, J., Layton, R., VanderPlas, J., Joly, A., Holt, B., Varoquaux, G.: API design for machine learning software: experiences from the scikit-learn project. CoRR, abs/1309.0238 (2013)
3. Chin, W.-S., Zhuang, Y., Juan, Y.-C., Wu, F., Tung, H.-Y., Yu, T., Wang, J.-P., Chang, C.-X., Yang, C.-P., Chang, W.-C., et al.: Effective string processing and matching for author disambiguation. J. Mach. Learn. Res. **15**(1), 3037–3064 (2014)
4. Culotta, A., Kanani, P., Hall, R., Wick, M., McCallum, A.: Author disambiguation using error-driven machine learning with a ranking loss function. In: 6th International Workshop on Information Integration on the Web (IIWeb-2007), Vancouver, Canada (2007)

5. Fan, R.-E., Chang, K.-W., Hsieh, C.-J., Wang, X.-R., Lin, C.-J.: Liblinear: a library for large linear classification. J. Mach. Learn. Res. **9**, 1871–1874 (2008)
6. Fellegi, I.P., Sunter, A.B.: A theory for record linkage. J. Am. Stat. Assoc. **64**, 1183–1210 (1969)
7. Ferreira, A.A., Gonçalves, M.A., Laender, A.H.: A brief survey of automatic methods for author name disambiguation. ACM SIGMOD Rec. **41**(2), 15–26 (2012)
8. Ferreira, A.A., Veloso, A., Gonçalves, M.A., Laender, A.H.: Effective self-training author name disambiguation in scholarly digital libraries. In: Proceedings of 10th Annual Joint Conference on Digital Libraries, pp. 39–48. ACM (2010)
9. Friedman, J.H.: Greedy function approximation: a gradient boosting machine. Ann. Stat. **29**, 1189–1232 (2001)
10. Gentil-Beccot, A., Mele, S., Holtkamp, A., O'Connell, H.B., Brooks, T.C.: Information resources in high-energy physics: Surveying the present landscape and charting the future course. J. Am. Soc. Inf. Sci. Technol. **60**(1), 150–160 (2009)
11. Guyon, I., Weston, J., Barnhill, S., Vapnik, V.: Gene selection for cancer classification using support vector machines. Mach. Learn. **46**(1–3), 389–422 (2002)
12. Han, H., Giles, L., Zha, H., Li, C., Tsioutsiouliklis, K.: Two supervised learning approaches for name disambiguation in author citations. In: Proceedings of 2004 Joint ACM/IEEE Conference on Digital Libraries, pp. 296–305. IEEE (2004)
13. Huang, J., Ertekin, S., Giles, C.L.: Efficient name disambiguation for large-scale databases. In: Fürnkranz, J., Scheffer, T., Spiliopoulou, M. (eds.) PKDD 2006. LNCS (LNAI), vol. 4213, pp. 536–544. Springer, Heidelberg (2006)
14. Kang, I.-S., Na, S.-H., Lee, S., Jung, H., Kim, P., Sung, W.-K., Lee, J.-H.: On co-authorship for author disambiguation. Inf. Process. Manag. **45**(1), 84–97 (2009)
15. Khabsa, M., Treeratpituk, P., Giles, C.L.: Large scale author name disambiguation in digital libraries. In: 2014 IEEE International Conference on Big Data (Big Data), pp. 41–42. IEEE (2014)
16. Lange, D., Naumann, F.: Frequency-aware similarity measures: why Arnold Schwarzenegger is always a duplicate. In: Proceedings of 20th ACM International Conference on Information and Knowledge Management, pp. 243–248. ACM (2011)
17. Levin, M., Krawczyk, S., Bethard, S., Jurafsky, D.: Citation-based bootstrapping for large-scale author disambiguation. J. Am. Soc. Inf. Sci. Technol. **63**(5), 1030–1047 (2012)
18. Liu, W., Islamaj Doğan, R., Kim, S., Comeau, D.C., Kim, W., Yeganova, L., Wilbur, W.J.: Author name disambiguation for pubmed. J. Assoc. Inf. Sci. Technol. **65**(4), 765–781 (2014)
19. Louppe, G., Wehenkel, L., Sutera, A., Geurts, P.: Understanding variable importances in forests of randomized trees. In: Advances in Neural Information Processing Systems, pp. 431–439 (2013)
20. Malin, B.: Unsupervised name disambiguation via social network similarity. In: Workshop on Link Analysis, Counterterrorism, and Security, vol. 1401, pp. 93–102 (2005)
21. Newman, M.E.: The structure of scientific collaboration networks. Proc. Natl. Acad. Sci. **98**(2), 404–409 (2001)
22. Pedregosa, F., Varoquaux, G., Gramfort, A., Michel, V., Thirion, B., Grisel, O., Blondel, M., Prettenhofer, P., Weiss, R., Dubourg, V., Vanderplas, J., Passos, A., Cournapeau, D., Brucher, M., Perrot, M., Duchesnay, E.: Scikit-learn: machine learning in Python. J. Mach. Learn. Res. **12**, 2825–2830 (2011)
23. Philips, L.: The double metaphone search algorithm. C/C++ Users J. **18**(6), 38–43 (2000)

24. Ruggles, S., Sobek, M., Fitch, C.A., Hall, P.K., Ronnander, C.: Integrated Public Use Microdata Series. Historical Census Projects, Department of History, University of Minnesota (2008)

25. Schulz, C., Mazloumian, A., Petersen, A.M., Penner, O., Helbing, D.: Exploiting citation networks for large-scale author name disambiguation. EPJ Data Sci. 3(1), 1–14 (2014)

26. Smalheiser, N.R., Torvik, V.I.: Author name disambiguation. Ann. Rev. Inf. Sci. Technol. 43(1), 1–43 (2009)

27. Song, Y., Huang, J., Councill, I.G., Li, J., Giles, C.L.: Efficient topic-based unsupervised name disambiguation. In: Proceedings of 7th ACM/IEEE-CS Joint Conference on Digital Libraries, pp. 342–351. ACM (2007)

28. Strotmann, A., Zhao, D.: Author name disambiguation: what difference does it make in author-based citation analysis? J. Am. Soc. Inf. Sci. Technol. 63(9), 1820–1833 (2012)

29. Taft, R.L.: Name search techniques. Technical report Special Report No. 1, New York State Identification and Intelligence System, Albany, NY, February 1970

30. The National Archives. The soundex indexing system, May 2007

31. Torvik, V.I., Smalheiser, N.R.: Author name disambiguation in medline. ACM Trans. Knowl. Disc. Data (TKDD) 3(3), 11 (2009)

32. Tran, H.N., Huynh, T., Do, T.: Author name disambiguation by using deep neural network. In: Nguyen, N.T., Attachoo, B., Trawiński, B., Somboonviwat, K. (eds.) ACIIDS 2014, Part I. LNCS, vol. 8397, pp. 123–132. Springer, Heidelberg (2014)

33. Treeratpituk, P., Giles, C.L.: Disambiguating authors in Academic Publications using random forests. In: Proceedings of 9th ACM/IEEE-CS Joint Conference on Digital Libraries, pp. 39–48. ACM (2009)

34. Treeratpituk, P., Giles, C.L.: Name-ethnicity classification and ethnicity-sensitive name matching. In: AAAI, Citeseer (2012)

35. Ward Jr., J.H.: Hierarchical grouping to optimize an objective function. J. Am. Stat. Assoc. 58(301), 236–244 (1963)

Towards Flexible K-Anonymity

Rima Kilany[1(✉)], Maria Sokhn[2], Hussein Hellani[3], and Shaban Shabani[2]

[1] Université Saint Joseph, B.P. 11-0514 Riad El Solh, Lebanon
rima.kilany@usj.edu.lb
[2] HES-SO Valais Wallis, Technopole 3, 3960 Sierre, Switzerland
maria.sokhn@hes-so.ch, shaban.shabani@hevs.ch
[3] Université Saint Joseph, Bir Hassna, B.P. 13-6007 Beirut, Lebanon
hussein.hellani@hotmail.com

Abstract. Data published online nowadays needs a high level of privacy to gain confidentiality as well as to maintain the privacy laws. The focus on k-anonymity enhancements along the last decade, allows this method to be elected as the starting point of any research. In this paper we focus on the external anonymization through a new method: the « Flexible k-anonymity » . It aims to anonymize external published data, by defining a semantic ontology that distinguishes between sparse and abundant quasi-identifiers, and describes aggregation levels relations, in order to achieve adequate k-blocks. For the validation of our proposal, we apply the aforementioned anonymization method to the Comiqual dataset. Comiqual (Collaborative measurement of internet quality), is a large-scale measurement platform for assessing the internet quality access of mobile and ADSL users by collecting mobility traces and private data related to internet metric values.

1 Introduction

In order to have an efficient and useful anonymization process, we should consider data sanitization and refinement at two levels: First, the internal level, where the threat is mainly linked to employees or intruders. Indeed, they have been entrusted with authorized access to the network and can easily reach data repositories and violate individuals' privacies. Second, the external level, which mainly addresses published data, and attacks from people outside the organization.

In this paper we focus on the external level challenges by proposing an approach based on the k-anonymity principle. With the k-anonymity principle, the records are made indistinguishable from at least k-1 other records [1]. For this purpose, quasi-identifiers are examined for each record and a k-block is constructed in order to release them to the public. Quasi-identifiers are fields which, when combined, make a record unique and identifiable. Two methods are used to achieve the k-anonymity: *generalization* which substitutes the values of a given attribute with more general values and *suppression* which is used to mask the given information totally by an asterisk "*" and to moderate the generalization process when tuples with less than k-blocks occur [3].

© Springer International Publishing Switzerland 2016
A.-C. Ngonga Ngomo and P. Křemen (Eds.): KESW 2016, CCIS 649, pp. 288–297, 2016.
DOI: 10.1007/978-3-319-45880-9_22

For the validation of our proposal we used the Comiqual dataset. Comiqual (Collaborative measurement of internet quality)[1], is a large-scale measurement platform for assessing the internet quality access of mobile and ADSL users by collecting mobility traces and private data related to internet metric values.

The main contribution of this paper may be summarized as follows:

- The definition of a semantic ontology which distinguishes between scanty and abundant quasi-identifiers, by defining different classes for those identifiers as well as aggregation level relations between them.
- Flexible k-anonymity: a new anonymization method to be applied at the external level, by inferring aggregation levels from the ontology in order to be able to use different k-anonymity values and build appropriate k-blocks.
- The proposition of a complete anonymization process from the receipt of the data until its publishing.

The rest of this paper is structured as follows: Sect. 2 presents the related works, Sect. 3 gives an overview of Comiqual, Sect. 4 details the external anonymization approach, Sect. 5 presents the sanitization process, and finally, Sect. 6 concludes the paper.

2   Related Work

Most existing work on privacy, considered only location privacy of published data or what is called "second use of data" and employed k-anonymity based methods. In this section we will briefly explain the k-anonymity method [1] and some of its variants.

2.1   K-Anonymity Enhancements

The k-anonymity privacy protection achieved by L. Sweeney since 2002 [1] using generalization and suppression tools, opened the door to many researchers to add more enhancements or propose new methods based on the k-condition. In fact, most researchers [1, 8, 9] assume a uniform case study such as the medical dataset which focuses on grouping the attributes into personal identifiers (e.g. name, email address, etc.), quasi-identifiers (e.g. age, zip code, etc.) and sensitive attributes (e.g. disease) to apply the k-anonymity where each record in the same quasi-identifier block is indistinguishable from at least (k-1) other records within the same block [1]. The larger the value of k, the better the privacy is protected [2].

K-anonymity alone does not ensure privacy when sensitive values in an equivalence class lack diversity, which is known as the homogeneity attack. L-diversity [13] method is found to bridge this gap and is composed of three progressive levels: (1) distinct l-diversity, where each equivalent class has at least L values for each sensitive attribute but this doesn't prevent the probabilistic inference attacks. (2) in entropy l-diversity, each equivalence class must not only have enough different sensitive values, but the

---

[1] http://comiqual.usj.edu.lb/.

different sensitive values must also be distributed evenly enough. (3) the recursive l-diversity, makes sure that the most frequent value does not appear too frequently, and the less frequent values do not appear too rarely. Obviously these methods are too restrictive and require a specific distribution of the data values. The main issue with l-diversity is that it does not consider semantic meanings of sensitive values. This leads to a less conservative notion of l-diversity. T-closeness [14] is a refinement of l-diversity and it aims to create equivalent classes that resemble the initial distribution of attributes in the table. An equivalence class is said to have t-closeness if the distance between the distribution of a sensitive attribute in this class and the distribution of the attribute in the whole table is no more than a threshold t. The case studies that are elected to apply the anonymization methods, fit the researchers' purpose and prove high privacy protection. But what if we change the case study, do we still obtain the same results?

Three main issues are in common with the aforementioned methods: (1) they do not consider semantic meanings of sensitive values. (2) they are applied on sensitive attributes only, hence in our case these methods cannot be useful as there are no sensitive attributes in the dataset. (3) it is very difficult to build appropriate blocks in a sparse environment with scanty attribute values.

2.2   External Protection

Prior to the release of the "second use of data" version, the rule of thumb is to group the records into k similar blocks in order to satisfy the k-anonymity. Generalization and suppression are used for this purpose. According to Sweeny, in some cases, the price of removing an isolated record would be less than the price to pay in terms of information precision loss when generalizing all its possibly related data records [2]. This is not the case of the Comiqual dataset where suppression is forbidden and users should be able to view every and each measurement record sent to the server on the public website, using any device and from any location. According to the aforementioned analysis concerned with anonymization methods applied on sensitive attributes, some kind of threats such as homogeneity attacks become impractical to be applied on trajectory datasets. This is because this type of data distribution has no sensitive attributes and the sensitivity is embedded within the quasi-identifiers themselves such as location attributes. On another hand, traceability attacks with some background knowledge are very potential with any LBS system, where individual movements are disclosed using time-referenced location information as quasi-identifiers. The adversary may already know some portion of the trajectory of an individual in the dataset and may be interested in the rest (e.g. adversary knows that a particular person lives in a particular house. He also knows that she leaves the house and comes back home at specified times, so he may be interested in finding the locations she visited.). Standard generalization could not achieve good enough result in collecting the best k tuples and deceive the attackers to single out an individual, in other words, generalizing data in a non-smart manner leads to traceability attacks and sometimes cannot succeed in building an appropriate k-block in a sparse data environment (e.g. five different device models).

Due to the weakness of the above k-anonymity's enhancements in fixing all the shortcomings, introducing a semantic ontology system becomes necessary to let artificial intelligence control the whole anonymization process. Ontologies allow us to model concepts, their relationships and properties as well as other more subtle aspects of a domain. The idea is to add an ontology layer on top of the k-anonymity method in order to create a more robust privacy-enforcing system [3]. Vocabulary k-anonymity method [6] perceived the extremely sparse data of web query logs, and proposed an algorithm to cluster vocabularies by semantic similarities. Such methods do not apply well on measurement applications like Comiqual, because these measurement platforms do not store any sensitive attributes.

3   Comiqual Overview

In Comiqual, users send measurement details periodically. These details include user-name, email address, machine type, installed operating system, battery status, cell info, cell id, GPS location, IP address, ISP provider, and network type. Comiqual mobile agent (MA) manages and controls the measurement process between the Comiqual server and the peer server that examines the measurement speed (e.g. michigan.mlab2.lca01.meas-urement-lab.org). MA sends the results back to the internal Comiqual database to be published on the website. Each participant's mobile has its own id represented by a combination of user's email address and mobile IMEI (International Mobile Station Equipment Identity). This unique id called Measurement Agent ID (MAID) is associated with each single measurement record generated by this user. Changing any of the mobile device or the email address by the user, leads to the generation of a new MAID.

In the use case of the Comiqual dataset, MAID, IMEI and email address can be considered as Personal Identifiable Information (PII). The quasi-identifiers are: GPS location, cell id, device model, ISP provider and network type. While username and email could be simply hashed when being stored on the internal servers, majority of the mentioned quasi-identifiers combined together along with the location attributes, could lead to single out an individual by detecting a certain mobility pattern or presence pattern and therefore should be wisely anonymized, on the internal and the external level.

Comiqual main constraints are: (1) the dataset does not include sensitive attributes. The sensitivity is embedded within the quasi-identifiers themselves as shown in Table 1 where for example the device model and the GPS location are the most two sensitive attributes in comparison to others. (2) Suppression is not allowed in any of the internal and external anonymization levels in order not to lose any collected measure-ment detail. (3) Generalization can be applied at the external level only in order not to lose accuracy of the collected data. It is important to mention that Comiqual dataset belongs to a family of broadband measurement applications as well as many similar applications like weather detection, prayer timings, etc. that are very trendy nowadays. A huge number of online participants are continuously sharing personal information on public sites (GPS locations, device model, etc.). Therefore the main purpose of this research is to foil the traceability of those users and protect them against many privacy attacks.

Table 1. Comiqual data representation

Quasi identifiers				Non-sensitive
Device model	GPS location	ISP	Net type	Result (Mbps)
Samsung-Galaxy	L1	ISP1	DSL	R1
IPhone 6	L2	ISP2	4G	R2
HTC one	L3	ISP3	3G	R3

Comiqual has sparse data represented by the "mobile model" which is very difficult to be grouped into k-block of similarity. Indeed there are different mobile series that fall under each brand. The rapid development and competency between mobile vendors is introducing more sparse data in the model since we have new brand series often. This added more challenges to the grouping of similar items during the anonymization process because of the existence of sparse data that does not satisfy the k-condition. Standard generalization could not achieve very good result in collecting the best k tuples and deceive the attackers to single out an individual. In other words, generalizing data via non-smart manner leads to traceability attacks and sometimes cannot succeed the k-block in a sparse data environment (e.g. five different device models). Due to the weakness of the k-anonymity's enhancements in fixing the sparse data issue, introducing a semantic ontology system becomes necessary to let the artificial intelligence control the whole anonymization process.

4   Process Overview

Simply removing the identifiers of individuals or replacing them by a pseudonym does not protect their privacy from inference attacks. Indeed several works, among which the one of Gambs et al. [15], demonstrate that a reverse analysis over the data may allow the identification of a person. Hence, it is important to combine the hashing aspect with the injection of some noise. The process is divided into two phases: the internal named the sanitization process, and the external process described in details in Sect. 5.

At each stage of the sanitization process (Fig. 1), we propose and apply an appropriate anonymization method in order to alleviate the specific attack risks associated to the data state at each level. The steps of the process are applied in the following order:

1. Hash the PII of each entry in order to avoid direct identification.
2. Encrypt the hashed PII used to filter the true data
3. Encrypted PII will be stored in a secure text file
4. → 7 Add fake records with fake PII - that are hashed and encrypted - until satisfying the k-anonymity (we are currently working on the enhancement of this series of steps that will be applied at the internal level)
8. The data that is composed from fake and real records will be published to the internal database and the temporary location will be cleared after a t time.
9. Filter the real records by means of true PII and preserve them in a temporary database

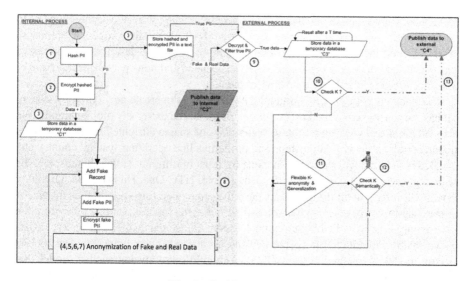

Fig. 1. Sanitization process

10. Check k satisfaction, publish immediately to external database if records' number satisfy k-condition, and the temporary location will be cleared after a t time

11. If the records' number is less than k, flexible k-anonymity will be applied to differentiate sparse and non-sparse attribute and enforce K_S and K simultaneously based on probability of sparse P(s). First generalize the normal and sparse attributes to satisfy k and k_S then use the ontology to fill the remaining records (k-k_S) based on their most common criterion

12. Ensure K and K_S satisfaction (using semantic ontology)

13. Publish the anonymized data to the external website

5   External Anonymization

Experiments show that constructing a k-block of multiple quasi-identifiers fails to succeed due to the nature of data. For example, our case study includes mobile model attributes such as: Samsung, iPhone, HTC, Huawei, etc. In addition, there are different mobile series that fall under each brand. The rapid development and competency between mobile vendors is introducing more sparse data in the model since we have new brand series every short time. This adds more challenges to the grouping of similar items during the anonymization process because of the existence of sparse data that does not satisfy the k-condition.

5.1   Semantic Ontology Model

To enhance the anonymization methods, many researchers such as the p-sensitivity [4], vocabulary k-anonymity [6], ontology k-anonymity [3] and ontological semantics technology [8] are based on semantic ontology. The hardness of applying k-anonymity in sparsely data environment and the eventual benefit of using semantics encouraged us to use a semantic ontology and change the core of the k-anonymity process. Contrary to the above cited methods, we propose a semantic ontology to describe the domain of quasi-identifiers in order to distinguish between sparse and non-sparse attributes and to maximize k-block in a sparse data environment.

Resolving sparsity of the Comiqual dataset, lead us to create new ontology system that has the role of providing best common criteria of the extremely mobile brands. We focus on mobile device model that is a sub-class of sparse attribute class. "Mobile" is encountered by a many relationships and properties like operating system, country etc. The idea is to find always a common name for a list of different mobile brands, e.g. the following devices are very sparse: Samsung Note, HTC One, Huawei, LG. Ontology system will infer that all these devices have the same operating system, thus instead of suppressing them all, ontology will alter their names to "android mobile" therefore the k-anonymity is achieved without too much losses. Our semantic model (Fig. 2) consists of two main hierarchies: the sparse and the normal attributes branches. The mobile concept model is inspired from MOKM, a mobile ontology knowledge model that supports information intercommunication among mobile applications and improves the cooperative ability of mobile users [7]. We extended this model by adding the distinction between sparse and non-sparse attributes.

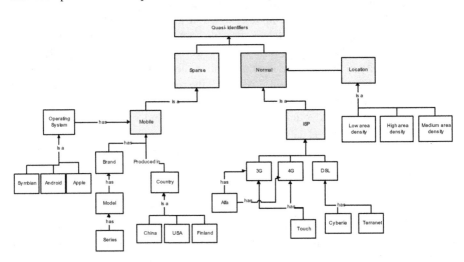

Fig. 2. Semantic ontology model for comiqual data

5.2 Flexible K-Anonymity

In order to select k similar mobiles, we scroll up the sparse attributes hierarchy to infer the aggregation level from the ontology that will enable us to build the appropriate k-block. The properties "is a", "produced in", "has OS", etc. with the joint/disjoint relations between subclasses, enables us to infer at least one common criterion between many devices. For instance, assume we encounter in our dataset the following device models: HTC One, Samsung S3, Huawei P8 and LG G2. The system infers that those devices have the same android operating system, and that they are produced in the same country, therefore their diverse names will be replaced by either "android mobile" or "Chinese mobile" in order to satisfy the k-anonymity and overcome the sparse data problem.

Keeping sparse attributes (e.g. "mobile model") out of consideration during the construction of k-anonymity blocks can lead to eventual individual identification. While dealing with such attributes, in the same way as for non-sparse ones, it could end up with the anonymization process failure. Flexible k-anonymity is based on splitting the quasi-identifiers attributes into sparse and normal classes towards having sparse data partially contribute in the creation of k-blocks, by k of sparse value ks, where ks < k. This method allows us to use different k values for the same dataset rather than fix a static one. The level of contribution depends mainly on the percentage of sparse data available within the dataset that is called sparse probability P(s) that indicates how much "sparse data" should contribute in k-block construction. The number of the remaining tuples to be filled into the dataset will be inferred from the ontology through the detection of an appropriate aggregation level, by going up in the hierarchy of sparse attribute classes as shown in Fig. 1. We will begin by some definitions, then explain how to apply the flexible k-anonymity approach in order to choose the k values in an optimal way for sparse and non-sparse attributes.

A: Attribute; A_S: Sparse Attribute; $P(s)$: sparse probability.
K_S: k-anonymity of sparse; K: k-anonymity of non-sparse

Let $RT(A_1,...,A_n)$ be a table, $QI_{RT}(A_i,...,A_j)$ the quasi-identifiers where $A_i,...,A_j \subseteq A_1,...,A_n$ and $A_S \subseteq A_i,...,A_j$. $K_S = P(s) \times K$ (with $P(s) > 0$)

RT is said to satisfy *flexible k-anonymity* if each sequence of values in $RT[QI_S]$ appears at least K_S occurrences in $RT[QI_S]$ and each sequence of values in $RT[QI_{RT}]$ appears with at least K occurrences in $RT[QI_{RT}]$. The complementary of K $(K-K_S)$ is semantically selected.

5.3   Optimal K-Anonymity and K of Sparse

Flexible k-anonymity approach is based on: (1) determine the sparse and non-sparse quasi-identifiers, i.e. "network type" and "ISP name" are non-sparse attribute while "mobile model" is a sparse attribute. (2) Assign the static k value in a way that enables normal attributes to realize k-anonymity. (3) Evaluate the sparse probability P(s) by conducting a heuristical study on the sparse data values within the dataset. In the Comiqual use case, we found that this probability for the "device mobile" attribute represents no more than 30 % of the whole dataset. (4) Calculate k of sparse value K_s by multiplying P(s) times k. P(s) represents the direct relation between K and k_s:

$$a)\ K_s = P(s) \times K \quad b)\ K_s := \begin{cases} K for P(s) \ll \\ otherwise; Ks < K \end{cases}$$

Generally, optimal k-anonymity is NP-hard and not easy to be evaluated [12]. In practice for *k* to be a small constant around 5 or 6, gives positive results [1]. In *flexible k-anonymity*, defining the *k* value is done intuitively by estimating the occurrences of similar or approximate records of the non-sparse attribute, whereas k of sparse K_S represents portion of this predefined k value, defined by sparse data proportion of the whole dataset. Semantic ontology provides the complementary of k by unifying the sparse data under the most common criterion name. For instance, assume that k = 6 mobiles and sparse probability P(s) = 50 %, the K_S value will be $6 \times 0.5 = 3$. This means that we should have at least 3 similar mobiles of the 6-blocks to satisfy the flexible k-anonymity. The remaining three sparse records are semantically altered into the most common criterion such as "mobile model name", e.g. the three mobile names might be turned into as "android mobile" or "Chinese mobile". Flexible k-anonymity usage, can be extended to enhance some non-sparse attribute like location, e.g. it can be used to distinguish between low, medium and high area density, then assign different k value for each class. Accordingly, a user with Samsung S6 could be generalized to Samsung mobile in a low area density but when moving to high area density, generalization could not be applied on same model's group that satisfy the k-anonymity. As a result of introducing flexible k-anonymity, we are going to have more accurate anonymized data, with a high level of privacy.

6 Conclusion

In this paper we showed that in order to achieve data anonymization, applying k-anonymity enhancements such as l-diversity, t-closeness and others, could be impractical for datasets holding spatio-temporal information of individuals and which do not contain any sensitive attributes, such as the Comiqual dataset. For such datasets, simply removing the identifiers of individuals or replacing them by a pseudonym does not protect their privacy from inference attacks. In this paper we demonstrated the need for a sanitization process, which should introduce a level of protection at both the internal and the external levels. Future work will detail a new anonymization algorithm to be applied at the internal level when data is at rest, based on hashing the PID, as well as the addition of noise, because deletion of sensible data is not always permitted, as in the case of the Comiqual dataset. For the external level, we proposed the *flexible k-anonymity* for the second use of data, to fight counter sparse data when constructing k-block, by using semantic ontology system that infers the common criteria for the sparse data.

As for future work, we would also like to extend our study to investigate how the continuous flow of data affects the proposed sanitization process and how the arrival of new collected entries can be synchronized with the anonymized data without affecting the efficiency of the process.

References

1. Sweeney, L.: k-anonymity: a model for protecting privacy. Int. J. Uncertainty Fuzziness Knowl. Based Syst. **10**(05), 557–570 (2002)
2. Lv, P.: Utility-based anonymization for continuous data publishing. In: Computational Intelligence and Industrial Application. Pacific-Asia Workshop (2008)
3. Omran, E., Bokma, A., Abu-Almaati, S.: A k-anonymity based semantic model for protecting personal information and privacy. In: 2009 IEEE (IACC 2009), Patiala, India, 6–7 2009
4. Xiao, Z., Meng, X.: p-sensitivity: a semantic privacy-protection model for location-based services. In: Mobile Data Management Workshops (2008 MDMW) (2008)
5. Daubert, J., Grube, T., Muhlhauser, M., Fischer, M.: Internal attacks in anonymous publish-subscribe P2P overlays. In: 2015 International Conference on NetSys (2015)
6. Liu, J., Wang, K.: Enforcing vocabulary k-anonymity by semantic similarity based clustering, in data mining. In: 2010 IEEE 10th International Conference on ICDM (2010)
7. Junwu, Z., Bin, L., Fei, W., Sicheng, W.: Mobile ontology. Int. J. Digit. Content **4**(5), 46–54 (2010)
8. Ringenberg, T., Taylor, J.: Semantic Anonymization of Medical Records. IEEE, San Diego (2014)
9. Bertino, E., Ooi, B., Yang, Y., Deng, R.: Privacy and ownership preserving of outsourced medical data. In: 2005 ICDE (2005)
10. You, T.-H., Peng, W.-C., Lee, W.-C.: Protecting moving trajectories with dummies. In: Proceedings of the 2007 International Conference on Mobile Data, pp. 278–282 (2007)
11. Kido, H., Yanagisawa, Y., Satoh, T.: An anonymous communication technique using dummies for location-based services. In: ICPS, pp. 88–97 (2005)
12. Meyeson, A., Williams, R.: On the complexity of optimal K-anonymity. In: PODS 2004, New York (2004)
13. Machanavajjhala, A., Gehrke, J., Kifer, D., Venkitasubramaniam, M.: l-diversity: privacy beyond K-anonymity. In: ICDE (2006)
14. Li, N., Li, T., Venkatasubramanian, S.: t-closeness: privacy beyond k-anonymity and l-diversity. In: ICDE 2007, pp. 106–115 (2007)
15. Gambs, S., Killijian, M.-O., Núñez, M., del Cortez, P.: De-anonymization attack on geolocated data. J. Comput. Syst. Sci. **80**(8), 1597–1614 (2014). http://dx.doi.org/10.1016/j.jcss.2014.04.024

Applications

Ontology-Based Collaborative Development of Domain Information Space for Learning and Scientific Research

Anton Anikin$^{(\boxtimes)}$, Dmitry Litovkin, Marina Kultsova, and Elena Sarkisova

Volgograd State Technical University, Volgograd, Russia
anton@anikin.name

Abstract. This paper is devoted to a problem of creating a domain information space for information retrieval and reuse in various subject domains. We proposed an ontology-based approach to the collaborative development of the domain information space using personal human cognitive spaces. In framework of the proposed approach an information space ontology was created, an algorithm for generation of the domain information space was developed on the basis of the personal cognitive spaces using the rule-based reasoning over ontology. We considered three scenarios of the information space construction: an individual and collaborative creation of the information space, as well as creation of the new information space on the basis of the existing spaces. Also we presented a software architecture for the construction of domain information space which includes two software applications: web-application for individual and collaborative creation of the domain information space and builder of personal electronic learning collections. The implementation of the developed approach was illustrated by the example of information space construction for the programming languages domain.

The proposed approach is consistent with the concept of Open Science and allows describing the domain structure, combining the distributed open scientific information resources into thematic collections and creating the domain information spaces using these resources.

The ontology-based approach and software tools can be used in the scientific research and learning process for the open educational resources retrieval and reuse, as well as for extending the personal cognitive spaces of learners, teachers and researchers.

Keywords: Semantic web · Ontology · OWL-DL · Open Science · Reasoning · Human cognitive space · Information space · Information retrieval · Collaborative knowledge construction

1 Introduction

The total number of various information resources available on the Internet is growing significantly in the recent years. These resources can be used by the person or group of persons in the investigation of some subject domain,

© Springer International Publishing Switzerland 2016
A.-C. Ngonga Ngomo and P. Křemen (Eds.): KESW 2016, CCIS 649, pp. 301–315, 2016.
DOI: 10.1007/978-3-319-45880-9_23

particularly, during learning or scientific research. A high accessibility of the huge number of learning resources which are focused on different learning goals and outcomes, and have different coverage of the subject domain, complexity level and representation, allows improving the efficiency of learning. In the scientific research the use of open information resources is important on the early stages of this process for study the state of the art in subject domain, identification of research problems and goals, statement of the specific research tasks and data collection [6,19]. Also the scientific research process, as opposed to the learning process, is closely associated with acquiring new knowledge and creating new information resources, which also should be available for other researchers, at that an important problem is the support of collaborative work of researchers.

In the learning process there are common educational goals for some learning groups that allows using a common set of Open Educational Resources (OER), considering at that some individual requirements. In the scientific researches it is possible to set some common or related goals, the achievement of these goals requires use of common set of information resources and interaction between the researchers for collaborative creation of new information resources.

The subject domain might have a rather complex structure and cover a huge set of concepts and relations between them, that complicates extremely the study of appropriate information resources by the person who is not familiar with the domain. In this case it is advisable to create the own domain model or use and extend the domain models created by other persons.

So, the important and urgent problem is implementation of the personalized information retrieval for the person or groups in accordance with their understanding of subject domain. In this paper we propose an ontology-based approach that allows collaborative setting the association relations between information resources and concepts of subject domain to retrieve the 'top' information resources for the persons or groups.

2   Related Work

Information support has a key role in the learning process and determines the quality and efficiency of this process. Competency-based learning is directed to the acquirement of appropriate competencies, knowledge and skills, information support of learning implies use of the appropriate information and educational resources. At that it is possible to use existing information and learning resources as well as create new resources for educational needs of individual learners and groups. The use of existing open information resources allows to meet the needs of the individual learners and groups, and it is substantially cheaper then creation of new educational content. The problem is to find the resources that achieve the learning objectives and meet the requirements of learner among a great number of available resources, and provide an effective personal learning trajectory to achieve required learning outcomes. The OpenCourseWare approach implies that learning content should be presented as a combination of the reusable and remixable learning objects. The total number of low-quality learning objects on

Internet grows rapidly, it makes the search and filtering of educational content rather challenging and labor-intensive task. The efficiency of learning resources retrieval can be improved using resource metadata annotation on the basis of the proper metadata standards and ontological models [3,7,11,20]. Other important step towards decreasing the costs while increasing the quality of the educational materials is application of collaborative techniques in content development process. This approach affects different aspects of the learning content development such as annotating, personalization and sharing [18].

Various electronic libraries (such as ACM Digital Library, IEEE Xplore Digital Library, SpringerLink, ISI Web of Sciences) are widely used in scientific research for access to high quality peer-reviewed scientific papers, but they provide access only on subscription base and do not support semantic search in subject domains. The OpenScience approach involves publishing the open research, campaigning for open access, encouraging scientists to practice open notebook science, and generally making it easier to publish and communicate scientific knowledge. The services like arXive.org allow publishing the papers and provide open access to them. The services like ShareLatex [28], Authorea [25] and Intech [26] provide the ability for collaborative writing and/or publishing the papers. Scientific social networks like ResearchGate [27], Academia [24] allow to share publications, connect and collaborate with specialists in the related fields and co-authors. SocioNet [14,15] service allows to organize the resource descriptions into collections and publish them. Some software tools (like Mendeley Desktop, Zotero) allow to create personal repositories of the information resources for use in research. During the scientific research process the domain models (conceptual models, mind maps, ontologies) are created using various software tools, some of these tools allow to associate the information resources with the components of the domain models, but none of these provides reuse of the resulting model and resources and search of the models and resources created and gathered by other researchers.

Thus, information support of research and learning processes involves thematic information retrieval and creation of the collections of information resources for some topic as a result of their retrieval and integration [5,12]. The specificity of the thematic information retrieval, both for learning and research processes, consists in the following: (a) in the beginning of the retrieval a person can not realize clearly his information needs and has only general idea - a topic, so he is not able to create correct search query for the search engine; (b) during the information retrieval a person redefines his information needs. In the capacity of the information retrieval results we can consider not only information resources as such, but also more precise identification of the personal information needs. The use of general-purpose web search engines based on vector space retrieval model for solving this task is complicated by the fact that these systems, as a rule, do not take into account the semantics of search query and document, that reduces the search quality. The great disadvantage of the original vector space model is that in theory the model assumes the independence of the terms, that is not the case in the reality of information retrieval [4,21]. One of convenient tools for thematic information retrieval are the thematic directories. These directories are focused on the creation of thematic collections of information resources with the convenient system of links and resource hierarchy. This tool allows the person to find required resource using directory navigation or specific retrieval mechanism of the directory. The significant disadvantage of the thematic directories is rigid structure of the directory that reflects the vision of some subject domain by some group of experts in some aspect.

According to [9,10], a cognitive space is the set of concepts and relations among them held by a human. The cognitive space can be individual as well as shared by a group of people. Using the modern software tools the cognitive space can be mapped into conceptual model represented as a mind map, topic map, concept map (conceptual diagram) or ontology. An information space [9,10] is the set of objects and relations among them held by information system. The components of the information space for the information retrieval task include concepts, documents, words, relations among words and documents. So the information space should be consistent with the cognitive space of particular humans or groups.

An essential and actual problem is creation of the domain information space which is relevant to the personal cognitive space of subject of information process.

The use of ontologies for the cognitive and information spaces representation is an appropriate and promising approach. For the information retrieval the ontology can provide the expressive terminology to describe the content, and the inferences sanctioned by ontology can be used to improve the quality of search. In the tasks of the learning process support [1,2,8,13] and in the domain modeling tasks [22,23] the domain model should provide the breadth and depth of knowledge and skills, granularity and scalability. The domain model should be modular and extensible so that it could cover the new subdomains, and the ontology meets all of these requirements.

Ontological representation of cognitive and informational spaces allows us to consider the information retrieval task as a known Ad-hoc Object Retrieval task [16], where: INPUT: (a) keyword query; (b) query type - "Type query": the intention of the query is to find entities of a particular class; (c) query intent - class "information resource"; (d) data graph - the cognitive and information spaces where the objects are the concepts of the domain and the information resources. OUTPUT: a ranked list of information resource identifiers from data graph.

3   An Ontology-Based Approach to the Information Space Development

In this paper we propose an approach to collaborative development of the domain information space which allows setting the relations and mapping between the cognitive spaces of different persons, as well as sharing and collaborative using of this information space.

The cognitive space of some subject domain $CognitiveSpace$ is defined as follows:

$$CognitiveSpace =< Concepts, IncludesRelations >, \qquad (1)$$

where: $Concepts$ - set of the concepts of the subject domain;
$IncludesRelations$ - set of the subsumption relations defined on the set of the concepts.
The information space $InformationSpace$ that is relevant to the cognitive space $CognitiveSpace$:

$$InformationSpace =< Objects, Relations, Rules >, \qquad (2)$$

where: $Objects$ - set of the objects of the subject domain held by the information system,
$Relations$ - set of the relations between these objects,
$Rules$ - set of the reasoning rules for setting the relations between the objects.
The set of the objects of the subject domain $Objects$ is defined as follows:

$$Objects =< Concepts, InformationResources >, \qquad (3)$$

where: $Concepts$ - set of the concepts of the subject domain;
$InformationResources$ - set of the information resources associated with the concepts of the subject domain.
The set of the relations $Relations$ includes two types of relations:

$$Relations =< ConceptRepresentationRelations, IRRelations >, \qquad (4)$$

where $ConceptRepresentationRelations$ - set of the association relations between the concepts of the subject domain and information resources:

$$Concepts \times InformationResources \rightarrow \{undefined, bad, good, excellent\}, \qquad (5)$$

where: $excellent$ - the resource describes the concept in full;
bad - the resource contains minimal information about the concept;
$good$ - intermediate value between the excellent and bad;
$undefined$ - the resource describes the concept with relevance which is not defined yet;
$IRRelations$ - set of the relations between the information resources.
The information space can be defined using following three techniques described below or their combinations (Fig. 1):

1. **Individual creation of the information space.** The person defines his own cognitive space $CognitiveSpace$, then associates the information resources with the concepts of the subject domain and judges their relevance (i.e. defines the relations $ConceptRepresentationRelations$). So, the information space $InformationSpace$ which is relevant to the person's cognitive space is created. Afterwards these spaces can be used by other persons.

2. **Collaborative creation of the information space.** The person creates the cognitive and information spaces, then provides to other persons possibility to extend and redefine the information space by defining the new information resources and relations between resources and concepts of the subject domain.

3. **Creating the information space on the basis of existing information spaces, created by other persons.** The person creates the cognitive space for some subject domain $CognitiveSpace\_1$, after that sets correspondences between concepts of $CognitiveSpace\_1$ and existing information space $CognitiveSpace\_2 : Concepts\_1 \times Concepts\_2 \rightarrow \{equivalence, compatible\}$, where: *equivalence* - equivalence relation between two concepts; *compatible* - less general or intersection relations between two concepts, that correspond to set-theoretic relations between classes and relations at an alignment between ontologies [17]. Thus, using the integration mechanism, the association relations are defined between the concepts of $CognitiveSpace\_1$ and information resources of the information space $InformationSpace\_2$. The person can redefine these relations later.

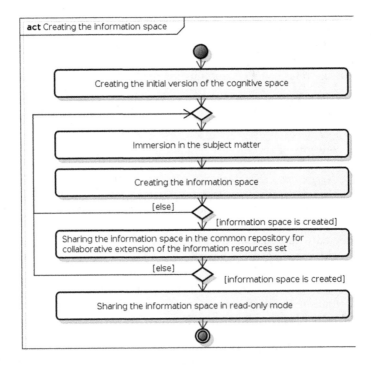

Fig. 1. Cognitive and information space creating process (UML activity diagram)

For formal representation of the cognitive and information spaces we propose to use the ontology described with OWL language (Fig. 2).

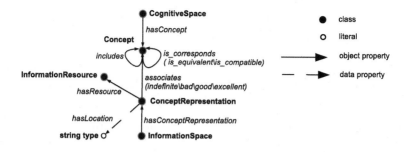

Fig. 2. The structure of the ontology of cognitive and information spaces

In terms of the proposed ontological model, creation of the information space for some subject domain reduced to creating the instance of the class *CognitiveSpace* and instances *concept$_i$* of the class *Concept* and defining the relations "includes" between these instances, as well as the creating one instance *cognitive_space* of the class *CognitiveSpace* and relations *hasConcept* between instance *cognitive_space* and all instances *concept$_i$*. Such approach allows describing the cognitive spaces of the individuals and groups within common ontological model, and also creating the information spaces on their basis by setting the relations between the concepts of different cognitive spaces. The creation of new relations between the objects of information space is carried out using logical inference over ontology. These relations can be used for inclusion in new information space the information resources which were previously included in other information spaces.

We applied the proposed approach to creation of the domain information space on the basis of cognitive space for the subject domain of programming languages. The fragment of cognitive space ontology "Operators in C" for domain "Programming Languages" is shown in Fig. 3.

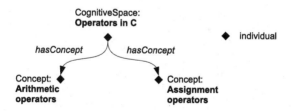

Fig. 3. The fragment of the cognitive space ontology for "Programming Languages" domain

In this fragment the cognitive space "Operators in C" is represented as an instance of class *CognitiveSpace* in corresponding ontology and includes the concepts of the subject domain "Arithmetic operators" and "Assignment operators", which are represented as the instances of the class *Concept* of the ontology, and the relationship "includes" between them.

The creation of the information space implies defining the instances *information_resources* of the class *InformationResource* and one instance *information_space* of the class *InformationSpace*. To define the relations "concept representation" between the concept *concept* ∈ *concepts* and the information resource *information_resource* ∈ *information_resources* it is necessary to:

1. define the instance *concept_representation* of the class *Concept Representation*;
2. define the relation *hasConceptRepresentation* between the instances *infor mation_space* and *concept_representation*;
3. define the relation *hasConceptRepresentation* between the instances *concept_representation* and *information_resource* ∈ *information_ resources*;
4. define the relation from the set *undefined, bad, good, excellent* between the instances *concept_representation* and *concept* ∈ *concepts*.

The ontology of the information space for "Programming Language C" domain is shown in Fig. 4.

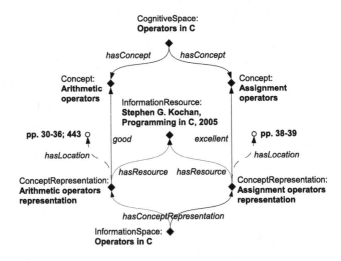

Fig. 4. The ontology of information space "Operators in C"

This fragment defines the information space "Operators in C" as an instance of the class *InformationSpace* of the ontology, at that the information space is the extension of the cognitive space defined above. The information resources are defined as instances of the class *InformationResource*, also the relations between the resources and appropriate representation of the concepts in the information space are specified, as well as the relations between the representations of the concepts in the information space (instances of the class

ConceptRepresentation) and the concepts of the cognitive space. The task of the determining the correspondences between the concepts of the different subject domains using the ontology representation boils down to the defining the relations from the set of *is_equivalent, is_compatible* between instances of the class *Concept* that belong to the different subject domains. To create the information space *InformationSpace_1* on the basis of the existent information space *InformationSpace_2* and the correspondences between the cognitive spaces *CognitiveSpace_1* and *CognitiveSpace_2* we developed the algorithm described below using the proposed ontology representation:

1. define the instance *information_space_1* of the class *InformationSpace* for the new information space;
2. for each information resource from the *InformationSpace_2* that is relevant for concepts of the cognitive space *CognitiveSpace_2*:
   2.1 define the instance *concept_representation_1* of the class *ConceptRepresentation* and
   2.2 set the relation *hasResource* between the instance *concept_representation_1* and the instance of the class *InformationResource*, belonging to the *InfromationSpace_2*;
3. for each instance *concept_representation_1* defined above:
   3.1 set the relation *hasConceptRepresentation* between the instances *informaton_space_1* and *concept_representation_1*;
   3.2 set the relation "concept representation" with one of the instances of the class *Concept* belonging to the *CognitiveSpace_1*.

To define the relevant information resources on the step 2.2 of the algorithm and the relations "concept representation" on the step 3.2, the set of the SWRL-rules was developed. These rules allow to take into consideration the correspondences between the concepts of the cognitive spaces *CognitiveSpace_1* and *CognitiveSpace_2* as well as the association relations between the concepts of the cognitive space *CognitiveSpace_2* and information resources of the information space *InformationSpace_2*. The SWRL-rules for the association of the two information spaces (Fig. 6) are defined as follows:

$$hasConceptRepresentation(?is1, ?cr1) \wedge hasConceptRepresentation(?is2, ?cr2)$$
$$\wedge hasResource(?cr1, ?res) \wedge good(?res, ?c1) \wedge is\_equivalent(?c2, ?c1)$$
$$\wedge sameAs(?cr1, ?c1) \wedge sameAs(?cr2, ?c2) \rightarrow$$
$$hasResource(?cr2, ?res) \wedge good(?res, ?c2);$$

$$(6)$$

$$hasConceptRepresentation(?is1, ?cr1) \wedge hasConceptRepresentation(?is2, ?cr2)$$
$$\wedge hasResource(?cr1, ?res) \wedge good(?res, ?c1) \wedge$$
$$is\_compatible(?c2, ?c1) \wedge sameAs(?cr1, ?c1) \wedge sameAs(?cr2, ?c2) \rightarrow$$
$$hasResource(?cr2, ?res) \wedge indefinite(?res, ?c2),$$

$$(7)$$

where $?is1, ?is2, ?c1, ?c2, ?cr1, ?cr2, ?res$ - the variables of the SWRL rules.

The example of the associating the cognitive spaces is shown in Fig. 5. The resulting information space is shown in Fig. 6. Thus, the relations *is_equivalent* and *is_compatible* are defined between the concepts "Assigment operators", and the concepts "Arithmetic operators" and "Increment and decrement operators" of the cognitive spaces "Operators in C" and "Lvalue in C", represented by the instances of the classes in the ontology (Fig. 5). In addition, the relations for inclusion of the information resources of the information space "Lvalue in C" into the information space "Operators in C" (Fig. 6) are defined as a result of reasoning over ontology using the SWRL-rules defined above.

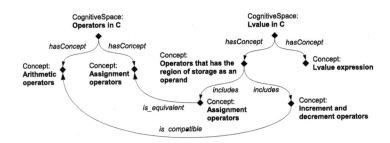

Fig. 5. The example of set of correspondences between two information spaces

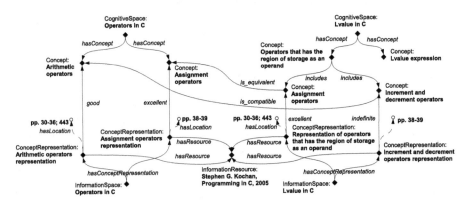

Fig. 6. The example of association of two information spaces

4   Software Tools for the Information Space Development

Software architecture (Fig. 7) and software tool for collaborative development of domain information space were designed, the Cognitive-information space editor was implemented as Java server application. It allows to create the cognitive space within domains as a set of concepts and relations (Fig. 8) and fill it with the annotations of the information resources to convert into the information space (Fig. 9).

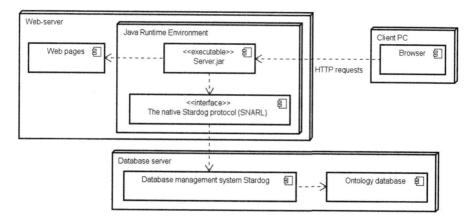

Fig. 7. Cognitive-information space editor architecture

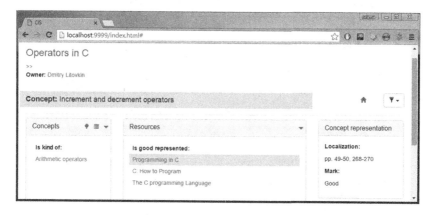

Fig. 8. Cognitive-information space editor: concepts

Also the software architecture and tool for creating the personal learning collections Collection Builder [1,2] were designed and implemented within framework of proposed knowledge-based approach as desktop application using C# language (Fig. 10).

It allows to create the personal learning collection (subset of the elements of the domain information space created by tutors) on the basis of learner profile and annotations of learning resources. For the query it uses the learning objective defined with domain concepts and competencies as well as the current student's competencies and knowledge as the prerequisites for the learning resources use. The collection is created using SWRL-rules and it is visualized in tree-structured form in accordance with domain structure defined in domain ontology.

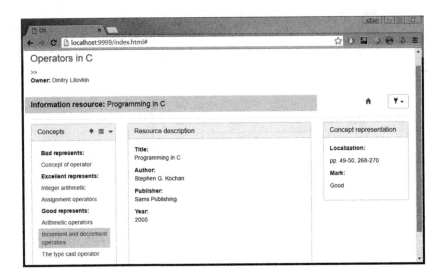

Fig. 9. Cognitive-information space editor: resources

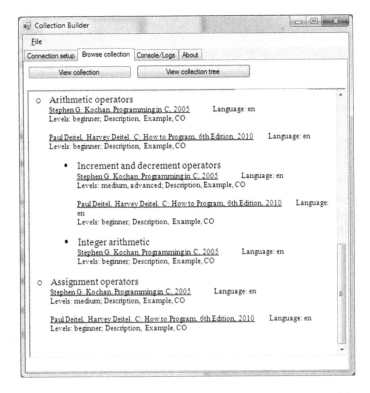

Fig. 10. Collection builder: tree-structured visualization of the personal learning collection

Developed software tools was applied for creation of personal collections for the course "Programming Languages" in Volgograd State Technical University (testing conditions: PC Intel Core i5-3337U, 4 GB of RAM; the number of learning resources in the repository - 50; the number of tutors - 4, the number of students - 20, the number of created collections - 20). Test results has shown that the average time of collection creation decreased almost by 99 %; automatically generated collection contains 100 % of learning resources obtained by the intersection of the collections created by tutors for each student, and 91 % of learning resources obtained by combining the tutors collections; the average value of collection recall increased by 29 %, precision - by 2,9 %, F-measure - by 16,3 % in comparison with non-automated process.

5 Conclusion and Future Work

The ontology-based approach to collaborative construction of the domain information space on the basis of the cognitive spaces of individuals or groups and the existing information spaces was proposed, that allows decreasing the time and increasing the efficiency of retrieval and reuse of the information resources which are relevant to the subject domain and the cognitive space of the information process subject.

This work is carried out in the framework of ongoing project, the main provisions of the proposed approach have been already used for the development of the software tool for distributed learning resources retrieval and creation of the personal learning collections [1,13]. The ontological model for knowledge representation was developed including ontologies of learning course domain, learning resource, learner's profile and personal learning collection. The last one includes the set of semantic rules for creating the personal learning collection. The new two-stage method for electronic learning resources retrieval and integration into personal learning collection was developed based on ontology reasoning rules. Developed models, method and software tool were successfully applied for creation of information space in the form of personal learning collection for the course "Programming Languages. C++" in Volgograd State Technical University.

Further evolution of the project implies the following activities: extending the set of SWRL-rules for the purpose of automatic generation of information space; implementation of the repository of information spaces; testing and evaluation of the proposed approach and developed software tools for information support of scientific research and learning process.

Acknowledgement. This paper presents the results of research carried out under the RFBR grant 15-07-03541 Intelligent support of decision making in management of large scale systems on the base of integration of different types of reasoning on ontological knowledge.

References

1. Anikin, A., Kultsova, M., Zhukova, I., Sadovnikova, N., Litovkin, D.: Knowledge based models and software tools for learning management in open learning network. In: Kravets, A., Shcherbakov, M., Kultsova, M., Iijima, T. (eds.) JCKBSE 2014. CCIS, vol. 466, pp. 156–171. Springer, Heidelberg (2014)
2. Anikin, A., Litovkon, D., Kultsova, M.: An ontology-based approach to collaborative development of domain information space. Economics and education. In: Proceedings of the 12th International Conference on Educational Technologies (EDUTE 2016). Proceedings of the 10th International Conference on Business Administration (ICBA 2016), Barcelona, Spain, 13–15 February 2016, pp. 13–19 (2016)
3. Brusilovsky, P.: Adaptive hypermedia for education and training. In: Durlach, P., Lesgold, A. (eds.) Adaptive Technologies for Training and Education, pp. 46–68. Cambridge University Press, Cambridge (2012)
4. Manning, C.D., Raghavan, P., Schütze, H.: Introductionto Information Retrieval. Cambridge University Press, Cambridge (2008)
5. Chugreev, V.L.: Model structurnogo predstavleniya informacii i metod ee tematichecskogo analiza na osnove chastotno-kontekstnoy klassifikacii: autoreferat k.t.n.: 05.13.01, Saint Petersburg State Electrotechnic University, Saint Petersburg, Russia (2003)
6. Creswell, J.W.: Educational Research: Planning, Conducting and Evaluating Quantitative and Qualitative Research, 3rd edn, pp. 8–9. Prentice Hall, Upper Saddle River (2008). ISBN 0-13-613550-1
7. Digital libraries in education, science and culture: analytical survey. UNESCO Institute for information technologies in education, Moscow (2007)
8. Sklavakis, D., Refanidis, I.: The MATHESIS meta-knowledge engineering framework: ontology-driven development of intelligent tutoring systems. Appl. Ontology 9(3–4), 237–265 (2014)
9. Newby, G.B.: Cognitive space and information space. J. Am. Soc. Inf. Sci. Technol. 52(12), 1026–1048 (2001)
10. Newby, G.B.: Metric multidimensional information space. In: Proceedings of TREC-5. The National Institute of Science and Technology, Gaithersburg (1996)
11. Kogalovsky, M.R., Parinov, S.I.: Technology for semantic enrichable scientific and educational digital libraries. In: Proceedings of the Economical Efficiency of Information Business-Systems, 16 April 2015. Moscow State University, Moscow (2015)
12. Kozlov, D.D.: Reshenie zadachi tematicheskogo informacionnogo poiska v runet: autoreferat k.t.n.: 05.13.11, Moscow State University, Moscow, Russia (2004)
13. Kultsova, M., et al.: Ontology-based learning content management system in programming languages domain. In: Kravets, A., Shcherbakov, M., Kultsova, M., Shabalina, O. (eds.) CIT&DS 2015. CCIS, vol. 535, pp. 767–777. Springer, Switzerland (2015)
14. Parinov, S., Kogalovsky, M.: Semantic linkages in research information systems as a new data source for scientometric studies. Scientometric 98(2), 927–943 (2014)
15. Parinov, S., Lyapunov, V., Puzyrev, R., Kogalovsky, M.: Semantically enrichable research information system SocioNet. In: Klinov, P., Mouromtsev, D. (eds.) KESW 2015. CCIS, vol. 518, pp. 147–157. Springer, Heidelberg (2015). doi:10.1007/978-3-319-24543-0_11
16. Pound, J., Mika, P., Zaragoza, H.: Ad-hoc object retrieval in the web of data. In: Proceedings of the 19th WWW, pp. 771–780 (2010)

17. Shvaiko, P., Euzenat, J.: Ontology matching: state of the art and future challenges. IEEE Trans. Knowl. Data Eng. **25**(1), 158–176 (2013). Institute of Electrical and Electronics Engineers

18. Tarasowa, Darya, Auer, S.: Collaborative authoring of OpenCourseWare: the best practices and complex solution. In: Mouromtsev, D., d'Aquin, M. (eds.) Open Data for Education. LNCS, vol. 9500, pp. 103–131. Springer, Heidelberg (2016). doi:10. 1007/978-3-319-30493-9_6

19. Trochim, W.M.K.: Research Methods Knowledge Base (2006). http://www. socialresearchmethods.net/kb/contents.php

20. Vasiliev, V., Kozlov, F., Mouromtsev, D., Stafeev, S., Parkhimovich, O.: ECOLE: an ontology-based open online course platform. In: Mouromtsev, D., d'Aquin, M. (eds.) Open Data for Education. LNCS, vol. 9500, pp. 41–66. Springer, Heidelberg (2016). doi:10.1007/978-3-319-30493-9_3

21. Stock, W.G., Stock, M.: Handbook of Information Science, p. 901. de Gruyter Saur, Berlin (2013)

22. Zaripova, V.M., Petrova, I.Y., Kravets, A., Evdoshenko, O.: Knowledge bases of physical effects and phenomena for method of energy-informational models by means of ontologies. Commun. Comput. Inf. Sci. **535**, 224–237 (2015)

23. Drug Interaction Knowledge Base Ontology. https://bioportal.bioontology.org/ ontologies/DIKB

24. https://www.academia.edu

25. http://www.authorea.com

26. http://www.intechopen.com

27. https://www.researchgate.net

28. https://www.sharelatex.com

Challenges of Implementation and Practical Deployment of Aviation Safety Knowledge Management Software

Peter Vittek(✉), Andrej Lališ, Slobodan Stojić, and Vladimír Plos

Department of Air Transport, Czech Technical University in Prague,
Horská 3, 128 03 Prague 2, Czech Republic
{xvittek,lalisand,stojislo,plos}@fd.cvut.cz

Abstract. This paper introduces practical issues of implementing and deploying safety knowledge management software to aviation safety. Domain specific intangible nature of the issues concerned is described and all factors which prevent successful application of any knowledge management system are documented. Aviation organizations are struggling to find any effective solution which would on one hand allow timely tracking of the dynamic knowledge and on the other to not limit or bias reporters or investigators within the knowledge gathering process. The article deals mainly with practical issues concerning the deployment of reporting software, which constitutes the interface between humans and ontology model behind the software, capable to address trade-offs between various conflicting design criteria as well as many aviation organization types, as the aviation safety knowledge is to be gathered in cooperation with the industry.

Keywords: Aviation safety · Reporting · Report forms · Knowledge management · Taxonomy

1 Introduction

Aviation safety is currently the subject of many efforts and research activities. One of the aspects of these efforts is to formalize all available knowledge and to encourage common and harmonized process for keeping it up to date [1]. The knowledge is then to be used backwards within the industry for safety management or safety oversight activities, providing respective authorities with timely and adequate suggestions on what to do. Given the complexity of the aviation industry, ever increasing safety standards as well as constant traffic growth, this process is a must and no single country or aviation organization can handle the issues completely on its own.

Because aviation became considerably globalized, the very same technology is being used all over the world. This entails not only more extensive experience with the technology itself, in terms of reliability and safety assessment, but also issues stemming from the cultural and maturity differences, especially as far as

© Springer International Publishing Switzerland 2016
A.-C. Ngonga Ngomo and P. Křemen (Eds.): KESW 2016, CCIS 649, pp. 316–327, 2016.
DOI: 10.1007/978-3-319-45880-9_24

the intangible safety of aviation system components and human interactions is concerned. Typically, hardware issues are defined using sensory data which are easy to quantify and usually based on mathematical or physical equations, providing very little room for bias or uncertainty. Owing to this, nowadays hardware solutions have reached excellent degree of reliability [2] and certainly they are rarely considered as unsafe. On the other hand, human and component interactions are still quite easily capable of generating unacceptable level of risk [3], but despite the fact that human factors are still subjected to specific research [4], there are no sensory data available nor can they be effectively obtained.

All recent accidents and incidents in the aviation exhibit presence of contributory factors, which are hardly manageable; they are often difficult to define or track, creating much more room for bias and uncertainty and frequently leading to inadequate corrective measures or oversimplified solutions [5]. This is because authorities are for various reasons rather prone to fix the consequences than learn about the true causes, mostly because of the lack of knowledge needed and the demanding nature of the learning process.

Accordingly, the industry demands adequate software solutions, which would enable effective knowledge gathering, processing and sharing among multiple aviation organization types and authorities. The solutions have to ensure these mechanisms be harmonized so that the learning process can provide for all its stakeholders desired degree of understanding both the issues and how to effectively control them. This article will describe existing efforts in this domain as well as the approach being developed within research projects at Czech Technical University in Prague.

2   Aviation Safety Knowledge

Because the hardware technology and its individual components have reached satisfactory degree of safety and reliability, this chapter will deal only with the persisting issue of component interactions and human presence in the system.

From the standpoint of system theory and safety engineering, the interactions concerned need to be first identified within the hazard analysis, either when new system is being designed or when existing system is to be modified [5]. Aviation industry rules and regulations do not allow hazard analysis to be left out prior to the operations, so this scenario is omitted. In other words, hazards are typically well-known and should always be there appropriately documented. Each of the hazards must be mitigated by the means of applying suitable safety constraints to make sure that the process controlled will remain under control at all times. Then, monitoring of the system, or more precisely controlled process, should allow for identification of any undesired behavior, usually utilizing safety key performance indicators [6]. Any violation of this designed safety control structure should be subjected to analysis to introduce adequate mitigation measures. Monitoring the system may also lead to identification of new hazards and subsequent modifications of the existing safety control structure.

Whilst from the standpoint of theory this is the best practice, dedicated systems are facing specific issues when being implemented and deployed. In the aviation, there are many different aviation organization types (such as airlines, flight schools, maintenance organizations, air navigation service providers etc.) which all operate within different part of the industry. Not only do they interact when it comes to the industry operations as a whole, but their internal safety control structures and technologies interact among themselves as well. Gathering information from all these interfaces, therefore, requires more than just interdepartmental cooperation; it is often about inter-organizational and international efforts.

To gather and track the interactions, individual interfaces must be monitored. Because the interfaces are mostly intangible, the only way how to achieve this is to observe them. In aviation, this typically requires someone to report anything he or she perceives as violation of the existing safety constraints, either originating directly from operational staff or indirectly from external or internal audits and occurrence investigation. In extreme cases, reporting may involve reporter's own mistakes or flawed actions. Depending on the way the information was achieved, these are called occurrence or investigation reports and audit findings. In some cases, semi-automated or automated processing is possible, but this is only in case where there is some hardware at least on one side of the interface.

Majority of the interfaces need to be observed so effective reporting forms are the key enabler towards the desired solutions. These must address the most common issue regarding any reporting forms: not to limit or bias the reporter in describing the occurrence or finding but, on the other hand, to guide him as much as possible to ensure highest quality of the data gathered. Processing and storing the data to knowledge management system database will then allow further decision-making about the system controlled. The more accurate the data stored, the better the mitigation measures.

3   Knowledge Management Tools

The essential tool used to manage aviation safety knowledge is the safety management system [7]. In its variations is can be applied to state level in form of a state safety plan or programme [2], but the principles remain always the same: it is a system of processes related to gathering and utilizing safety information in order to identify hazards and manage risks. The system has to be well documented in terms of roles and responsibilities, hazard analysis and risk assessment, data processing and privacy ensuring, and activities related to the follow-up of the information gathered, first of all safety assurance and safety promotion. These principles are all documented in the domain-related literature [8,9] and are not subject of this article.

In practice, room for improvements of knowledge management process still exists in aviation safety in its very fundamentals: the reporting process and data quality assurance. European Union (EU) has established joint repository for data gathered throughout the European mandatory and voluntary reporting system ECCAIRS, named as European Central Repository (ECR). The repository was then used to derive industry safety knowledge, subsequently represented and formalized in the so-called "risk portfolios" produced by European Aviation Safety Agency (EASA) [10]. They present basic dependencies also named as "safety issues", i.e. patterns describing what are the key contributory factors of the most frequently observed incidents or accidents. Because this knowledge is derived from central repository based on data from all Member States, it is highly relevant and cannot be obtained so easily from individual safety management systems. It is to be implemented into local safety management systems afterwards and utilized for both investigation as well as mitigation measures.

After years of its existence and more than one million records, the room for improvements was recognized as the data in ECR demonstrated various levels of quality, relevance and completeness and suggested that there are different reporting cultures among individual EU Member States [11]. Whilst this tells a lot about maturity, cultural differences and attitude towards safety of the reporting entity, it also points out the effectiveness and usability of the reporting system itself. It has a very complicated hierarchical structure of terms it uses and from the standpoint of user, an extensive training to use the system is a must. Despite the training, many users did not recognize potential of many features of the system, such as Analysis Tools, Safety Performance Indicators tool, Server Side Services or Application Programming Interfaces [11]. Together with the complicated structure of the terms, the system certainly needs refinements.

With regard to this, European Commission adopted new safety reporting framework for aviation, enforced throughout Commission Regulation (EU) No 376/2014. This reporting framework sets new processes for safety reporting and also introduces attributes and contextual information to be gathered. It narrows the reporting process and provides reporting entities with some guidance on what and how to report. EU Member States and individual aviation organizations are encouraged to assure compatibility with their local software solutions and reporting systems. This framework aims to resolve many recognized deficiencies of the previous reporting scheme with providing guidance, ensuring higher degree of user-friendliness and finally higher level of data consistency. .

As one of the examples of the progress, the so-called "smart forms" developed by EASA can be mentioned. The Agency recently introduced this concept aimed to simplify the reporting process by removing some burden from the reporter by limiting the fields in the form. The core of the concept is a simple hierarchical modeling of the form's sections. Each section has its context defined by the new EU legislation, e.g. aircraft section has attributes such as aircraft type, serial number or operator. Every reporter has to select his background and then is given a set of basic questions to answer (Fig. 1). Based upon them the system filters irrelevant sections with their data fields. This should lead to the reporter having more time to focus on relevant data fields and not to spend too much time filling the form.

Apart from that, Reduced Interface Taxonomy (RIT) is currently under development at EU level to support usage of the ECCAIRS system itself. Not only is this taxonomy expected to reduce the vast amount of terms used by ECCAIRS (more than 4000) but it should also reduce the complicated structure

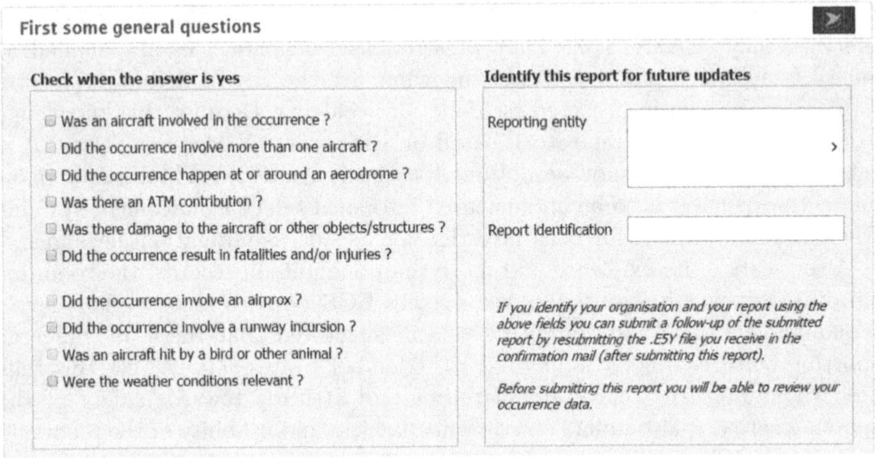
First some general questions

Check when the answer is yes

- Was an aircraft involved in the occurrence ?
- Did the occurrence involve more than one aircraft ?
- Did the occurrence happen at or around an aerodrome ?
- Was there an ATM contribution ?
- Was there damage to the aircraft or other objects/structures ?
- Did the occurrence result in fatalities and/or injuries ?
- Did the occurrence involve an airprox ?
- Did the occurrence involve a runway incursion ?
- Was an aircraft hit by a bird or other animal ?
- Were the weather conditions relevant ?

Identify this report for future updates

Reporting entity

Report identification

If you identify your organisation and your report using the above fields you can submit a follow-up of the submitted report by resubmitting the .E5Y file you receive in the confirmation mail (after submitting this report).

Before submitting this report you will be able to review your occurrence data.

Fig. 1. EASA reporting form [12]

of the original ICAO ADREP taxonomy [13] behind ECCAIRS. The structure of the system and the taxonomy is organized in a complex multilevel hierarchy (Fig. 2), suggesting that the structure was apparently not designed to allow searching terms. Very few aviation experts are really familiar with the structure and terms thus this shortcoming needs to be addressed too.

Another example is that ADREP was intended to distinguish between for instance event types and contributory factors, whilst contributory factors were divided into explanatory and descriptive. It depended on some internal logic of the taxonomy, which distributed the terms into those two categories and limited users in the way they could be used. Whilst the logic may seem to be clear for the authors, it wasn't so much for the users from the industry. Due to this, explanatory and descriptive factors are now being abandoned in the very ECCAIRS system.

EU Member States and aviation organizations have established own reporting systems, which are all in line with the applicable Regulation. The EASA smart forms are, however, the most advanced solution available to date, despite the fact that they don't use any advanced modeling. Even compared to ECCAIRS, smart forms provide more user-friendly way how to record some safety occurrence. They assure higher quality of aviation safety data and in return more accurate risk portfolios etc.

4 Addressing Practical Issues with Ontology

Due to its ability to cover all objects and relations between them, ontological modeling emerge as a powerful tool for systematic approach to aviation safety management. Ontology brings explicit specification of conceptualization [14], which offers great potential to address the issues with ECCAIRS and even to

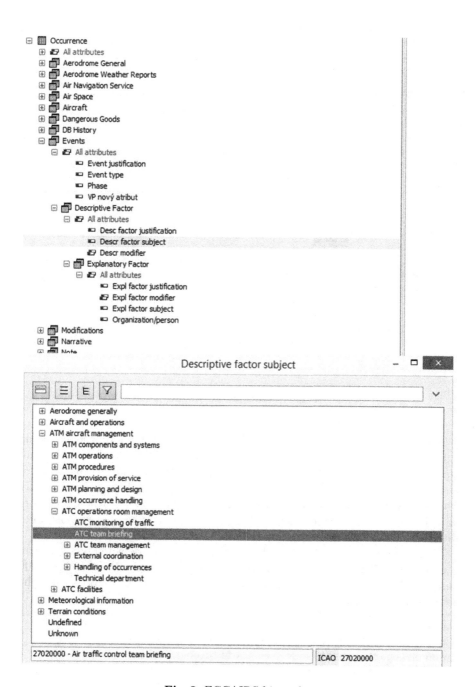

Fig. 2. ECCAIRS hierarchy

surpass EASA's smart forms. Ontology is formal, shared and explicit [15] and in its core well defines meanings in comprehensible structure. It is logic-based and easily processed, it covers whole domain and enable understanding of all its aspects. It was defined as shared understanding of a domain [16].

For research purposes a special ontology, related to the domain of aviation was developed. A domain model defines objects, events and relations that capture similarities and regularities in a given domain of discourse [17,18].

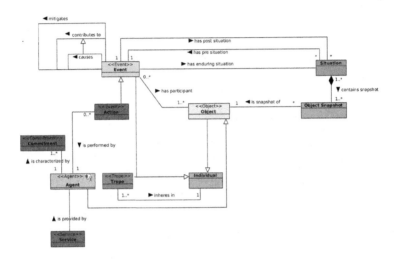

Fig. 3. Model of safety event

The primary task was answering the question how to approach modeling of terms used by ECCAIRS. EU legislation mandates EU Member States and aviation organizations to have ECCAIRS and ADREP compatible solutions for aviation safety reporting. The idea was to break down the ECCAIRS structure into its very terms and to model them using OWL, creating a new structure which finally allows advanced features. RIT taxonomy reduction was also integrated so that, where possible, only reduced vocabulary is used.

To exemplify the use of OWL, the very same term as depicted on Fig. 2 can be used, i.e. "27020000 - Air traffic control team briefing". Number 27020000 is an ECCAIRS identification, which can be preserved using OWL too, but the core difference is how the very term is modeled:

```
Briefing
and has_location some Operations_room
and has_participant some Air_traffic_controller
and has_participant min 2
```

The modeling is based on the core term, in this case it is "Briefing". It means that all other briefings are modeled likewise and so they can be all easily found and the differences between them identified. Links to other terms as "Operations room" and "Air traffic controller" are set using specific relations which allow effective searching of this particular "Briefing".

Ontology as such enables its application for many purposes and it depends on the application which needs are to be met. Formal description of the events (Fig. 3) is an useful feature enabling creation of ontology-based safety management tools such as tool for safety indicators. This new approach to safety knowledge tools allows generation of ontology-based smart forms which can utilize many different relations (as outlined above) than just simple hierarchical modeling as EASA's smart forms or ECCAIRS are based on. A whole approach to making reporting more user-friendly and appearing is make it easy and logical.

Approach chosen within newly developed safety reporting tool includes ontology-based smart form concept. The idea lies in finding mutual connections between individual contributory factors, relevant safety events, model objects and relations in order to create clear structure of occurrences. Relations are used in a way that assures not only effective use of advanced tools and features, but also the option to search for desired term.

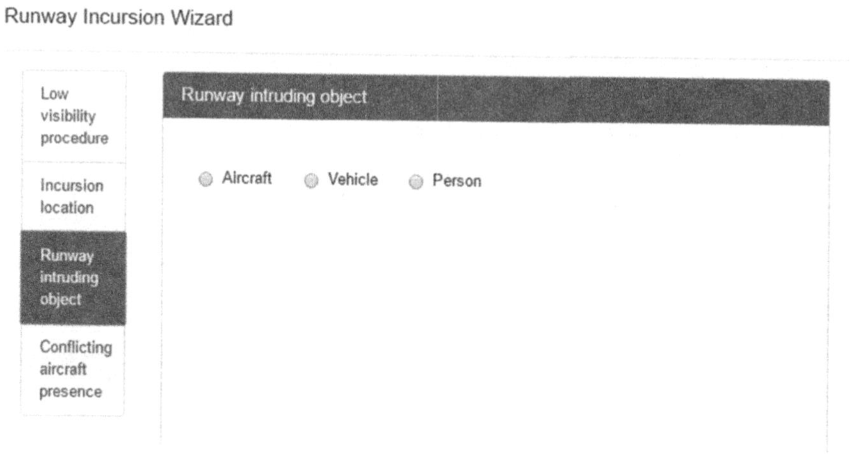

Runway Incursion Wizard

Low visibility procedure	Runway intruding object
Incursion location	◉ Aircraft ◉ Vehicle ◉ Person
Runway intruding object	
Conflicting aircraft presence	

Fig. 4. Smart reporting form - first level

The objective of wizard is to help a reporter by reducing the list of attributes only to those related and relevant for specific event type. Presented form on the previous figure (Fig. 4) shows a first step of interactive report submission, where form as such is reduced according to chosen fields. On the following picture (Fig. 5) presented form is generated according to the chosen object involved in reported occurrence. Generation of these forms uses legislation requirements, expert assessment, available safety studies and advanced modeling of both the structure of the terms as well as their meaning. Especially the modeling plays key role in form generation; each filled information narrows the rest set of data fields to be filled.

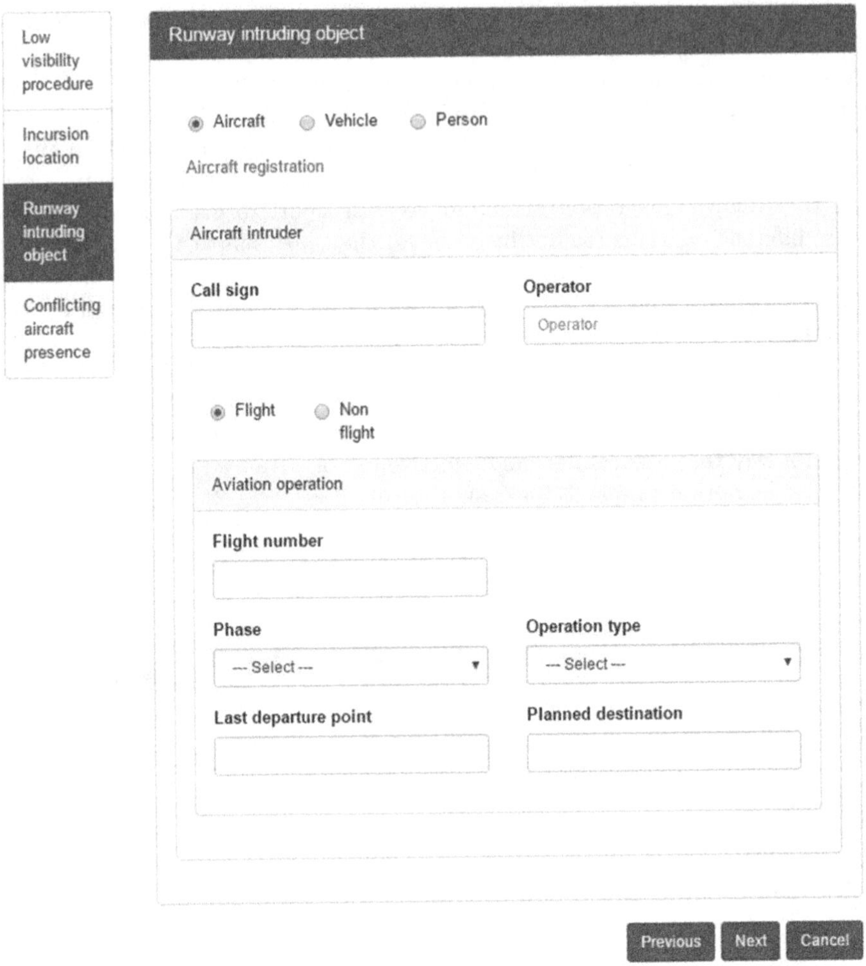

Fig. 5. Smart reporting tool - second level

Such approach is applied on all relevant event types. Its main purpose is to guide reporter correctly and logically through the reporting process in order to catch and emphasize all important facts regarding given safety event. Reporter is guided to submit comprehensible event description including underlining the contributing factors and circumstances. The tool allow using contributory factors but in line with the newest approach, i.e. using them as event types preceding the occurrence and not as special terms to be selected from separate lists. The interface of the tool is appealing and provides reporter with clear information and instructions.

Additionally, the model allows searching the most appropriate term to be used for event type, contributing factor or circumstance. Using controlled OWL language and then full-text search, one can search the entire vocabulary (Fig. 6). Analytic and statistical features of the tool bring additional possibilities for effective safety management. Different kind of data representation, statistics, query builders, etc., covers organization's needs and could support their productivity. Such easy-to-use software should decrease reluctance to such initiatives. Benefits brought through system implementation are numerous, including their compatibility and up-to-date format in relation to current legislation.

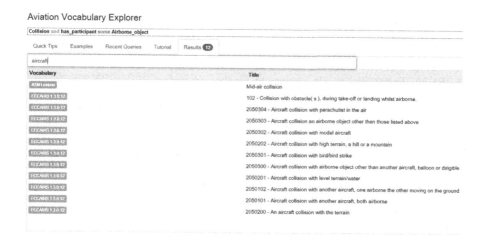

Fig. 6. Aviation vocabulary explorer

5   Conclusions

Development of safety knowledge systems is understood as support and encouraging step leading to effective safety management. It brings different kinds of benefits and covers whole domain. Current solutions have their strengths and weaknesses, but their main advantage is the fact that they are being used for quite some time, influencing their users to express quite strong reluctance toward implementation and deployment of the new, more advanced solutions.

Different strategies could be chosen in order to overcome this kind of problems. The focus should be placed on making the solution more understandable and easy to use, with clear expression of advantages that such system could provide. Also, the solution must fulfill the requirements for all kinds of organization, regardless of their current size of maturity level.

Ontological modeling is recognized as powerful tool for reaching expected result in the domain of aviation safety. It covers all aspects adequately and enables required level of occurrence structure description. Features that are brought as a result of an ontology implementation are numerous. Smart reporting forms developed through projects at the Czech Technical University defines the direction of the future safety management development.

Acknowledgement. This paper was supported by the Technology Agency of the Czech Republic, grant No. TA04030465.

References

1. ICAO - International Civil Aviation Organization: Paper A38-WP/85 on Consolidated Aviation Safety Knowledge Management: An Enabler of Improved Operational Safety for the 38th ICAO Assembly in September/October 2013, Montréal, Quebec, Canada (2013)
2. ICAO - International Civil Aviation Organization: Safety Management Manual (SMM): Doc 9859, 3rd edn., Montréal, Quebec, Canada (2013). ISBN 978-92-9249-214-4
3. Socha, V., Schlenker, J., Kalavsky, P., Kutilek, P., Socha, L., Szabo, S., Smrcka, P.: Effect of the change of flight, navigation and motor data visualization on psychophysiological state of pilots. In: SAMI 2015 - IEEE 13th International Symposium on Applied Machine Intelligence and Informatics, Proceedings, Herlany, Slovakia, 22–24 January 2015, pp. 339–344 (2015). ISBN 978-1-4799-8221-9
4. Novák, L., Němec, V., Soušek, R.: Effect of normobaric hypoxia on psychomotor pilot performance. In: The 18th World Multi-conference on Systemics, Cybernetics and Informatics. International Institute of Informatics and Systemics, Orlando, Florida, vol. II, pp. 246–250 (2014). ISBN 978-1-941763-05-6
5. Leveson, N.: Engineering a Safer World: Systems Thinking Applied to Safety. MIT Press, Cambridge (2011)
6. TRADE - Training Resources, Data Exchange. A Handbook of Techniques and Tools: How to Measure Performance. U.S. Department of Energy, Washington, DC (1995)
7. Stolzer, A.J., Halford, C.D., Goglia, J.J.: Implementing safety management systems in aviation. In: Ashgate Studies in Human Factors for Flight Operations. Ashgate, Burlington (2011). ISBN 978-1-4094-0165-0
8. Waring, A.: Safety Management Systems, 1st edn. Chapman & Hall, London (1996)
9. Gabbar, H.A.: The Design of a Practical Enterprise Safety Management System. Kluwer Academic Publishers, Dordrecht (2004)
10. EASA - European Aviation Safety Agency. Annual Safety Review 2014, Cologne, Germany (2015). ISBN 978-92-9210-196-1
11. Post, W.: ECCAIRS Survey. Presentation at ECCAIRS Steering Committee Meeting, Brussels, 26–27 October 2015. Joint Research Centre, Brussels, Belgium (2015). http://eccairsportal.jrc.ec.europa.eu/index.php/Documents/39/0/
12. JRC - Joint Research Centre. Aviation Safety Reporting (2016). http://www.aviationreporting.eu/
13. International Civil Aviation Organisation. ADREP 2000 taxonomy (2000)
14. Gruber, T.R.: A translation approach to portable ontology specification. Knowl. Acquisition **5**(2), 199–220 (1993)
15. Studer, R., Benjamins, V.R., Fensel, D.: Knowledge engineering: principles and methods. Data Knowl. Eng. **25**(1–2), 161–197 (1998)

16. Gomez-Perez, A., Fernandez-Lopez, M., Corcho, O.: Ontological Engineering. Springer, Heidelberg (2005)
17. Guizzardi, G.: Ontological Foundations for Structural Conceptual Models. University of Twente, Enschede (2005)
18. Ledvinka, M., Křemen, P.: JOPA: accessing ontologies in an object-oriented way. In: Proceedings of the 17th International Conference on Enterprise Information Systems, pp. 212–221. SciTePress - Science and Technology Publications, Porto (2015). ISBN 9789897580963

Requirements to Modern Semantic Search Engine

Ricardo Usbeck[1][(✉)], Michael Röder[1], Peter Haase[2], Artem Kozlov[2],
Muhammad Saleem[1], and Axel-Cyrille Ngonga Ngomo[1]

[1] AKSW Group, University of Leipzig, Leipzig, Germany
{usbeck,roeder,ngonga}@informatik.uni-leipzig.de
[2] metaphacts GmbH, Walldorf, Germany
{ph,ak}@metaphacts.com

Abstract. Since the introduction of computing machines into companies and industries, searching large enterprise data is an open challenge including diverse and distributed datasets, missing alignment of vocabularies within divisions as well as data isolated in format silos. In this article, we report the requirements of commercial enterprises to the next generation of semantic search engine for large, distributed data. We describe our elicitation process to gather end user requirements, the challenges arising for real-world use cases as well as how such an implementation of this paradigm can be benchmarked. In the end, we present the design of the DIESEL search engine, which aims to implement the requirements of commercial enterprise to semantic search.

1   Introduction

Computing machines have been used for decades in companies and the amount of work that is supported by information technology is still growing. However, the problem of searching in large enterprise data is still an open problem due to grown data infrastructure of many companies. Several different software solutions are used for different tasks leading to a distributed and diverse landscape of data silos comprising different formats. The integration of all these data silos to enable enterprise wide search is a complex and time consuming task.

On the other hand, Linked Data technologies, built upon the Resource Description Framework (RDF), already showed their strength to extract, interlink and search over a variety of datasets. Still, the main disadvantage of RDF hindering its proliferation in the area of companies is its high entry barrier. Normal users need to be trained to cope with RDF data. Additionally, large parts of the existing software and database landscape would have to be adapted.

The main idea of the DIESEL project[1] is to use the strengths of Linked Data technologies to create an RDF-based layer on top of the diverse data silos. This layer enables the development of a search engine over all company and open data islands that makes use of the advantages of RDF without (a) its

---

[1] http://www.diesel-project.eu.

© Springer International Publishing Switzerland 2016
A.-C. Ngonga Ngomo and P. Křemen (Eds.): KESW 2016, CCIS 649, pp. 328–343, 2016.
DOI: 10.1007/978-3-319-45880-9_25

drawbacks and (b) without huge integration efforts. The consortium, which is composed of two companies, a German and a Swiss-based company, as well as the Leipzig University, strives to provide an extensible open-source framework. In this article, we are going to present current requirements to modern semantic search engines. Our contributions in this article are as follows:

- First, we describe our use cases within the DIESEL project and to which extent these already point to novel research requirements.
- We present six up-to-date requirements analyses towards semantic-driven search over large, distributed enterprise as well as open data.
- We describe the influence of these requirements on our project and homogenize the raised issues to gain a concise overview and start outlining the DIESEL architecture.
- Consequently, we generate and collect matching benchmark definitions and datasets and present them to spark community contributions to the DIESEL platform.

To the best of our knowledge, we present the first systematic and industry-related as well as open data-based requirements specification for a large scale, semantic search engine.

2   Use Cases

In the following, we describe our three use cases within the DIESEL project.

Querying Enriched Encyclopedic Knowledge. DBpedia, YAGO as well as other open projects collect and systematize the world's knowledge and store it in a mostly structured way. However, accessing more complex knowledge from these sources remains a difficult task [19]. Thus, we will develop a generic search engine for encyclopedic knowledge that exceeds the capabilities of existing, text-based information retrieval approaches and truly understands the user intention while querying. This use case will make use of keywords and phrases and provide the user with state-of-the-art input comfort like auto-completion for query formulation. Furthermore, we will extend this knowledge by using unstructured streams (e.g. Twitter), tabular data (e.g. WHO) and other data sources which are not yet structured. Finally, we ensure that users can use any available knowledge base no matter of its location by implementing a federation layer—to query multiple RDF-based knowledge bases per input query—based on W3C standards like RDF and SPARQL 1.1.

Wikidata. This use case focuses on how enterprises search can be supported by utilizing Wikidata. In contrast to the previous use case, the focus here is not limited to Wikidata itself, but rather its integration with enterprise data sources. Wikidata provides entity descriptions for approximately 17 million entities covering a variety of domains [22].[2] While wikidata is not an enterprise

---

[2] https://www.wikidata.org/wiki/Special:Statistics.

search data source per se, its potential for supporting and enriching enterprise search is immense. This is largely due to the fact that Wikidata develops more and more towards a hub of identifiers for any kind of entities that enables the interlinking between disparate sources. In this context, the use case aims at enriching, contextualizing and integrating enterprise data with an open knowledge graph. Built on top of the Wikidata Knowledge Graph, the Wikidata Query Service[3] exposes this knowledge to the community and third-party developers through a scalable Web-based SPARQL endpoint, enabling queries such as "How did the population of Berlin develop over time", "Which countries are run by a female president", or "What are the most notable works displayed in the British Museum". The main focus will be, to allow customers to query the wealth of Wikidata in a way similiar to modern Web search engines without additional effort.

Medium-Large Enterprise Search and Knowledge Graph. Finally, we aim to leverage enterprise search by introducing a single point that enables querying all of a companies data as well as open data sources. This use case has as objective to ameliorate the search experience of employees in their everyday work. Here, we will extend the DIESEL engine to integrate the information from different sources where a semantically enriched federated search will be used as a single global search. Additionally, users will be able to visualise results in a structured and accessible graphical interface.

2.1   Elicitation Strategies

For our three use cases, we followed different elicitation strategies to account for the environmental situation of their future deployment.

Querying Enriched Encyclopedic Knowledge. First, we gathered academic requirements pertaining to search on encyclopedic data. Thus, we compiled the set of requirements behind the challenges Question Answering over Linked Data (QALD)[4] and Open Knowledge Base and Question-Answering (OKBQA)[5] by collecting:

- The motivation behind the benchmarks they provide. For example, to unify and extend existing as well as newly created datasets to compare the performance of systems.
- The current weaknesses of existing systems which took or are taking part in these challenges by analysing their internal structure and the corresponding publications.
- The drawbacks of existing datasets.

Overall, the elicitation led to the conclusion that current encyclopedic knowledge bases (YAGO, DBpedia) contain a large yet incomplete amount of valuable

---

[3] https://query.wikidata.org/.

[4] http://qald.sebastianwalter.org/.

[5] http://www.okbqa.org/.

knowledge. However, a completion using knowledge gathered from data sources of another structure (in particular text and tables) would improve (1) the spectrum of queries that can be answered and (2) the completeness of the answers.

Wikidata. Second, our German partner company used its community engagement in the Wikidata project, where it supported the development of the Wikidata Query Service based on the Blazegraph graph database.[6] As part of these activities, as well as through interviews with customers, which are anonymized, interested in using Wikidata and continued analysis of requirements expressed by the community (e.g. on the Wikidata mailing list), we elicited requirements for the Wikidata in enterprises use case.

Medium-Large Enterprise Search and Knowledge Graph. Finally, we analysed the requirements for enterprises which want to leverage knowledge graphs and search in their business environment. The collection of requirements for this use case was output of:

– Several informal interviews to the Swiss companies customers: this interviews had the objective to understand the organization's needs at management level, and to identify up to which extent they perceive the importance of integrating information within the company as well as with the German companies customers and leads.
– One formal interview with controlling personnel from a engineering contractor for rotating machines which seeks to manage and search its internal technical documentation: This interview focused on more technical approach, and had as objective to gather a single customer perspective on the information extraction aspects and the design of a user interface that could fulfil the needs.
– One formal interview with a customer of the German partner company, a large Swiss bank. The discussed potential use case involves analysts in a financial research department, who rely on search for their day-to-day analysis tasks, e.g., the task of predicting the development of inflation rates.
– A public online survey: This survey was designed with the objective of capturing most common requirements among different organisation, and also to understand how important it would be to have a global search among company data. The survey was created using Google forms, and is available at https://t.co/8eBakzLLPK[7]. It was distributed via our twitter account[8], through our LinkedIn contacts, and in the Search Engine Land LinkedIn group. We plan to leave the survey open considering that any future input can still be relevant for us.

2.2   Use Case-Driven Requirements

Querying Enriched Encyclopedic Knowledge. The requirements for this use case focus on the data life cycle from extraction to distributed querying

---

[6] http://metaphacts.com/wikidata.
[7] At the 22nd April, we gathered 13 responses. In the online version you can also see preliminary reports.
[8] https://twitter.com/project_diesel.

Table 1. Requirements for an enhanced encyclopaedic search.

ID	Title	Description
1–1	Knowledge Extraction from unstructured sources	It must be possible to extract supplementary triples from text and use them to extend existing knowledge bases
1–2	Knowledge Extraction from semi-structured sources	It must be possible to extract supplementary triples from tables and XML documents and use them to extend existing knowledge bases
1–3	Federation of Queries	It must be possible to use knowledge from distributed RDF stores within one query within reasonable time
1–4	Runtime Efficiency	Search system for RDF data should be able to return answers within 3s
1–5	Quality	Generated answers should have a high precision rather than a high recall to support trustworthy decision making
1–6	Verbalization of Answers	Users should see more than URIs

via SPARQL, quality of service and verbalization of answers. Table 1 lists the requirements from the QALD and OKBQA challenge in detail.

Wikidata. The second use case focuses on the deployment of Wikidata in an corporate environment to enhance readily available data. Thus, the requirements evolve around interfacing between existing and future modules to enrich a company's workflow. Table 2 presents 9 requirements elicited from customer interviews.

Medium-Large Enterprise Search and Knowledge Graph. At the core of the DIESEL project is the design of a semantic search engine for large enterprise data. Thus, the last requirement collection aimed at covering as much aspects as possible in a concise way. For the sake of readability, we separated this use case requirement into four parts to account for the different elicitation methodologies. First, we present the results of the informal interviews with customers from our partner companies, see Table 3.

Although there is a strong need for web-search, DIESEL will not focus on searching the Web since this is a different task than enterprise search. However, DIESEL will support searching web knowledge stored in internal data repositories which will have been transformed to RDF via our internal tools.

Second, we present the requirements from interviews with the rotary-machine producing companies controlling personal. Table 4 summaries the requirements. This customers also requires searching multiple documents, stored in various places and formats. Furthermore, this elicitation points out the importance of a context-aware knowledge extraction to reduce noise as well as the issue of corporate access control.

Table 2. Requirements for using Wikidata within company search.

ID	Title	Description
2–1	Wikidata as an entity hub	Linking existing data with Wikidata identifiers to query existing data and knowledge graphs and link sofar isolated data sources
2–2	Structured and unstructured data	Link together structured and unstructured data (e.g. Wikidata with the Wikipedia articles)
2–3	Elasticsearch	Due to the important role of entity search and the requirement to bridge with unstructured data, rich keyword search is essential. Of particular interest is support for Elasticsearch, as this is the search engine of choice of the Wikimedia projects
2–4	Multilingualism	Use Wikidata as a useful resource for enabling multilingual search
2–5	Contextualizing enterprise data	Linking enterprise data sources with Wikidata. Enable effective contextualization. Might require the combination of enterprise data with open data
2–6	Wikidata Ontology	Consider usage of Wikidata ontology while developing search interfaces
2–7	Taxonomies	Besides formal ontologies, much of the knowledge is represented in hierarchical classification schemes and taxonomies. The use of these structures in search is essential
2–8	Hybrid searches	Search over Wikidata is very diverse, involving e.g. the following:- Entity searches- Structured queries ("What are the largest cities with a female mayor")- Property paths of unknown length (e.g. ancestor relations, territorial structures, taxonomic structures, part of relationships)- Discovery of links/paths between entities- Image Searches- Temporal and spatial data
2–9	Provenance	Management of trust and provenance is very important. Related aspects include the management of evidences, references, annotations etc.

Third, we describe the requirements for a search system in the financial industry. Here, a Swiss bank details similar but also different aspects of enterprise search engines, see Table 5. These requirements introduce the issue of reusing existing thesauri and schemata, which is hard due to diverse matching problems. Moreover, financial research also relies on sentiment analysis of people,

Table 3. Requirements for search in knowledge graphs of SMEs and larger enterprises via informal interviews with customers.

ID	Title	Description
3–1	Search for internal office documents and emails	Support DOC, PPT, XLS, PDF and MS[a] Exchange email Server
3–2	Search in web	Support search in wiki, blogs and supplier web pages for guidelines, instructions and technical manuals
3–3	Search by vocabulary	Support a common vocabulary with synonyms
3–4	Datastore	Support MS SharePoint, file servers and the Web
3–5	Search UI	Simple search slit using natural language and auto-completion
3–6	Enhanced Search	Filter by vocabulary terms (concepts)
3–7	Filtering	Intuitive filtering by data type, time, owner, concepts
3–8	Enrichment	Connect search results with other company (SAP ERP, CRM) and web data
3–9	MashUp	Possibility to add mashup widgets, e.g., location-based information such as bus time table, events, point of interest
3–10	Trust	Possibility to specify only trusted sources

[a] MS stands for Microsoft

documents and market news. Thus an additional sentiment module would be advantageous. Finally, to enhance usability of the system it requires to represent results in a contextual and visual way to speed up understanding of the result.

At last, we introduce the results from our public survey which collected requirements from eight different companies. Table 6 presents the six derived requirements. Note here, that accessing already existing data in wiki-like systems, calendars and emails is at the core of user requirements. Again, we will not focus on search the Web, see above.

3   System Design and Requirement Implications

In the following, we summarize and order the requirements to be able to derive a concise general architecture for our open source implementation. Moreover, we introduce possible existing frameworks and approaches to tackle the raised challenges. An overview of the DIESEL architecture is depicted in Fig. 1.

Knowledge Extraction for Enterprise Data. First, a modern semantic search engine and respectively the DIESEL project needs to support the requested data sources. According to customers, each use case reports that

Table 4. Requirements for search in knowledge graphs of SMEs and larger enterprises via an interview with controlling personal from an engineering contractor for rotatary machines.

ID	Title	Description
3–11	Office documents	Support PDF documents
3–12	None-text PDFs	Customer documents (old) are provided as scanned files in PDF format, thus, a OCR is required
3–13	MS Word documents	Customer documents can be provided in MS Word
3–14	MS Excel documents	Customer documents can be provided as MS Excel
3–15	Automatic meta-information extraction	Extract document name, location, metadata or document header and footer data
3–16	Search based on a taxonomy	A taxonomy of concepts with different labels is required.[a]
3–17	Multilingualism	Support documents in English and German
3–18	Headers and footers	Since some of the project or document information is repeated in headers and footers, it creates critical noise that hinders to find detailed information fast
3–19	Data source	Support WebDAV file system
3–20	Data source	Support MS SharePoint
3–21	Full text search	Resources should be searchable like Web search engines already do
3–22	Faceted search	By using the taxonomy and metainformation, configurable facets should be provided
3–23	Nearby terms	Support span searches within paragraphs
3–24	Result preview	Enable search result snippets. (Answer verbalisation)
3–25	Result detail view	The user should have the possibility to click and open the document directly from the result view
3–26	Access Control	Support Active Directory

[a] Note, the generation of the taxonomy takes a lot of time that nobody wants to invest.

DIESEL must be able to extract and search classical formats next to existing data sources:

- Support open data formats: XML, CSV, TSV, RDF, HTML.
- Support proprietary formats: MS Word, MS Excel.
- Support PDF which may require OCR technologies.
- Able to access existing data bases, mail servers, calendars, MS Share Point and media wiki.

Table 5. Requirements gathered during an specification discussion from a Swiss bank.

ID	Title	Description
3–27	Search over unstructured data	Support PDF and HTML documents containing e.g., speeches about economic developments
3–28	Concept and entity extraction	Effective search requires a precise identification of specific entities (such as places, organisation, persons) and concepts (financial domain concepts, topics)
3–29	Taxonomies and glossaries	Support existing glossaries, thesauri and taxonomies of relevant concepts and entities
3–30	Sentiment analysis	The system requires to understand the sentiment of particular search terms
3–31	Search over structured data	Support company internal sources with numerical data about economic indicators, such as inflation rates
3–32	Contextualized result visualization	The result view should contain contextual information, e.g., how documents are related, their sentiment, covered topics etc. using, e.g., through graphs, heat maps, tag clouds
3–33	Multilingualism	Support German and English

Table 6. Requirements derived from public survey. Users entered these in a comment field.

ID	Title	Description
3–34	Separate security spaces	Account for the security level of the user
3–35	Search on email	Support email search
3–36	Search on media wiki	Support search on companies media wiki installation
3–37	Calendars	Support search in various calendar formats
3–38	Popular search engines	The first source of information in companies used is Web search engines
3–39	Time	Reduce search time (less that 10 min for a global search)

Next to a semantic access to existing data sources, DIESEL needs to be able to truly understand the data at hand. Thus, the following criteria specify the capabilities of the data extraction modules:

– Extraction of domain-specific terminology (e.g. financial).
– Include metadata extraction (e.g. authors, creation dates) about the documents.

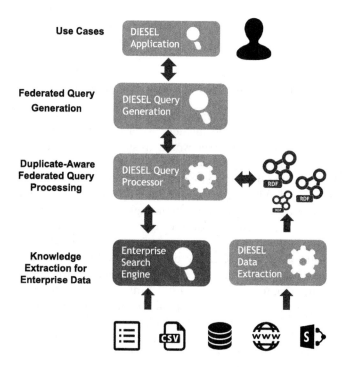

Fig. 1. Top-level architecture of the DIESEL search engine.

– Multilingual search capabilities per deployment for English and German.
– Extract entities such as places, organisation, persons, and also specific domain concepts (financial domain concepts, topics).

   We will tackle these criteria by deploying data wrappers to RDF[9] such as Google Refine or RDF123 and will extend the systems where needed. Furthermore, we will make use of our own FOX-Framework [16], which is based on AGDISTIS [20], to extract RDF entities from unstructured data as well as deploying REX [2], a web-scale extraction system for heavily templated websites such as media wiki info boxes.

DIESEL Search Engine Core. After introducing a RDF layer on top of existing enterprise data, we need to develop a search algorithm that abides the following:

– The DIESEL search engine should provide high quality answers instead of a large number possibly irrelevant data.
– DIESEL should show the diverse interpretations of user input.
– The system has to reuse domain specific vocabularies and taxonomies.

---

[9] https://www.w3.org/wiki/ConverterToRdf.

– Multilingual queries: one DIESEL instance is required to process each language.
– The system should extend result with similar concepts.
– The system should provide distance between concepts provided in the search.
– The system should be able to rank results w.r.t. user intention.

As can be seen in many requirements, we will focus on a high precision instead of a high F-measure which would include also a high recall. In industrial environments a clear, expressive and trustworthy results seems to be more important than to no every possible correct answer. Our algorithm will be based on the underlying RDF data and leverage its graph-shape by using an algorithm similar to our own SESSA [8]. This algorithm is a high precision spread-activation algorithm over RDF graphs. Based on its high-quality result set, we will make use of existing Least-General-Generalization algorithms [6] to increase the answer coverage. Further, we will deploy a novel version of SPARQL2NL [10] to enable end-users to understand the internal representation of their input. Additionally, we will use make use of Ginseng[10], a modular semantic search engine to satisfy the need for similar concepts, proper result ranking and related concepts based on a similarity function. We will develop all DIESEL components with multilingualism in mind, with an initial focus on the languages required for the use cases, specifically German and English. Note, that the core DIESEL modules will not focus on query generation. This will be part of the use case implementations and the respective industrial exploitations.

DIESEL Search over Distributed Data. Although, we transformed existing data into RDF and our algorithm works on any RDF graph data, we need to be able to plug-in all data sources and query all endpoints at once. Thus, we need to implement a SPARQL federation layer, which satisfies the following user needs:

– Support querying different information silos within one query.
– Ensure access right restriction by enrolling a policy layer to the federation engine.
– Efficient execution.
– The system should be able to search in Solr.
– The system should be able to search in Elasticsearch.

We will extend the QUETSAL [12] federation engine which provides time efficient and effective source selection and ranking algorithms. QUETSAL already provides means to execute queries only on a subset of policy-restricted data sources via its SAFE extension [5]. The extension will consider time-efficient joins for federated queries and a duplicate aware federation. Finally, we will add capabilities to query state-of-the-art full-text search engines.

---

[10] http://ginseng.aksw.org/.

Use Case Specific Criteria. Finally, there are requirements which only apply to certain use cases.

– Ability to include results from enterprise search engines.
– The search interface should provide a faceted search based in a taxonomy.
– The allows filtering by vocabulary terms (concepts).
– The system should also search into the Google Knowledge Graph

These extension will be built as extension to the previous two DIESEL modules, i.e., the query generation core and the query federation layer, to include other data sources, special filters or extend the Ginseng-based interface with novel widgets. Thus, DIESEL will remain modular and extensible.

Required Benchmarks. To ensure a stable and guided development process, the development of a large-scale semantic search engine requires target key performance indicators (KPIs). Our survey yielded a number of measurements and values from which we used the maximal boundaries to push the limits of the DIESEL search engine further. Table 7 details the chosen modules and their respective KPIs.

4   Related Work

To develop a fully-fledged semantic search engine for large enterprise data tackles many fields. Here, we will focus on the field of keyword-based search as it is at the core of this survey.

Semplore [24] is the first known hybrid search engine by IBM. It combines existing information retrieval index structures and functions to index RDF data as well as textual data. Semplore focuses on scalable algorithms and is evaluated on an early Question Answering over Linked Data (QALD[11]) dataset.

Bhagdev et al. [1] describe an approach to hybrid search combining keyword searches, Semantic Web inferencing and querying. The proposed K-Search outperforms both keyword search and pure semantic search strategies. Additionally, a user study reveals the acceptance of the Hybrid Search paradigm by end users.

A personalized hybrid search implementing a hotel search service as use case is presented in [23]. By combining rule-based personal knowledge inference over subjective data, such as expensive locations, and reasoning, the personalized hybrid search has been proven to return a smaller amount of data thus resulting in more precise answers.

SINA [15] aims at answering a keyword question using different datasets. First, simultaneous disambiguation and segmentation is performed using Hidden Markov Models (HMM) and the Hyperlink-Induced Topic Search (HITS) algorithm. The resources found are used to construct an Incomplete Query Graph (IQG) constisting of disjoint sub-graphs. To build the federated SPARQL query

---

[11] http://greententacle.techfak.uni-bielefeld.de/~cunger/qald/.

Table 7. Target benchmark modules and performance indicators.

Benchmark	Method	Estimation/Measurement for Evaluation
Indexing	Collection of real-world log-files from DIESEL prototypes. Measuring the performance of single look-ups	<50ms per request
Similarity	Measure time-efficient implementation of similarity measures [4,9]	<20ms per request
Auto-completion	Developing a benchmark for autocompletion	Survey on user satisfaction
Query Expansion	Reuse and extension of benchmarks [14]	>0.65 F-measure
Ranking	Extend existing benchmarks to match DIESEL use cases [7]	>0.6 Accuracy
Query generation benchmark	Evaluation of query generation using QALD and other datasets [18]. Extraction, extension and evaluation of federated queries from [11]	>0.5 F-measure
Verbalization benchmark	Reuse and extend existing benchmarks from [10]	User study
Federation benchmark	Reuse benchmarks from [13]	<120 ms overall query runtime
Knowledge Extraction from Unstructured Data	Existing benchmarks based on the GERBIL platform [21]	>0.7 F-measure A2KB
Scalable Knowledge Extraction from Structured Data	Reuse existing benchmarks [2]	>0.9 Precision

that retrieves the results, the IQG's are connected using a Minimum Spanning Tree approach inspired by Prim's algorithm.

The work of Tran et al. [17] tackles the problem of keyword search over RDF data. More specifically, their work is concerned with mapping keywords to a list of ranked conjunctive queries, with a special focus on efficient inference of implied connections. To accomplish this, a top-k algorithm is proposed that computes the best query interpretations of the keyword query using bidirectional graph exploration. The interpretations are then scored and mapped to conjunctive queries.

Optique [3] aims at developing end-user oriented, semantic query interfaces over enterprise data sources by applying the techniques of OBDA (Ontology-based Data Access). With OBDA, end users can express queries over an ontology as conceptualization of the domain; mappings are employed to relate the ontology to the schemas of the underlying data sources. Optique provides an end-to-end

OBDA platform, from end-user oriented query formulation to the actual query execution, but with a clear focus on relational data sources.

To the best of our knowledge, there is no open-soure semantic search engine satisfying a large subset of the requirements elicited earlier. Note, we are not focusing on exploring related commercial work.

5   Conclusion

In this article, we presented a concise requirements specification as well as benchmarking measures to develop an open data using, enterprise, semantic search engine over large data. We collected information from various sources, such as businesses, communities and academia. With these instruments at hand, we are going to develop the open-source DIESEL search engine together with our partners. Our intuition is that we cover enough use cases and domains to implement DIESEL in a way that will make it easy to generalize the project prototype to an industry-ready application.

In the future, we want to develop a first prototype and elaborate on the requirements to speed-up a close-to-reality implementation of our platform. We will integrate upcoming user requirements from interested parties in our development cycle and strive for an European dissemination of our efforts.

Acknowledgements. This work has been supported by Eurostars projects DIESEL (E!9367) and QAMEL (E!9725) as well as the European Union's H2020 research and innovation action HOBBIT (GA 688227).

References

1. Bhagdev, R., Chapman, S., Ciravegna, F., Lanfranchi, V., Petrelli, D.: Hybrid search: effectively combining keywords and semantic searches. In: Bechhofer, S., Hauswirth, M., Hoffmann, J., Koubarakis, M. (eds.) ESWC 2008. LNCS, vol. 5021, pp. 554–568. Springer, Heidelberg (2008)
2. Bühmann, L., Usbeck, R., Ngonga Ngomo, A.-C., Saleem, M., Both, A., Crescenzi, V., Merialdo, P., Qiu, D.: Web-scale extension of RDF knowledge bases from templated websites. In: Mika, P., et al. (eds.) ISWC 2014, Part I. LNCS, vol. 8796, pp. 66–81. Springer, Heidelberg (2014)
3. Giese, M., Soylu, A., Vega-Gorgojo, G., Waaler, A., Haase, P., Jiménez-Ruiz, E., Lanti, D., Rezk, M., Xiao, G., Özgür, L.Ö., Rosati, R.: Optique: zooming in on big data. IEEE Comput. **48**(3), 60–67 (2015)
4. Hoffart, J., Altun, Y., Weikum, G.: Discovering emerging entities with ambiguous names. In: Proceedings of the 23rd International Conference on World Wide Web, WWW 2014, pp. 385–396. ACM, New York (2014)
5. Khan, Y., Saleem, M., Iqbal, A., Mehdi, M., Hogan, A., Hasapis, P., Ngonga Ngomo, A.-C., Decker, S., Sahay, R.: SAFE: policy aware SPARQL query federation over RDF data cubes. In: Semantic Web Applications and Tools for Life Sciences (SWAT4LS) (2014)

6. Lehmann, J., Bühmann, L.: AutoSPARQL: let users query your knowledge base. In: Antoniou, G., Grobelnik, M., Simperl, E., Parsia, B., Plexousakis, D., De Leenheer, P., Pan, J. (eds.) ESWC 2011, Part I. LNCS, vol. 6643, pp. 63–79. Springer, Heidelberg (2011)

7. Lopez, V., Nikolov, A., Fernandez, M., Sabou, M., Uren, V., Motta, E.: Merging and ranking answers in the semantic web: the wisdom of crowds. In: Gómez-Pérez, A., Yu, Y., Ding, Y. (eds.) ASWC 2009. LNCS, vol. 5926, pp. 135–152. Springer, Heidelberg (2009)

8. Lukovnikov, D., Ngonga-Ngomo, A.-C.: Sessa - keyword-based entity search through coloured spreading activation. In: NLIWoD@ISWC (2014)

9. Mendes, P.N., Jakob, M., García-Silva, A., Bizer, C.: DBpedia spotlight: shedding light on the web of documents. In: Proceedings of the 7th International Conference on Semantic Systems, pp. 1–8. ACM (2011)

10. Ngonga Ngomo, A.-C., Bühmann, L., Unger, C., Lehmann, J., Gerber, D.: SPARQL2NL - Verbalizing SPARQL queries. In: Proceedings of WWW 2013 Demos, pp. 329–332 (2013)

11. Nikolov, A., Schwarte, A., Hütter, C.: FedSearch: efficiently combining structured queries and full-text search in a SPARQL federation. In: Alani, H., et al. (eds.) ISWC 2013, Part I. LNCS, vol. 8218, pp. 427–443. Springer, Heidelberg (2013)

12. Saleem, M., Ali, M.I., Verborgh, R., Ngonga Ngomo, A.-C.: Federated query processing over linked data. In: Tutorial at ISWC (2015)

13. Saleem, M., Ngonga Ngomo, A.-C., Xavier Parreira, J., Deus, H.F., Hauswirth, M.: DAW: duplicate-aware federated query processing over the web of data. In: Alani, H., et al. (eds.) ISWC 2013, Part I. LNCS, vol. 8218, pp. 574–590. Springer, Heidelberg (2013)

14. Shekarpour, S., K. Höffner, J. Lehmann, Auer, S.: Keyword query expansion on linked data using linguistic and semantic features. In: 7th IEEE International Conference on Semantic Computing, 16–18 September 2013, Irvine, California, USA (2013)

15. Shekarpour, S., Ngonga Ngomo, A.-C., Auer, S.: Question answering on interlinked data. In: Proceedings of the 22nd International Conference on World Wide Web, pp. 1145–1156. International World Wide Web Conferences Steering Committee (2013)

16. Speck, R., Ngonga Ngomo, A.-C.: Ensemble learning for named entity recognition. In: Mika, P., et al. (eds.) ISWC 2014, Part I. LNCS, vol. 8796, pp. 519–534. Springer, Heidelberg (2014)

17. Tran, T., Wang, H., Rudolph, S., Cimiano, P.: Top-k exploration of query candidates for efficient keyword search on graph-shaped (rdf) data. In: IEEE 25th International Conference on Data Engineering, ICDE 2009, pp. 405–416. IEEE (2009)

18. Unger, C., Forascu, C., Lopez, V., Ngomo, A.N., Cabrio, E., Cimiano, P., Walter, S.: Question answering over linked data (QALD-5). In: CLEF (2015)

19. Usbeck, R.: Combining linked data and statistical information retrieval. In: Presutti, V., d'Amato, C., Gandon, F., d'Aquin, M., Staab, S., Tordai, A. (eds.) ESWC 2014. LNCS, vol. 8465, pp. 845–854. Springer, Heidelberg (2014)

20. Usbeck, R., Ngonga Ngomo, A.-C., Röder, M., Gerber, D., Coelho, S.A., Auer, S., Both, A.: AGDISTIS - graph-based disambiguation of named entities using linked data. In: Mika, P., et al. (eds.) ISWC 2014, Part I. LNCS, vol. 8796, pp. 457–471. Springer, Heidelberg (2014)

21. Usbeck, R., Röder, M., Ngonga Ngomo, A.-C., Baron, C., Both, A., Brümmer, M., Ceccarelli, D., Cornolti, M., Cherix, D., Eickmann, B., Ferragina, P., Lemke, C., Moro, A., Navigli, R., Piccinno, F., Rizzo, G., Sack, H., Speck, R., Troncy, R., Waitelonis, J., Wesemann, L.: GERBIL - general entity annotation benchmark framework. In: 24th WWW Conference (2015)

22. Vrandečić, D., Krötzsch, M.: Wikidata: a free collaborative knowledgebase. Commun. ACM **57**(10), 78–85 (2014)

23. Yoo, D.: Hybrid query processing for personalized information retrieval on the semantic web. Knowl. Base Syst. **27**, 211–218 (2012)

24. Zhang, L., Liu, Q., Zhang, J., Wang, H., Pan, Y., Yu, Y.: Semplore: an IR approach to scalable hybrid query of semantic web data. In: Aberer, K., et al. (eds.) ASWC 2007 and ISWC 2007. LNCS, vol. 4825, pp. 652–665. Springer, Heidelberg (2007)

Medical Knowledge Representation for Evaluation of Patient's State Using Complex Indicators

Mikhail Lushnov, Vyacheslav Kudashov, Alexander Vodyaho,
Maxim Lapaev, Nataly Zhukova[✉], and Denis Korobov

Information Systems Department,
Saint-Petersburg State Technical University, Saint Petersburg, Russia
nazhukova@mail.ru

Abstract. In the paper a hierarchy of formalized models for data, information and knowledge representation in medical domain is proposed. The models allow solving problems of calculation, evaluation and analysis of complex indicators of patient's organism state. The model hierarchy is composed of raw objective numerical data description, the description of the outputs of statistical and intelligent processing and analyses procedures. To build the model a set of transformations are defined according to JDL fusion model adapted for medical objective data. Models are implemented as a system of ontologies. Experimental research of the models and transformations was conducted on historical data of Almazov Medicine research center (Saint-Petersburg, Russia).

Keywords: Knowledge representation · Medical scenarios · Measurement data processing and analysis

1 Introduction

The study of property sets of various phenomena has led scientists to the need for a systemic approach. The need for such approach in study of the entire organism was felt by researchers for a long time. There are a lot of different definitions of the term "system". One of them suggested in 1973 by P. N. Anokhin that is often used in medical domain is the following "a system is a set of components involved in a process of interaction during which they assist each other with the aim to receive useful result". Functional system is a unit of the integration of the whole organism, composed dynamically to achieve adaptive features of the organism. It is always based on the cyclical relationship selectively combining special central-peripheral formations.

System approach in medicine and biology is defined by the following main entities: (i) set of interdependent elements; (ii) mechanism of interaction with environment; (iii) inclusion of the investigated system as an element of higher level of a system (organ, tissue and whole organism); (iv) possibility of consideration of a system elements as systems of lower level [1].

For medical practice the main point of interest is internal interaction of functional system elements, and first of all collaborative and correlation-cooperative interactions

A.-C. Ngonga Ngomo and P. Křemen (Eds.): KESW 2016, CCIS 649, pp. 344–359, 2016.
DOI: 10.1007/978-3-319-45880-9_26

which result appearance of phenomena of self-organization in an organism and its subsystems in healthy and unhealthy state.

From the point of view of data processing and analysis definition of the term system is the following: a system is a set of elements, characterized by connections with each other and an additional feature - a function that does not match or is not characterized by any of the properties of its individual elements. System functional state is a result of interaction of environment and native properties of a subject i.e. a set attributes, properties, features which directly or indirectly characterize an activity.

For solving problems of such kind system analysis in particular multi dimension statistics and methods of optimization are used. For mining general rules of development of objects of different nature synergetic methods are used.

Identification of internal interactions of system elements needs deep knowledge of medicine and statistics and takes a lot of time. This limits possibility of receiving and usage information about internal interaction between system elements while maintaining a patient. The problem can be solved by means of usage of hierarchical formalized information models for describing medical data and results of its processing. The number of levels in the model must be agreed with the number of steps of data processing. Transitions between levels have to match processing steps.

Usage of formalized models gives following advantages:

- allows make systematization and specification of operations, in particular, define order of their execution, necessary preconditions, etc.;
- allows use machine processing on the defined steps.

In the Sect. 2 of the paper known approaches to data, information and knowledge presentation in medicine domain are discussed. Sections 3 and 4 provide descriptions of data and processes to be formalized. In the Sects. 5 and 6 descriptions of suggested hierarchical models are presented. In the Sect. 7 description of the model in the form of ontology is presented and an example of its usage is given.

2 Background

Data Representation Approaches. For medicine domain standards of description medical data, dictionaries, classifications are developed, multiple models of medicine informatics are suggested, general and special purpose ontologies are defined.

Ready solutions are well described in free publications. Number of publications for last 5 years is more than 12000 articles according to Springer Link and Elsevier Science Direct [4]. Systematization of publications in frame of PubMed project is realized. PubMed is an English language text data base of medical and biological publications created by National Center of Biotechnological Information (NCBI) on the base of USA National Medical Library (NLM) [2]. It is a free version of MEDLINE data base [3]. Detailed overview of the state of the art of the domain one can find in [4, 5]. It is necessary to mention also International Conference on Harmonization of Technical Requirements for Registration of Pharmaceuticals for Human Use [6], Consolidated Standards of Reporting Trials [7].

Meta ontologies of medicine domain are: global medical thesaurus Unified Medical Language System(UMLS) Metathesaurus [8]; integrator of data sources and meta ontologies SNOMED Clinical Terms [9]; Technical Implementation Guide [10]; Medical Subject Headings [11]; Medical Dictionary for Regulatory Activities [12]. International Classification of Diseases [13]; ICD10Data.com [14]; normalized names of medicines RxNorm [15] one can conceder as domain ontologies. Extended analyses of the ontologies is given in [5]. Nowadays in Russian medical centers ontologies are not used. Only ICD10 is used as a classifier.

Data Processing Approaches. JDL model is the de facto standard model of data processing and analyses. This model is a general model that is not directly connected with any subject domain. The model was initially developed at the beginning of the 1990s by the American Joint Directors of Laboratories and was modified many times. The JDL model is a functional multilevel model that defines "what is to be done" at each level but not "how it is to be done". The model is supposed to be implemented for each subject domain or a group of tasks. It divides the functions applicable to data into four levels. Below a brief description of JDL model levels according to [16] is given (Fig. 1).

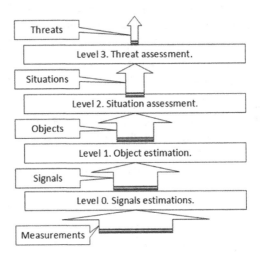

Fig. 1. JDL model levels

The core of the JDL model refers to the levels L0–L3. Level 0 is aimed to calculate and to estimate the characteristics of the separate signals as well as to solve the tasks of signals assessment and prediction. At Level 1 estimation of separate objects characteristics is fulfilled. The main goal of this level is the assessment and prediction of values of the separate objects parameters and the state of the analyzed objects. At Level 2 situation assessment is performed. The main objective of this level is the assessment and prediction of the structure and the properties of the relations between natural and technical objects of some part of the real world and their influence on each other and the environment. At this level the functions oriented on obtaining information on the analyzed situations in a certain context in terms of objects, the relations between objects and possible events are considered. Level 3 is the threat assessment level at which the situations and the dynamic of their changes are analyzed, assumptions about actions of external objects, probable threats and possible vulnerabilities are estimated [17].

3 Medical Processes Specification

Medical Processes Specification. For analysis of internal interactions of elements of functional biosystems methods of optimization which can give statistical functional estimations of the system (set of biological system parameters) are used. These methods allow calculate correlation estimations. Biological sense of such models is based on the ideas formulated by Yu. N. Shanin and coauthors [18] about maximum of correlation couples in normal state of organism and presence of misbalance in case of presence of pathology. In order to describe biosystem with the help of statistical methods one can use an integral estimation functional which describes the set of attributes of a number of biological object in a fixed moment of time. Functional can be conceded as complex measure. General algorithm of its calculation is described below. It is based on searching of variants of splitting of a set of objects to not crossed classes. Each class is a set of biological parameters of a functional subsystem, which give local maximum to a functional - sum of "internal" correlation couples mines some threshold value [19]. The result of calculation of the functional is integral indicator (II). It can be estimated using various algorithms.

Complex indicators are calculated for different systems. Using data, information and knowledge from the database of Federal Almazov North-West Medical Research Centre (Saint-Petersurg, Russia) (Center) we have allocated 4 sets of the parameters, most frequently used to analyze the patients with cardiovascular diagnoses. They are: biochemical parameters (BCP), peripheral blood system - red blood cells and leukocyte formula, acid-base status of the blood (the acid-alkaline balance - AAB), urine indicators.

One of the methods to observe dynamics of patient's state, that is called Botkin list, was developed by famous Russian clinician Sergey Botkin. It is based on time series of parameters that are measured at certain time periods, for example, once a day. This method uses only objective data, obtained from different sources: blood tests, urin analyses, acid-alkaline balance indicators and others to calculate integral indicators.

The method is based on calculation of integral indicator, also known as electrolyte balance functional. The common idea of calculating this indicator is shown below.

Algorithm Description. The set of objects (biochemical parameters, ions of blood, etc.) $R = (R_1, R_2, .., R_m)$ is divided into disjoint classes – sets of functional subsystem of physical parameters, delivering local maximum of the functional. Functional is described by sum of correlation links between parameters subtracted by a certain threshold, characterizing significance of the correlation links, according to:

$$F(a, R) = \sum_s \sum_{i,j \in R_s} (a_{ij} - a)$$

where a is a link threshold ($a_{ij} > a$ means the link is significant, $a_{ij} < a$ – link is not significant), a_{ij} is link indicator between i and j parameters ($a_{ij} = a_{ji}$ and a_{ii} are not considered and researched), $i, j \in R_s$ means parameters i and j are included into R_s class.

The procedure of data preliminary processing and calculating F(a, R) is given below.

Step 1.   Convert raw data to the standard form. The output is a matrix, where elements are time series that are results of tests

Step 2.   Analyze relations between time series using correlation theory. Correlations of time series are investigated separately for each physiological system

Step 3.   Estimate level of correlations and remove extra parameters

Step 4.   Estimate conditions and requirements to functional calculation

Step 5.   Calculate the functional F (a, R) separately for each system. Functional F (a, R) is a time series

Complex Parameters Estimation and Presentation. The result of the algorithm execution is an integral indicator of a patient state that is calculated using results of measurements. It can be compared with ones calculated for other patients in similar conditions to get some statistics and to make conclusions.

Fluctuation of the indicator can point at some changes in patient's condition. Using it can help doctors to find unexpected process and predict possible consequences for near next days. The fluctuations can also point on rhythmical properties of the internal processes.

Botkin list assumes representation of dynamics of integral indicator changing and timetable with plots of significant parameters. An expert needs to view different parameters at one time. Separate question is a handy output that prevents plots to superimpose on each other.

4   Medical Data Specification

Raw Data. The initial data are the results of various medical tests. Medical tests are divided into different groups depending on the accessories to physiological systems. Each specific analysis is characterized by two parameters: type of analysis and date. The number of tests of the same type defines a time series.

There are some problems with primary data processing. The main problem is that data is incomplete. We can't know values of required parameters at all moments of time. Commonly the parameters are measured to regularly. It is only possible to interpolate them to restore the dynamics of their changes. Another problem is that contradictions can be noticed between different sources of information. Data can be uncoordinated with illness history or with subjective opinion of the doctor.

Physiological System Parameters for Botkin List. Parameters of 4 human physiological systems that are mentioned above are considered in Botkin list.:

1. Functional leukogram (FLG) - hematocrit, hemoglobin, white blood cells, lymphocytes, neutrophils, and stab-nuclear and segment-nuclear neutrophils;

2. Functional AAR (FKSCHR) - ABE (vein), Ca2 + (Vienna), Cl- (vein), Glu (vein), HCO3- (P) (vein), Lac (vein), Na + (vein), cthb (vein), ctO2 (vein), the pCO_2 (vein), pO_2 (vein), K + (vein), osmolarity (vein);
3. Functional BHP (FBHP) - the Glu (vein), of Na + (vein), ALT, AST, Alb-min, Amylase (syv), Bilirubin total, Glucose, K + (vein), Creatinine, Uric acid, urea, total protein, CRP, trailning, total cholesterol;
4. Indicators of urine - specific gravity.

Parameters are measured 1–3 times a day. Integral indicator is calculated over whole parameters set excluding repeating parameters and parameters measured in different ways.

Modelling Base. Database of the Center is maintained since 2000 year to the present. The total number of patients exceeds 100 thousands. Historical data can be used to estimate integral indicators. System functional is calculated based on data about treating all the patients with cardiovascular diagnoses according to ICD-10 (I20-I22).

5 Medical Data Modeling Views

Problem solving is carried out according to the JDL model. As described above the model has a 4-level implementation of the functionality - the level of initial data, objects, situations, assessing situations (threats assessment). For medical domain initial data are results of diagnostic tests that the patent passes. Signals are preprocessed measurements. Object is a patient himself. Situation is the observed state of the patient described by integral parameters. The state can be estimated using graphical user interface with plots and tables of initial and calculated parameters values. The projection of JDL model on the problem of the indicators calculation is shown on Fig. 2.

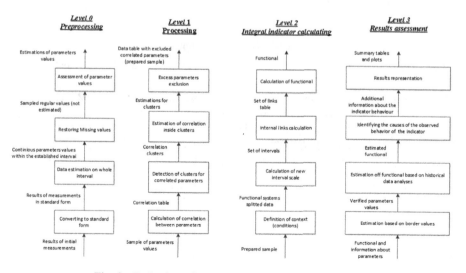

Fig. 2. Projection of JDL model for indicator calculation

Each level of JDL model is divided into 4 steps that define functions for data transformation. Specification of JDL model is not enough for solving processing tasks. To implement it in information systems it is necessary to define the input-output data and algorithms for each step. To describe them Informational JDL model and Logic (Algorithmic) JDL model were developed. In the next sections models of the levels are presented in details. In practice the models and their parts can be used separately or in combinations.

6   Medical Data Modeling Views Specification

The level of preprocessing depends on the characteristics of the input format, frequency and regularity of measurements, physical characteristics of the parameters and solved tasks. At the level of preprocessing there is a need to use specialized algorithms and define their parameters. In a number of the most difficult cases, experts are involved. After the preprocessing stage data has the following characteristics: (i) it is interpretable at the machine level; (ii) it is organized as a linked data; (iii) it can be used as input data for machine learning algorithms; (iv) semantic tools are applicable to data.

Processing level is based on the use of static methods, such as correlation methods, methods of cluster analysis and designed to convert data into a form required for solving applied and intelligent tasks. The level of calculation of integral indicators can be considered as the level of solving user tasks. A set of algorithms is defined according to the tasks and subject area. Results assessment level is for preparing additional information and views required for understanding and interpreting the results. Applied methods of intellectual analysis, semantic methods, methods of cognitive graphics, etc. are used at the level.

6.1   Preprocessing (Level 0)

(See Table 1).

Table 1. Preprocessing.

Level	Name	Description	Basic algorithms	Input	Output
$L^1 0-1$	Converting to standard form	Fill missing values with '-1'; convert matrix to the form with no negative value at first or last rows	Described at 6.5	Matrix	Matrix
$L^1 0-2$	Data estimation on whole interval	Interpolate parameters values	Cubic splines, Lagrange polynomial	Vector	Function; interpolated values
$L^1 0-3$	Restoring missing values	Festore missing values using function from previous level	Described at 6.5	Vector	Vector

(Continued)

Table 1. (*Continued*)

Level	Name	Description	Basic algorithms	Input	Output
$L^1$0– 4	Assessment of parameter values	Define table from vectors with filled values and estimate values statistically	Described at 6.5	List of vectors	Matrix

6.2 Processing (Level 1)

(See Table 2).

Table 2. Processing.

Level	Name	Description	Basic algorithms	Input	Output
L11– 1	Calculation of correlation between parameters	Consider parameters as independent random samples; calculate correlation between them	Correlation function	List of parameter vectors	Correlation matrix
L11– 2	Detection of clusters for correlated parameters	Split the set of parameters in clusters	k-means, FOREL	Correlation matrix	Clusters
L11– 3	Estimation of correlations inside clusters	Define extra parameters with correlation more than threshold	Logical operations	Cluster of parameter	List of extra parameter
L11– 4	Extra parameters exclusion	Define table excluding extra	Logical operations	List of extra parameter	Matrix

6.3 Integral Indicator Calculating (Level 2)

(See Table 3).

Table 3. Integral indicator calculating.

Level	Name	Description	Basic algorithms	Input	Output
L12 – 1	Definition of contexts (conditions)	Split parameters according to physiological systems	Hash set search	One matrix	Individual matrix for each system

(*Continued*)

Table 3. (*Continued*)

Level	Name	Description	Basic algorithms	Input	Output
L12–2	Calculation of new interval scale	Determine a new scale; split data into intervals	Logical operations	One matrix	Individual matrix for each interval
L12–3	Internal links calculation	Consider parameters as independent random samples; calculate correlation between them inside each interval	Correlation formula for samples	List of parameter vectors	Correlation matrix
L12–4	Calculation of functional	Calculate functional according to function from Sect. 3	Functional	Correlation matrix	Value of
	functional				

6.4   Results Assessment (Level 3)

(See Table 4)

Table 4. Data based assessment.

Level	Name	Description	Basic algorithms	Input	Output
L13–1	Estimation based on border values	Compare values with min and max thresholds	Logical operations	Matrix of parameter	Boolean matrix
L13–2	Estimation of functional based on historical data analyses	Compare calculated functional to functionals of other patients	Arithmetic operations	Value of functional	Value of difference
L13–3	Identifying the causes of the observed	Use intelligent methods, semantic techniques or	Intelligent methods, semantic queries,	Parameter and functional dynamics	Text

(*Continued*)

Table 4. (*Continued*)

Level	Name	Description	Basic algorithms	Input	Output
	behavior of the indicator	medical experts conclusions	manual inspection		
L13–4	Results representation	Represent results in user-friendly form	Graphical output	Parameter and functional dynamics	Tables, graphs, plots, text

6.5   Level Specification

For all levels of the models detailed specification was developed. Due to the high sensitivity of preprocessing stage detailed specification of this stage is given below. Processing stage specification is based on the formal statements of problems of correlation analysis and cluster analysis [20, 21], processing stage is based on procedures of functional calculation, results assessment is based on intelligent, in particular data mining techniques [22] and semantic technologies.

$L^1 0$–1: *Converting to Standard Form.* Let $D_0 = \{C_I\}_{i=1}^N$ is set of initial data. N - number of analyzes. Each analysis C_i is represented as a time series $C_i = (c_i(t_{i1}), c_i(t_{i2}), ..., c_i(t_{ip}))$. Double subscript t indicates that the tests were carried out at different times p. Index, generally speaking, depends on the number i. Let M - the number of days a patient is treated. T_{ij} -the time of analysis that is the number of days from the start of the treatment. The times t_{ij} satisfy the inequalities: $1 \leq tij \leq M$, thee are ordered: $t_{i1} < t_{i2} < ... < t_{ip}$.

Each time series $C_i = (c_i(t_{i1}), c_i(t_{i2}), ..., c_i(t_{ip}))$, $i = 1, ..., N$, can be written as $V'_i = (d'_{1i}, d'_{2i}, ..., d'_{Mi})$, moreover $d'_{ki} = c_i(t_{ir})$, if there exists a point in time t_{ir} equal to k, and $d'_{ki} = -1$ otherwise. If any component d_{ki} of vector V'_i is negative, at k day i analyses were not carried out. Set of column vectors V'_i forms a table P' with size $= M \times N$. Dropping, if necessary, extra lines, provide that in P first and last lines do not contain negative elements. Let m be a number of rows in table P. The resulting maping $(c_i(t_{i1}), c_i(t_{i2}), ..., c_i(t_{im})) \rightarrow (d_{1i}, d_{2i}, ..., d_{mi})$ is denoted by f_1. The function f_1 for every analysis C_i assigns m-dimension column vector $V_i = (d_{1i}, d_{2i}, ..., d_{mi})$.

$L^1 0$–2: *Data Estimation on Whole Interval.* There is a sufficient number of recovery methods. We focus on the cubic spline interpolation. Consider an arbitrary column vector $V_i = (d_{1i}, d_{2i}, ..., d_{mi})$. On the interval $[1; m]$ build a free-boundary conditions cubic spline (natural cubic spline) $S_i(t)$, passing through the points (k, d_{ki}), for which $d_{ki} > 0$. Thus, $S_i(k) = d_{ki}$, if $d_{ki} > 0$, and $S_i''(1) = S_i''(m) = 0$. Now we have results estimation on whole interval $[1; m]$. Exactly for any moment of time t from interval $[1; m]$ value $S_i(t)$ is an estimation for i analyses at time moment t.

L10–3: Restoring Missing Values. Assessment of Parameter Values. Let D' is the corrected table of data. Let $dki = Vi(k)$. Now the table has no negative elements and is used as an output for preprocessing level.

L10–4: Standard statistical procedures are applied

7 System of Ontologies for Medical Models Representation

The ontology for our model is presented in Fig. 3. This ontology presents basic concepts of medical data and entities associated with the data. There are two types of humans: patients and doctors. Doctors writes dairies, epicrises and conclusions about patients and maintains episodes, prescribing some medicine and analyses. Episode is the part of illness, describing a certain period usually starting at the first or further consultations or after a surgery operation. Patient is the main actor, who participates in an episode. Different kind of data about him is placed in medical database. The main type of data is the results of diagnostic tests, that can be subjective (doctor's epicrises) or objective (results of analyses). Each analysis consists of one or several tests. Test can be numeric or organoleptic. Botkin list method uses only numeric data. Tests correspond to a certain physiological system of the organism. There are many types of systems in human's body, but according to the Botkin list problem, we are interested in only 4 of them. Each system is characterized by a set of parameters and an integral functional indicator that is calculated for the set of parameters. Indicator can be calculated differently for each system.

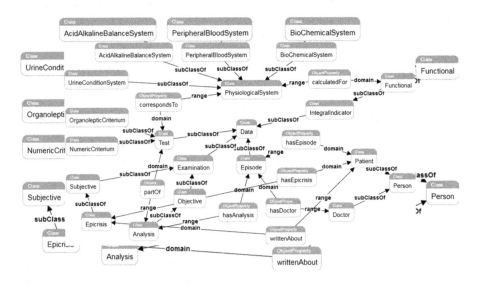

Fig. 3. Domain ontology

On Fig. 4 ontology of peripheral blood system is given. It describes parameters that correspond to this system and describes its properties. Its parameters have measurement units: percentage, milliliters, pieces. Also each parameter has permissible range that depends on gender and age of the patient. Ontology shows that all parameters belong to peripheral blood system and hemogram functional is calculated for this system.

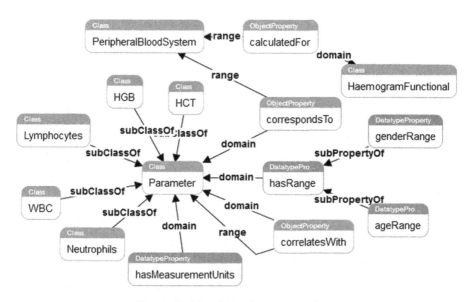

Fig. 4. Peripheral blood system ontology

8 Case Study

Prototype Description. A service for integral indicator calculation was implemented in SMDA (semantic medical data analysis) system, that is developed at ISST Laboratory, ITMO university [23]. The purpose of this system is processing operational and historical data. It is based on technologies of object-oriented databases (Cache, InterSystems [24], methods of intelligent analysis [25], semantic technologies, and graph knowledge representation and reasoning, (blazegraph and metaphacts, Metaphacts Inc) [26]. The service has been used for experimental researches over data, given by Federal Almazov North-West Medical Research Centre.

Data Description. Two patients, that were threated in 2014, are considered. Patient 1 was 45 years old, he or she applied to medical center at 10.06.14 with chronic ischemic heart disease (ICD I25). Drug treatment and surgery (coronary artery bypass surgery with plastic 2 valves with cardiopulmonary bypass) was hold. Patient 1 died at 22.07.2014. Patient 2 was 37 years old, applied to medical center at 24.02.2014 with other forms of angina pectoris diagnosis (ICD I20.8). Surgery (multivessel coronary angioplasty) was hold. Disease outcome: improvement, discharged at 27.05.2014.

Processing and Analyses. The main 7 parameters and their characteristics (for patient 1 sample) are represented in Table 5. In the Table 6 example of raw data for the first patient is presented. Rows contain time stamps (33 measures). Columns contain different parameters: HCT – Hematocrit, HGB – Hemoglobin, MCH - mean corpuscular volume, Eosinophil – eosinophil parameter measured in percentage (30 parameters at all). The identical table is defined for the second patient.

Table 5. Parameters for peripheral blood system.

Parameter	Full name	Min	Max	Mean	Std. dev.
HCT	Hematocrit	19.6	40.4	28.58	4.19
HGB	Hemoglobin	70.1	134.3	97.13	12.88
WBC	Leukocytes	1.7	24.8	14.23	4.984
Lymphocytes, %	Lymphocytes	1.5	20.2	6.827	4.59
Neutrophils, %	Neutrophils	74.1	95.8	88.50	5.28
Stab neutrophils, %	Stab-core neutrophils	2	37	8.39	6.945
Segment neutrophils, %	Segment-core neutrophils	58	90	81.15	7.07

Table 6. Raw data for patient 1.

Date/time	HCT	HGB	WBC	...	Neutrophils, %
18.06.14 08:18	37.1	120.9	22.5		94.2
19.06.14 08:22	35.9	118.2	24.8		94.7
20.06.14 08:38	29.4	97	11.7		91.3
21.06.14 09:44	26	90	15.1		79.8
...					
22.07.14 08:36	40.4	134.3	1.7		76.6

The dynamics of these parameters are shown on Fig. 5. We divided measures into 11 equal intervals (3 measures in each interval) and calculated integral indicator at each interval. Table 7 represents intervals division and numeric value of functional on each interval for the first patient. Similar calculations were made for the second patient. The dynamics of integral indicator is shown at Fig. 6.

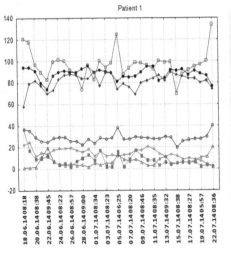

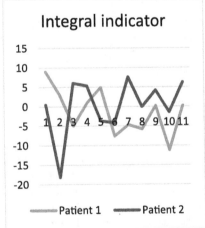

Fig. 5. Dynamics of blood parameters   **Fig. 6.** Dynamics of integral indicator

Table 7. Intervals for patient 1.

Interval no	Start	End	F
1	18.06.14 08:18	20.06.14 08:38	9.003937
2	21.06.14 09:44	23.06.14 08:06	3.10657
3	24.06.14 08:22	26.06.14 08:57	–4.96492
4	27.06.14 09:03	30.06.14 08:31	0.950779
5	01.07.14 08:34	03.07.14 08:23	4.909288
6	04.07.14 08:37	06.07.14 10:01	–7.61186
7	07.07.14 08:20	09.07.14 08:46	–4.69866
8	10.07.14 09:04	12.07.14 08:07	–5.81267
9	13.07.14 09:32	15.07.14 08:38	0.213378
10	16.07.14 09:48	18.07.14 08:26	–11.212
11	19.07.14 05:57	22.07.14 08:36	0.238856

Results Estimation. Examining the plot, we can see that the functional of the first patient falls gradually. This can point deterioration of his condition. On the contrary functional of the second patient felt sharply on the second interval and then began to rise gradually. The fall of it can point at immediate extreme condition in those 3 days. The true reason of this should be established by medical experts. But we can say presumably that it is caused by leukocyte reduction in blood after operation. And the rise can be interpreted as the condition improvement. SMDA interface allows to display information about separate parameters used by medical experts to understand reason of such behavior.

Benefits of the Model. Using a priori information about the physical nature of the parameters allowed at the preprocessing stage to select the algorithms for interpolation and recovery of missing values and to set initial parameters for them.

At the stage of processing descriptions of relations between the parameters allowed take into account correlations, in particular, made it possible to perform checking for correlation dependencies and exclude correlated parameters.

When solving users' tasks using information about relation between parameters and the systems of the organism allowed form samples of parameters to calculate the functional. The transition to an interval scale is made using statistical data about the dynamics of the behavior of the parameters of the earlier treated patients.

At the assessment stage the meta information about parameters allowed perform tests on the parameters taking into account patient's features, to identify possible causes of deviations (considering the parameters that were used to calculate the functional), to assess the similarity of the observed functional with functionals calculated in similar cases.

A formal presentation of the results allows multiple use of the results of pre-processing stage, which does not depend on calculated functionals.

Description of models in OWL language in the form of ontologies allows modify and expand the model, and thus the processing logic, without changes in the program code.

9 Conclusion

The main goal was to development of information models for calculating integral indicators based on measurements. The models are specified and implemented.

For the purpose of modeling systems of the body of patients with cardiovascular diagnoses according to ICD-10 (I20-I22) from the database of Federal Almazov North-West Medical Research Centre more than ten thousand diagnostic tests of 2014 year were unloaded. We used records about over 1000 patients.

Suggested information model was used by experts from Federal Almazov North-West Medical Research Centre. Received results proved its correctness and relevance. Calculated values for 98 % of patients were correct.

The hierarchy of formalized models:

– opens perspective of practical usage of calculation of complex indicators in medical domain;
– opens perspective of application of methods of machine learning and semantic technologies in medicine;
– allows accumulate medical knowledge and distribute it in medical community.

Further development of suggested approach is planned in directions of development of new methods of results interpretation using historical data, cognitive graphics, etc. Separate activity is to be realized in the direction of creation of context sensitive models which can take into account variations of parameters in specific conditions.

References

1. Petlenko V.P., Popov A.S.: Philosophical problems of medicine (1978)
2. USA National Medical Library. https://www.nlm.nih.gov/
3. MEDLINE. https://www.nlm.nih.gov/bsd/pmresources.html
4. Neznanov, A.A., Starichkova, Y.V.: Development of classification of clinical diagnoses in medical information systems, Business Informatics (2015)
5. Neznanov, A.A.: Modern mathematical models of medical informatics: the statistics to mining (2016)
6. ICH. http://www.ich.org
7. CONSORT. http://www.consort-statement.org
8. Metathesaurus. http://www.nlm.nih.gov/research/umls/quickstart.html
9. SNOMED CT. http://www.ihtsdo.org/snomed-ct
10. Technical Implementation Guide. http://ihtsdo.org/fileadmin/user_upload/doc/en_us/tig.html
11. MeSH. http://www.nlm.nih.gov/mesh/meshhome.html
12. MedDRA. http://www.meddra.org
13. ICD. http://www.who.int/classifications/icd/en
14. ICD10Data. http://www.icd10data.com
15. RxNorm. http://www.nlm.nih.gov/research/umls/rxnorm/index.html
16. Steinberg, A.N., Bowman, C.L., White, F.E.: Revisions to the JDL data fusion model. In: The Joint NATO/IRIS Conference, Quebec (1998)

17. Zhukova, N.A., Pankin, A.V.: Principles of managing the processing and analysis of multi-dimensional measurements in IGIS. In: Proceedings of the Information Technologies in Man-Agement, Saint-Petersburg, 9–11 October (2012)
18. Shanin Yu., N.: Postoperative intensive therapy (1978)
19. Mirkin, B.G., Kupershtok, B.L.: Amount of internal relations classification as an indicator of quality (1976)
20. Mandel ID.: Cluster analysis. Moscow, Finance and Statistics (1988)
21. Multivariate statistical analysis: Timashevicha, V.N. (ed.). Moscow, UNITY (1999)
22. Piatetsky-Shapiro, G.: From Data Mining to Knowledge Discovery in Databases (1996)
23. ISST. http://isst.ifmo.ru/en/
24. InterSystems Cache. http://www.intersystems.com/our-products/cache/cache-overview/
25. Witten, I.H., Frank, E., Hall, M.A.: Data Mining: Practical Machine Learning Tools and Techniques (2011)
26. Metaphacts. http://www.metaphacts.com/

Author Index

Printed in the United States
By Bookmasters